INFORMATION
RANDOMNESS
&
INCOMPLETENESS
PAPERS ON ALGORITHMIC INFORMATION THEORY

World Scientific Series in Computer Science

Forthcoming titles:

Series in Computer Science — Vol. 8

INFORMATION RANDOMNESS & INCOMPLETENESS

PAPERS ON ALGORITHMIC INFORMATION THEORY

Gregory J Chaitin

IBM Thomas J Watson Research Center
Yorktown Heights, New York

World Scientific
Singapore • New Jersey • Hong Kong

Published by

World Scientific Publishing Co. Pte. Ltd.
P.O. Box 128, Farrer Road, Singapore 9128

U. S. A. office: World Scientific Publishing Co., Inc.
687 Hartwell Street, Teaneck NJ 07666, USA

Library of Congress Cataloging-in-Publication Data

Chaitin, Gregory J.
    Information, randomness & incompleteness.

    1. Machine theory.   2. Computational complexity.
3. Stochastic processes.   I. Title.
QA267.C49   1987   511.3   87-27418
ISBN 9971-50-479-0
ISBN 9971-50-480-4 (pbk.)

Printed in Singapore by Kyodo-Shing Loong Printing Industries Pte Ltd.

The author and the publisher are grateful to the following for permission to reprint the papers included in this volume.

Academic Press, Inc. (*Adv. Appl. Math.*);
Association for Computing Machinery (*J. ACM, ACM SICACT News*);
Cambridge University Press (*Algorithmic Information Theory*);
Elsevier Science Publishers (*Theor. Comput. Sci.*);
IBM (*IBM J. Res. Dev.*);
IEEE (*IEEE Trans. Info. Theory*);
N. Ikeda, Osaka University (*Osaka J. Math.*);
John Wiley & Sons, Inc. (*Encyclopedia of Statistical Sci., Commun. Pure Appl. Math.*);
MIT Press (*The Maximum Entropy Formalism*);
Pergamon Journals Ltd. (*Comp. Math. Applic.*);
Plenum Publishing Corp. (*Int. J. Theor. Phys.*);
I. Prigogine, Universite Libre de Bruxelles (*Proc. 1985 Solvay Conference*);
Springer-Verlag (*Open Problems in Communication and Computation*);
Verlag Kammerer & Unverzagt (*The Universal Turing Machine – A Half Century Survey*);
W. H. Freeman and Company (*Sci. Am.*).

PREFACE

God not only plays dice in quantum mechanics, but even with the whole numbers! The discovery of randomness in arithmetic is presented in my book *Algorithmic Information Theory* recently published by Cambridge University Press. There I show that to decide if an algebraic equation in integers has finitely or infinitely many solutions is in some cases absolutely intractable. I exhibit an infinite series of such arithmetical assertions that are random arithmetical facts, and for which it is essentially the case that the only way to prove them is to assume them as axioms. This extreme form of Gödel incompleteness theorem shows that some arithmetical truths are totally impervious to reasoning.

The papers leading to this result were published over a period of more than twenty years in widely scattered journals, but because of their unity of purpose they fall together naturally into the present book, intended as a companion volume to my Cambridge University Press monograph. I hope that it will serve as a stimulus for work on complexity, randomness and unpredictability, in physics and biology as well as in metamathematics.

GREGORY CHAITIN

CONTENTS

Part VI — Technical Papers on Turing Machines

Part I—Introductory/Tutorial/Survey Papers

RANDOMNESS AND MATHEMATICAL PROOF

Scientific American 232 (May 1975), pp. 47-52.

Although randomness can be precisely defined and can even be measured, a given number cannot be proved to be random. This enigma establishes a limit to what is possible in mathematics.

by Gregory J. Chaitin

Almost everyone has an intuitive notion of what a random number is. For example, consider these two series of binary digits:

$$01010101010101010101$$
$$01101100110111100010$$

The first is obviously constructed according to a simple rule; it consists of the number 01 repeated ten times. If one were asked to speculate on how the series might continue, one could predict with considerable confidence that the next two digits would be 0 and 1. Inspection of the second series of digits yields no such comprehensive pattern. There is no obvious rule governing the formation of the number, and there is no rational way to guess the succeeding digits. The arrangement seems haphazard; in other words, the sequence appears to be a random assortment of 0's and 1's.

The second series of binary digits was generated by flipping a coin 20 times and writing a 1 if the outcome was heads and a 0 if it was tails. Tossing a coin is a classical procedure for producing a random number, and one might think at first that the provenance of the series alone would certify that it is random. This is not so. Tossing a coin 20 times can produce any one of 2^{20} (or a little more than a million) binary series, and each of them has exactly the same probability. Thus it should be no more surprising to obtain the series with an obvious pattern than to obtain the one that seems to be random; each represents an event with a probability of 2^{-20}. If origin in a probabilistic event were made the sole criterion of randomness, then both series would have to be considered random, and indeed so would all others, since the same mechanism can generate all the possible series. The conclusion is singularly unhelpful in distinguishing the random from the orderly.

Clearly a more sensible definition of randomness is required, one that does not contradict the intuitive concept of a "patternless" number. Such a definition has been devised only in the past 10 years. It does not consider the origin of a number but depends entirely on the characteristics of the sequence of digits. The new definition enables us to describe the properties of a random number more precisely than was formerly possible, and it establishes a hierarchy of degrees of randomness. Of perhaps even greater interest than the capabilities of the definition, however, are its limitations. In particular the definition cannot help to determine, except in very special cases, whether or not a given series of digits, such as the second one above, is in fact random or only seems to be random. This limitation is not a flaw in the definition; it is a consequence of a subtle but fundamental anomaly in the foundation of mathematics. It is closely related to a famous theorem devised and proved in 1931 by Kurt Gödel, which has come to be known as Gödel's incompleteness theorem. Both the theorem and the recent discoveries concerning the nature of randomness help to define the boundaries that constrain certain mathematical methods.

Algorithmic Definition

The new definition of randomness has its heritage in information theory, the science, developed mainly since World War II, that studies the transmission of messages. Suppose you have a friend who is visiting a planet in another galaxy, and that sending him telegrams is very expensive. He forgot to take along his tables of trigonometric functions, and he has asked you to supply them. You could simply translate the numbers into an appropriate code (such as the binary numbers) and transmit them directly, but even the most modest tables of the six functions have a few thousand digits, so that the cost would be high. A much cheaper way to convey the same information would be to transmit instructions for calculating the tables from the underlying trigonometric formulas, such as Euler's equation $e^{ix} = \cos x + i \sin x$. Such a message could be relatively brief, yet inherent in it is all the information contained in even the largest tables.

Suppose, on the other hand, your friend is interested not in trigonometry but in baseball. He would like to know the scores of all the major-league games played since he left the earth some thousands of years before. In this case it is most unlikely that a formula could be found for compressing the information into a short message; in such a series of numbers each digit is essentially an independent item of information, and it cannot be predicted from its neighbors or from some underlying rule. There is no alternative to transmitting the entire list of scores.

In this pair of whimsical messages is the germ of a new definition of randomness. It is based on the observation that the information embodied in a random series of numbers cannot be "compressed," or reduced to a more compact form. In formulating the actual definition it is preferable to consider communication not with a distant friend but with a digital computer. The friend might have the wit to make inferences about numbers or to construct a series from partial information or from vague instructions. The computer does not have that capacity, and for our purposes that deficiency is an advantage. Instructions given the computer must be complete and explicit, and they must enable it to proceed step by step without requiring that it comprehend the result of any part of the operations it performs. Such a program of instructions is an algorithm. It can demand any finite number of mechanical manipulations of numbers, but it cannot ask for judgments about their meaning.

The definition also requires that we be able to measure the information content of a message in some more precise way than by the cost of sending it as a telegram. The fundamental unit of information is the "bit," defined as the smallest item of information capable of indicating a choice between two equally likely things. In binary notation one bit is equivalent to one digit, either a 0 or a 1.

We are now able to describe more precisely the differences between the two series of digits presented at the beginning of this article:

$$01010101010101010101$$
$$01101100110111100010$$

The first could be specified to a computer by a very simple algorithm, such as "Print 01 ten times." If the series were extended according to the same rule, the algorithm would have to be only slightly larger; it might be made to read, for example, "Print 01 a million times." The number of bits in such an algorithm is a small fraction of the number of bits in the series it specifies, and as the series grows larger the size of the program increases at a much slower rate.

For the second series of digits there is no corresponding shortcut. The most economical way to express the series is to write it out in full, and the shortest algorithm for introducing the series into a computer would be "Print 01101100110111100010." If the series were much larger (but still apparently patternless), the algorithm would have to be expanded to the corresponding size. This "incompressibility" is a property of all random numbers; indeed, we can proceed directly to define randomness in terms of incompressibility: A series of numbers is random if the smallest algorithm capable of specifying it to a computer has about the same number of bits of information as the series itself.

This definition was independently proposed about 1965 by A. N. Kolmogorov of the Academy of Science of the U.S.S.R. and by me, when I was an undergraduate at the City College of the City University of New York. Both Kolmogorov and I were then unaware of related proposals made in 1960 by Ray J. Solomonoff of the Zator Company in an endeavor to measure the simplicity of scientific theories. During the past decade we and others have continued to explore the meaning of randomness. The original formulations have been improved and the feasibility of the approach has been amply confirmed.

Model of Inductive Method

The algorithmic definition of randomness provides a new foundation for the theory of probability. By no means does it supersede classical probability theory, which is based on an ensemble of possibilities, each of which is assigned a probability. Rather, the algorithmic approach complements the ensemble method by giving precise meaning to concepts that had been intuitively appealing but that could not be formally adopted.

The ensemble theory of probability, which originated in the 17th century, remains today of great practical importance. It is the foundation of statistics, and it is applied to a wide range of problems in science and engineering. The algorithmic theory also has important implications, but they are primarily theoretical. The area of broadest interest is its amplification of Gödel's incompleteness theorem. Another application (which actually preceded the formulation of the theory itself) is in Solomonoff's model of scientific induction.

Solomonoff represented a scientist's observations as a series of binary digits. The scientist seeks to explain these observations through a theory, which can be regarded as an algorithm capable of generating the series and extending it, that is, predicting future observations. For any given series of observations there are always several competing theories, and the scientist must choose among them. The model demands that the smallest algorithm, the one consisting of the fewest bits, be selected. Stated another way, this rule is the familiar formulation of Occam's razor: Given differing theories of apparently equal merit, the simplest is to be preferred.

Thus in the Solomonoff model a theory that enables one to understand a series of observations is seen as a small computer program that reproduces the observations and makes predictions about possible future observations. The smaller the program, the more comprehensive the theory and the greater the degree of understanding. Observations that are random cannot be reproduced by a small program and therefore cannot be explained by a theory. In addition the future behavior of a random system cannot be predicted. For random data the most compact way for the scientist to communicate his observations is for him to publish them in their entirety.

Defining randomness or the simplicity of theories through the capabilities of the digital computer would seem to introduce a spurious element into these essentially abstract notions: the peculiarities of the particular computing machine employed. Different machines communicate through different computer languages, and a set of instructions expressed in one of those languages might require more or fewer bits when the instructions are translated into another language. Actually, however, the choice of computer matters very little. The problem can be avoided entirely simply by insisting that the randomness of all numbers be tested on the same machine. Even when different machines are employed, the idiosyncrasies of various languages can readily be compensated for. Suppose, for example, someone has a program written in English and wishes to utilize it with a computer that reads only French. Instead of translating the algorithm itself he could preface the program with a complete English course written in French. Another mathematician with a French program and an English machine would follow the opposite procedure. In this way only a fixed number of bits need be added to the program, and that number grows less significant as the size of the series specified by the program increases. In practice a device called a compiler often makes it possible to ignore the differences between languages when one is addressing a computer.

Since the choice of a particular machine is largely irrelevant, we can choose for our calculations an ideal computer. It is assumed to have unlimited storage capacity and unlimited time to complete its calculations. Input to and output from the machine are both in the form of binary digits. The machine begins to operate as soon as the program is given it, and it continues until it has finished printing the binary series that is the result. The machine then halts. Unless an error is made in the program, the computer will produce exactly one output for any given program.

Minimal Programs and Complexity

Any specified series of numbers can be generated by an infinite number of algorithms. Consider, for example, the three-digit decimal series 123. It could be produced by an algorithm such as "Subtract 1 from 124 and print the result," or "Subtract 2 from 125 and print the result," or an infinity of other programs formed on the same model. The programs of greatest interest, however, are the smallest ones that will yield a given numerical series. The smallest programs are called minimal programs; for a given series there may be only one minimal program or there may be many.

Any minimal program is necessarily random, whether or not the series it generates is random. This conclusion is a direct result of the way we have defined randomness. Consider the program P, which is a minimal program for the series of digits S. If we assume that P is not random, then by definition there must be another program, P', substantially smaller than P that will generate it. We can then produce S by the following algorithm: "From P' calculate P, then from P calculate S." This program is only a few bits longer than P', and thus it must be substantially shorter than P. P is therefore not a minimal program.

The minimal program is closely related to another fundamental concept in the algorithmic theory of randomness: the concept of complexity. The complexity of a series of digits is the number of bits that must be put into a computing machine in order to obtain the original series as output. The complexity is therefore equal to the size in bits of the minimal programs of the series. Having introduced this concept, we can now restate our definition of randomness in more rigorous terms: A random series of digits is one whose complexity is approximately equal to its size in bits.

The notion of complexity serves not only to define randomness but also to measure it. Given several series of numbers each having n digits, it is theoretically possible to identify all those of complexity $n - 1, n - 10, n - 100$ and so forth and thereby to rank the series in decreasing order of randomness. The exact value of complexity below which a series is no longer considered random remains somewhat arbitrary. The value ought to be set low enough for numbers with obviously random properties not to be excluded and high enough for numbers with a conspicuous pattern to be disqualified, but to set a particular numerical value is to judge what degree of randomness constitutes actual randomness. It is this uncertainty that is reflected in the qualified statement that the complexity of a random series is *approximately* equal to the size of the series.

Properties of Random Numbers

The methods of the algorithmic theory of probability can illuminate many of the properties of both random and nonrandom numbers. The frequency distribution of digits in a series, for example, can be shown to have an important influence on the randomness of the series. Simple inspection suggests that a series consisting entirely of either 0's or 1's is far from random, and the algorithmic approach confirms that conclusion. If such a series is n digits long, its complexity is approximately equal to the

RANDOMNESS AND MATHEMATICAL PROOF

logarithm to the base 2 of n. (The exact value depends on the machine language employed.) The series can be produced by a simple algorithm such as "Print 0 n times," in which virtually all the information needed is contained in the binary numeral for n. The size of this number is about $\log_2 n$ bits. Since for even a moderately long series the logarithm of n is much smaller than n itself, such numbers are of low complexity; their intuitively perceived pattern is mathematically confirmed.

Another binary series that can be profitably analyzed in this way is one where 0's and 1's are present with relative frequencies of three-fourths and one-fourth. If the series is of size n, it can be demonstrated that its complexity is no greater than four-fifths n, that is, a program that will produce the series can be written in $4n/5$ bits. This maximum applies regardless of the sequence of the digits, so that no series with such a frequency distribution can be considered very random. In fact, it can be proved that in any long binary series that is random the relative frequencies of 0's and 1's must be very close to one-half. (In a random decimal series the relative frequency of each digit is, of course, one-tenth.)

Numbers having a nonrandom frequency distribution are exceptional. Of all the possible n-digit binary numbers there is only one, for example, that consists entirely of 0's and only one that is all 1's. All the rest are less orderly, and the great majority must, by any reasonable standard, be called random. To choose an arbitrary limit, we can calculate the fraction of all n-digit binary numbers that have a complexity of less than $n - 10$. There are 2^1 programs one digit long that might generate an n-digit series; there are 2^2 programs two digits long that could yield such a series, 2^3 programs three digits long and so forth, up to the longest programs permitted within the allowed complexity; of these there are 2^{n-11}. The sum of this series $(2^1 + 2^2 + \cdots + 2^{n-11})$ is equal to $2^{n-10} - 2$. Hence there are fewer than 2^{n-10} programs of size less than $n - 10$, and since each of these programs can specify no more than one series of digits, fewer than 2^{n-10} of the 2^n numbers have a complexity less than $n - 10$. Since $2^{n-10}/2^n = 1/1,024$, it follows that of all the n-digit binary numbers only about one in 1,000 have a complexity less than $n - 10$. In other words, only about one series in 1,000 can be compressed into a computer program more than 10 digits smaller than itself.

A necessary corollary of this calculation is that more than 999 of every 1,000 n-digit binary numbers have a complexity equal to or greater than $n - 10$. If that degree of complexity can be taken as an appropriate test of randomness, then almost all n-digit numbers are in fact random. If a fair coin is tossed n times, the probability is greater than .999 that the result will be random to this extent. It would therefore seem easy to exhibit a specimen of a long series of random digits; actually it is impossible to do so.

Formal Systems

It can readily be shown that a specific series of digits is not random; it is sufficient to find a program that will generate the series and that is substantially smaller than the series itself. The program need not be a minimal program for the series; it need only be a small one. To demonstrate that a particular series of digits is random, on the other hand, one must prove that no small program for calculating it exists.

It is in the realm of mathematical proof that Gödel's incompleteness theorem is such a conspicuous landmark; my version of the theorem predicts that the required proof of randomness cannot be found. The consequences of this fact are just as interesting for what they reveal about Gödel's theorem as they are for what they indicate about the nature of random numbers.

Gödel's theorem represents the resolution of a controversy that preoccupied mathematicians during the early years of the 20th century. The question at issue was: "What constitutes a valid proof in mathematics and how is such a proof to be recognized?" David Hilbert had attempted to resolve the controversy by devising an artificial language in which valid proofs could be found mechanically, without any need for human insight or judgement. Gödel showed that there is no such perfect language.

Hilbert established a finite alphabet of symbols, an unambiguous grammar specifying how a meaningful statement could be formed, a finite list of axioms, or initial assumptions, and a finite list of rules of inference for deducing theorems from the axioms or from other theorems. Such a language, with its rules, is called a formal system.

A formal system is defined so precisely that a proof can be evaluated by a recursive procedure involving only simple logical and arithmetical manipulations. In other words, in the formal system there is an algorithm for testing the validity of proofs. Today, although not in Hilbert's time, the algorithm could be executed on a digital computer and the machine could be asked to "judge" the merits of the proof.

Because of Hilbert's requirement that a formal system have a proof-checking algorithm, it is possible in theory to list one by one all the theorems that can be proved in a particular system. One first lists in alphabetical order all sequences of symbols one character long and applies the proof-testing algorithm to each of them, thereby finding all theorems (if any) whose proofs consist of a single character. One then tests all the two-character sequences of symbols, and so on. In this way all potential proofs can be checked, and eventually all theorems can be discovered in order of the size of their proofs. (The method is, of course, only a theoretical one; the procedure is too lengthy to be practical.)

Unprovable Statements

Gödel showed in his 1931 proof that Hilbert's plan for a completely systematic mathematics cannot be fulfilled. He did this by constructing an assertion about the positive integers in the language of the formal system that is true but that cannot be proved in the system. The formal system, no matter how large or how carefully constructed it is, cannot encompass all true theorems and is therefore incomplete. Gödel's technique can be applied to virtually any formal system, and it therefore demands the surprising and, for many, discomforting conclusion that there can be no definitive answer to the question "What is a valid proof?"

Gödel's proof of the incompleteness theorem is based on the paradox of Epimenides the Cretan, who is said to have averred, "All Cretans are liars" [see "Paradox," by W. V. Quine; *Scientific American,* April, 1962]. The paradox can be rephrased in more general terms as "This statement is false," an assertion that is true if and only if it is false and that is therefore neither true nor false. Gödel replaced the concept of truth with that of provability and thereby constructed the sentence "This statement is unprovable," an assertion that, in a specific formal system, is provable if and only if it is false. Thus either a falsehood is provable, which is forbidden, or a true statement is unprovable, and hence the formal system is incomplete. Gödel then applied a technique that uniquely numbers all statements and proofs in the formal system and thereby converted the sentence "This statement is unprovable" into an assertion about the properties of the positive integers. Because this transformation is possible, the incompleteness theorem applies with equal cogency to all formal systems in which it is possible to deal with the positive integers [see "Gödel's Proof," by Ernest Nagel and James R. Newman; *Scientific American,* June, 1956].

The intimate association between Gödel's proof and the theory of random numbers can be made plain through another paradox, similar in form to the paradox of Epimenides. It is a variant of the Berry paradox, first published in 1908 by Bertrand Russell. It reads: "Find the smallest positive integer which to be specified requires more characters than there are in this sentence." The sentence has 114 characters (counting spaces between words and the period but not the quotation marks), yet it supposedly specifies an integer that, by definition, requires more than 114 characters to be specified.

As before, in order to apply the paradox to the incompleteness theorem it is necessary to remove it from the realm of truth to the realm of provability. The phrase "which requires" must be replaced by "which can be proved to require," it being understood that all statements will be expressed in a particular formal system. In addition the vague notion of "the number of characters required to specify" an integer can be replaced by the precisely defined concept of complexity, which is measured in bits rather than characters.

The result of these transformations is the following computer program: "Find a series of binary digits that can be proved to be of a complexity greater than the number of bits in this program." The program tests all possible proofs in the formal system in order of their size until it encounters the first one proving that a specific binary sequence is of a complexity greater than the number of bits in the program. Then it prints the series it has found and halts. Of course, the paradox in the statement from which the program was derived has not been eliminated. The program supposedly calculates a number that no program its size should be able to calculate. In fact, the program finds the first number that it can be proved incapable of finding.

The absurdity of this conclusion merely demonstrates that the program will never find the number it is designed to look for. In a formal system one cannot prove that a particular series of digits is of a complexity greater than the number of bits in the program employed to specify the series.

A further generalization can be made about this paradox. It is not the number of bits in the program itself that is the limiting factor but the number of bits in the formal system as a whole. Hidden in the program are the axioms and rules of inference that determine the behavior of the system and provide the algorithm for testing proofs. The information content of these axioms and rules can be measured and can be designated the complexity of the formal system. The size of the entire program therefore exceeds the complexity of the formal system by a fixed number of bits c. (The actual value of c depends on the machine language employed.) The theorem proved by the paradox can therefore be stated as follows: In a formal system of complexity n it is impossible to prove that a particular series of binary digits is of complexity greater than $n + c$, where c is a constant that is independent of the particular system employed.

Limits of Formal Systems

Since complexity has been defined as a measure of randomness, this theorem implies that in a formal system no number can be proved to be random unless the complexity of the number is less than that of the system itself. Because all minimal programs are random the theorem also implies that a system of greater complexity is required in order to prove that a program is a minimal one for a particular series of digits.

The complexity of the formal system has such an important bearing on the proof of randomness because it is a measure of the amount of information the system contains, and hence of the amount of information that can be derived from it. The formal system rests on axioms: fundamental statements that are irreducible in the same sense that a minimal program is. (If an axiom could be expressed more compactly, then the briefer statement would become a new axiom and the old one would become a derived theorem.) The information embodied in the axioms is thus itself random, and it can be employed to test the randomness of other data. The randomness of some numbers can therefore be proved, but only if they are smaller than the formal system. Moreover, any formal system is of necessity finite, whereas any series of digits can be made arbitrarily large. Hence there will always be numbers whose randomness cannot be proved.

The endeavor to define and measure randomness has greatly clarified the significance and the implications of Gödel's incompleteness theorem. That theorem can now be seen not as an isolated paradox but as a natural consequence of the constraints imposed by information theory. In 1946 Hermann

Weyl said that the doubt induced by such discoveries as Gödel's theorem had been "a constant drain on the enthusiasm and determination with which I pursued my research work." From the point of view of information theory, however, Gödel's theorem does not appear to give cause for depression. Instead it seems simply to suggest that in order to progress, mathematicians, like investigators in other sciences, must search for new axioms.

Bibliography

- *A Profile of Mathematical Logic.* Howard DeLong. Addison-Wesley, 1970.

- *Theories of Probability: An Examination of Foundations.* Terrence L. Fine. Academic Press, 1973.

- *Universal Gambling Schemes and the Complexity Measures of Kolmogorov and Chaitin.* Thomas M. Cover. Technical Report No. 12, Statistics Department, Stanford University, 1974.

- "Information-Theoretic Limitations of Formal Systems." Gregory J. Chaitin in *Journal of the Association for Computing Machinery,* Vol. 21, pages 403-424; July, 1974.

(a)
10100 → COMPUTER → 11111111111111111111

(b)
01101100110111100010 → COMPUTER → 01101100110111100010

Figure 1.   Algorithmic Definition of Randomness: Algorithmic definition of randomness relies on the capabilities and limitations of the digital computer.   In order to produce a particular output, such as a series of binary digits, the computer must be given a set of explicit instructions that can be followed without making intellectual judgments.   Such a program of instructions is an algorithm.   If the desired output is highly ordered (a), a relatively small algorithm will suffice; a series of twenty 1's, for example, might be generated by some hypothetical computer from the program 10100, which is the binary notation for the decimal number 20.   For a random series of digits (b) the most concise program possible consists of the series itself.   The smallest programs capable of generating a particular program are called the minimal programs of the series; the size of these programs, measured in bits, or binary digits, is the complexity of the series.   A series of digits is defined as random if series' complexity approaches its size in bits.

Alphabet, Grammar, Axioms, Rules of Inference

→ COMPUTER →

Theorem 1, Theorem 2, Theorem 3, Theorem 4, Theorem 5, ...

Figure 2.   Formal Systems: Formal systems devised by David Hilbert contain an algorithm
that mechanically checks the validity of all proofs that can be formulated in the
system.   The formal system consists of an alphabet of symbols in which all
statements can be written; a grammar that specifies how the symbols are to be
combined; a set of axioms, or principles accepted without proof; and rules of
inference for deriving theorems from the axioms. Theorems are found by writing
all the possible grammatical statements in the system and testing them to
determine which ones are in accord with the rules of inference and are therefore
valid proofs.   Since this operation can be performed by an algorithm it could be
done by a digital computer.   In 1931 Kurt Gödel demonstrated that virtually all
formal systems are incomplete: in each of them there is at least one statement that
is true but that cannot be proved.

Observations: 0101010101

Predictions: 0101010101010101010101
Theory: Ten repetitions of 01
Size of Theory: 21 characters

Predictions: 01010101010000000000
Theory: Five repetitions of 01 followed by ten 0's
Size of Theory: 42 characters

Figure 3.   Inductive Reasoning: Inductive reasoning as it is employed in science was
analyzed mathematically by Ray J. Solomonoff.   He represented a scientist's
observations as a series of binary digits; the observations are to be explained and
new ones are to be predicted by theories, which are regarded as algorithms
instructing a computer to reproduce the observations.   (The programs would not
be English sentences but binary series, and their size would be measured not in
characters but in bits.)   Here two competing theories explain the existing data;
Occam's razor demands that the simpler, or smaller, theory be preferred.   The task
of the scientist is to search for minimal programs.   If the data are random, the
minimal programs are no more concise than the observations and no theory can
be formulated.

Illustration is a graph of number of n-digit sequences as a
function of their complexity. The curve grows exponentially from
approximately 0 to approximately 2^n as the complexity
goes from 0 to n.

Figure 4.   Random Sequences: Random sequences of binary digits make up the majority
of all such sequences. Of the 2^n series of n digits, most are of a complexity that
is within a few bits of n. As complexity decreases, the number of series diminishes
in a roughly exponential manner. Orderly series are rare; there is only one, for
example, that consists of n 1's.

RUSSELL PARADOX
Consider the set of all sets that are not members of themselves.
Is this set a member of itself?

EPIMENIDES PARADOX
Consider this statement: "This statement is false."
Is this statement true?

BERRY PARADOX
Consider this sentence: "Find the smallest positive integer
which to be specified requires more characters than there are
in this sentence."
Does this sentence specify a positive integer?

Figure 5.   Three Paradoxes: Three paradoxes delimit what can be proved. The first,
devised by Bertrand Russell, indicated that informal reasoning in mathematics can
yield contradictions, and it led to the creation of formal systems. The second,
attributed to Epimenides, was adapted by Gödel to show that even within a formal
system there are true statements that are unprovable. The third leads to the
demonstration that a specific number cannot be proved random.

(a) This statement is unprovable.

(b) The complexity of 01101100110111100010 is greater than 15 bits.

(c) The series of digits 01101100110111100010 is random.

(d) 10100 is a minimal program for the series 11111111111111111111.

Figure 6.   Unprovable Statements: Unprovable statements can be shown to be false, if
they are false, but they cannot be shown to be true. A proof that "This statement
is unprovable" (a) reveals a self-contradiction in a formal system. The assignment
of a numerical value to the complexity of a particular number (b) requires a proof
that no smaller algorithm for generating the number exists; the proof could be
supplied only if the formal system itself were more complex than the number.
Statements labeled c and d are subject to the same limitation, since the
identification of a random number or a minimal program requires the
determination of complexity.

ON THE DIFFICULTY OF COMPUTATIONS

IEEE Transactions on Information Theory IT-16 (1970), pp. 5-9.

GREGORY J. CHAITIN[1]

Abstract

Two practical considerations concerning the use of computing machinery are the amount of information that must be given to the machine for it to perform a given task and the time it takes the machine to perform it. The size of programs and their running time are studied for mathematical models of computing machines. The study of the amount of information (i.e., number of bits) in a computer program needed for it to put out a given finite binary sequence leads to a definition of a random sequence; the random sequences of a given length are those that require the longest programs. The study of the running time of programs for computing infinite sets of natural numbers leads to an arithmetic of computers, which is a distributive lattice.

Section I

The modern computing machine sprang into existence at the end of World War II. But already in 1936 Turing and Post had proposed a mathematical model of computing machines (Figure 7 on page 14).[2] The mathematical model of the computing machine that Turing and Post proposed, commonly referred to as the Turing machine, is a black box with a finite number of internal states. The box can read and write on an infinite paper tape, which is divided into squares. A digit or letter may be written on each square of the tape, or the square may be blank. Each second the machine performs one of the following actions. It may stop, it may shift the tape one square to the right or one square to the left, it may erase the square on which the read-write head is positioned, or it may write a digit or letter on the square on which the read-write head is positioned. The action it performs is determined solely by the internal state of the black box at the moment, and the current state of the black box is determined solely by its previous internal state and the character read on the square of the tape on which its read-write head was positioned.

Incredible as it may seem at first, a machine of such primitive design can multiply numbers written on its tape, and can write on its tape the successive digits of π. Indeed, it is now generally accepted that any calculation that a modern electronic digital computer or a human computer can do, can also be done by such a machine.

Section II

How much information must be provided to a computer in order for it to perform a given task? The point of view we will present here is somewhat different from the usual one. In a typical scientific application, the computer may be used to analyze statistically huge amounts of data and produce a

---

[1]   Manuscript received May 5, 1969; revised July 3, 1969. This paper was presented as a lecture at the Pan-American Symposium of Applied Mathematics, Buenos Aires, Argentina, August 1968.

The author is at Mario Bravo 249, Buenos Aires, Argentina.

[2]   Their papers appear in Davis [1]. As general references on computability theory we may also cite Davis [2]-[4], Minsky [5], Rogers [6], and Arbib [7].

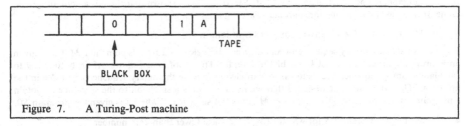

Figure 7.   A Turing-Post machine

brief report in which a great many observations are reduced to a handful of statistical parameters. We would view this in the following manner. The same final result could have been achieved if we had provided the computer with a table of the results, together with instructions for printing them in a neat report. This observation is, of course, ridiculous for all practical purposes. For, had we known the results, it would not have been necessary to use a computer. This example, then, does not exemplify those aspects of computation that we will emphasize.

Rather, we are thinking of such scientific applications as solving the Schrödinger wave equation for the helium atom. Here we have no data, only a program; and the program will produce after much calculation a great deal of printout. Or consider calculating the apparent positions of the planets as observed from the earth over a period of years. A small program incorporating the very simple Newtonian theory for this situation will predict a great many astronomical observations. In this problem there are no data – only a program that contains, of course, a table of the masses of the planets and their initial positions and velocities.

Section III

Let us now consider the problem of the amount of information that it is necessary to provide to a computer in order for it to calculate a given finite binary sequence. A computing machine is defined for these purposes to be a device that accepts as input a program, performs the calculations indicated to it in the program, and finally puts out the binary sequence it has calculated. In line with the mathematical theory of information, it is natural for the program to be viewed as a sequence of bits or 0's and 1's. Furthermore, in computer engineering all programs and data are represented in the machine's circuits in binary form. Thus, we may consider a computer to be a device that accepts one binary sequence (the program) and emits another (the result of the calculation).

$$011001001 \rightarrow Computer \rightarrow 111111001000110010100$$

As an example of a computer we would then have an electronic digital computer that accepts programs consisting of magnetized spots on magnetic tape and puts out its results in the same form. Another example is a Turing machine. The program is a series of 0's and 1's written on the machine's tape at the start of the calculation, and the result is a sequence of 0's and 1's written on its tape when it stops. As was mentioned, the second of these examples can do anything that the first can.

Section IV

We are interested in the amount of information that must be supplied to a computer M in order for it to calculate a given finite binary sequence S. We may now define this as the size or length of the smallest binary sequence that causes the machine M to calculate S. We denote the length of the shortest program for M to calculate S by $L(M, S)$. It has been shown that there is a computing machine M that has the following three properties.[3]

- $L(M, S) \leq k + 1$ for all binary sequences S of length k.

In other words, any binary sequence of length k can be calculated by this computer M if it is given an appropriate program at most $k + 1$ bits in length. The proof is as follows. If no better way to calculate a binary sequence occurs to us, we can always include the binary sequence as a table in the program. This computer is so designed that we need add only a single bit to the sequence to obtain a program for computing it. The computer M emits the sequence S when it is given the program $S0$.

- Those binary sequences S for which $L(M, S) < j$ are fewer than 2^j in number.

Thus, most binary sequences of length k require programs of about the same length k, and the number of sequences that can be computed by smaller programs decreases exponentially as the size of the program decreases. The proof is as follows. There are only $2^j - 2$ binary sequences less than j in length. Thus, there are fewer than 2^j programs less than j in length, for each program is a binary sequence. At best, a program will cause the computer to calculate a single binary sequence. At worst, an error in the program will trap the computer in an endless loop, and no binary sequence will be calculated. As each program causes the computer to calculate at most one binary sequence, the number of sequences calculated must be smaller than the number of programs. Thus, fewer than 2^j binary sequences can be calculated by means of programs less than j in length.

- For any other computer M' there exists a constant $c(M')$ such that for all binary sequences S, $L(M, S) \leq L(M', S) + c(M')$.

In other words, this computer requires shorter programs than any other computer, or more exactly it does not require programs much longer than those required by any other computer. The proof is as follows. The computer M is designed to interpret the circuit diagrams of any other computer M'. Given a program for M' and the circuit diagrams of M', the computer M proceeds to calculate how M' would behave, i.e., it proceeds to simulate M'. Thus, we need only add a fixed number of bits to any program for M' in order to obtain a program that enables M to calculate the same result. This program for M is of the form $PC1$.

The 1 at the right end of the program indicates to the computer M that this is a simulation, C is a fixed binary sequence of length $c(M') - 1$ giving the circuit diagrams of the computer M', which is to be imitated, and P is the program for M'.[4]

Section V

Kolmogorov [9] and the author [11], [12] have independently suggested that computers such as those previously described be applied to the problem of defining what is meant by a random or patternless finite binary sequence of 0's and 1's. In the traditional foundations of the mathematical theory of probability, as expounded by Kolmogorov in his classic [10], there is no place for the concept of an individual random sequence of 0's and 1's. Yet it is not altogether meaningless to say that the sequence

$$11001011111001100101110000010$$

is more random or patternless than the sequences

---

[3]   Solomonoff [8] was the first to employ computers of this kind.
[4]   How can the computer M separate PC into P and C? C has each of its bits doubled, except the pair of bits at its left end. These are unequal and serve as punctuation separating C from P.

$$1111111111111111111111111111111$$
$$0101010101010101010101010101,$$

for we may describe these last two sequences as thirty 1's or fifteen 01's, but there is no shorter way to specify the first sequence than by just writing it all out.

We believe that the random or patternless sequences of a given length are those that require the longest programs. We have seen that most of the binary sequences of length k require programs of about length k. These, then, are the random or patternless sequences. Those sequences that can be obtained by putting into a computer a program much shorter than k are the nonrandom sequences, those that possess a pattern or follow a law. The more possible it is to compress a binary sequence into a short program calculation, the less random is the sequence.

As an example of this, let us consider those sequences of 0's and 1's in which 0's and 1's do not occur with equal frequency. Let p be the relative frequency of 1's, and let $q = 1 - p$ be the relative frequency of 0's. A long binary sequence that has the property that 1's are more frequent than 0's can be obtained from a computer program whose length is only that of the desired sequence reduced by a factor $H(p, q) = -p \log_2 p - q \log_2 q$. For example, if 1's occur approximately $3/4$ of the time and 0's occur $1/4$ of the time in a long binary sequence of length k, there is a program for computing that sequence with length only about $H(3/4, 1/4) k = 0.80 k$. That is, the program need be only approximately 80 percent the length of the sequence it computes. In summary, if 0's and 1's occur with unequal frequencies, we can compress such sequences into programs only a certain percentage (depending on the frequencies) of the size of the sequence. Thus, random or incompressible sequences will have about as many 0's as 1's, which agrees with our intuitive expectations.

In a similar manner it can be shown that all groups of 0's and 1's will occur with approximately the expected frequency in a long binary sequence that we call random; 01100 will appear $2^{-5}k$ times in long sequences of length k, etc.[5]

Section VI

The definition of random or patternless finite binary sequences just presented is related to certain considerations in information theory and in the methodology of science.

The two problems considered in Shannon's classical exposition [15] are to transmit information as efficiently and as reliably as possible. Here we are interested in examining the viewpoint of information theory concerning the efficient transmission of information. An information source may be redundant, and information theory teaches us to code or compress messages so that what is redundant is eliminated and communications equipment is optimally employed. For example, let us consider an information source that emits one symbol (either an A or a B) each second. Successive symbols are independent, and A's are three times more frequent than B's. Suppose it is desired to transmit the messages over a channel that is capable of transmitting either an A or a B each second. Then the channel has a capacity of 1 bit per second, while the information source has entropy 0.80 bits per symbol; and thus it is possible to code the messages in such a way that on the average $1/0.80 = 1.25$ symbols of message are transmitted over the channel each second. The receiver must decode the messages; that is, expand them into their original form.

In summary, information theory teaches us that messages from an information source that is not completely random (that is, which does not have maximum entropy) can be compressed. The definition of randomness is merely the converse of this fundamental theorem of information theory; if lack of randomness in a message allows it to be coded into a shorter sequence, then the random

---

[5]   Martin-Löf [14] also discusses the statistical properties of random sequences.

messages must be those that cannot be coded into shorter messages. A computing machine is clearly the most general possible decoder for compressed messages. We thus consider that this definition of randomness is in perfect agreement and indeed strongly suggested by the coding theorem for a noiseless channel of information theory.

Section VII

This definition is also closely related to classical problems of the methodology of science.[6]

Consider a scientist who has been observing a closed system that once every second either emits a ray of light or does not. He summarizes his observations in a sequence of 0's and 1's in which a 0 represents "ray not emitted" and a 1 represents "ray emitted." The sequence may start

0110101110 ...

and continue for a few million more bits. The scientist then examines the sequence in the hope of observing some kind of pattern or law. What does he mean by this? It seems plausible that a sequence of 0's and 1's is patternless if there is no better way to calculate it than just by writing it all out at once from a table giving the whole sequence. The scientist might state:

My Scientific Theory: 0110101110 ...

This would not be considered an acceptable theory. On the other hand, if the scientist should hit upon a method by which the whole sequence could be calculated by a computer whose program is short compared with the sequence, he would certainly not consider the sequence to be entirely patternless or random. The shorter the program, the greater the pattern he may ascribe the sequence.

There are many parallels between the foregoing and the way scientists actually think. For example, a simple theory that accounts for a set of facts is generally considered better or more likely to be true than one that needs a large number of assumptions. By "simplicity" is not meant "ease of use in making predictions." For although general relativity is considered to be the simple theory par excellence, very extended calculations are necessary to make predictions from it. Instead, one refers to the number of arbitrary choices that have been made in specifying the theoretical structure. One is naturally suspicious of a theory whose number of arbitrary elements is of an order of magnitude comparable to the amount of information about reality that it accounts for.

Section VIII

Let us now turn to the problem of the amount of time necessary for computations.[7] We will develop the following thesis. Call an infinite set of natural numbers perfect if there is no essentially quicker way to compute infinitely many of its members than computing the whole set. Perfect sets exist. This thesis was suggested by the following vague and imprecise considerations.[8]

One of the most profound problems of the theory of numbers is that of calculating large primes. While the sieve of Eratosthenes appears to be as quick an algorithm for calculating all the primes as is possible, in recent times hope has centered on calculating large primes by calculating a subset of the primes, those that are Mersenne numbers. Lucas's test can decide the primality of a Mersenne number with rapidity far greater than is furnished by the sieve method. If there are an infinity of Mersenne primes, then it appears that Lucas has achieved a decisive advance in this classical problem of the theory of numbers.

---

[6] Solomonoff [8] also discusses the relation between program lengths and the problem of induction.
[7] As general references we may cite Blum [16] and Arbib and Blum [17]. Our exposition is a summary of that of [13].
[8] See Hardy and Wright [18], Sections 1.4 and 2.5 for the number-theoretic background of the following remarks.

ON THE DIFFICULTY OF COMPUTATIONS

An opposing point of view is that there is no essentially better way to calculate large primes than by calculating them all. If this is the case, it apparently follows that there must be only finitely many Mersenne primes.

These considerations, then, suggested that there are infinite sets of natural numbers that are arbitrarily difficult to compute, and that do not have any infinite subsets essentially easier to compute than the whole set. Here difficulty of computation refers to speed. Our development will be as follows. First, we define computers for calculating infinite sets of natural numbers. Then we introduce a way of comparing the rapidity of computers, a transitive binary relation, i.e., almost a partial ordering. Next we focus our attention on those computers that are greater than or equal to all others under this ordering, i.e., the fastest computers. Our results are conditioned on the computers having this property. The meaning of "arbitrarily difficult to compute" is then clarified. Last, we exhibit sets that are arbitrarily difficult to compute and do not have any subset essentially easier to compute than the whole set.

Section IX

We are interested in the speed of programs for generating the elements of an infinite set of natural numbers. For these purposes we may consider a computer to be a device that once a second emits a (possibly empty) finite set of natural numbers and that once started never stops. That is to say, a computer is now viewed as a function whose arguments are the program and the time and whose value is a finite set of natural numbers. If a program causes the computer to emit infinitely many natural numbers in size order and without any repetitions, we say that the computing machine calculates the infinite set of natural numbers that it emits.

A Turing machine can be used to compute infinite sets of natural numbers; it is only necessary to establish a convention as to when natural numbers are emitted. For example, we may divide the machine's tape into two halves, and stipulate that what is written on the right half cannot be erased. The computational scratchwork is done on the left half of the tape, and the successive members of the infinite set of natural numbers are written on the nonerasable squares in decimal notation, separated by commas, with no blank spaces permitted between characters. The moment a comma has been written, it is considered that the digits between it and the previous comma form the numeral representing the next natural number emitted by the machine. We suppose that the Turing machine performs a single cycle of activity (read tape; shift, write, or erase tape; change internal state) each second. Last, we stipulate that the machine be started scanning the first nonerasable square of the tape, that initially the nonerasable squares be all blank, and that the program for the computer be written on the first erasable squares, with a blank serving as punctuation to indicate the end of the program and the beginning of an infinite blank region of tape.

Section X

We now order the computers according to their speeds. $C \geq C'$ is defined as meaning that C is not much slower than C'.

What do we mean by saying that computer C is not much slower than computer C' for the purpose of computing infinite sets of natural numbers? There is a computable change of C's time scale that makes C as fast as C' or faster. More exactly, there is a computable function $f(n)$ (for example $n!$ or $n^{n^{n^{...}}}$ with n exponents) with the following property. Let P' be any program that makes C' calculate an infinite set of natural numbers. Then there exists a program P that makes C calculate the same set of natural numbers and has the additional property that every natural number emitted by C' during the first t seconds of calculation is emitted by C during the first $f(t)$ second of calculation, for all but a finite number of values of t. We may symbolize this relation between the computers C and C' as $C \geq C'$, for it has the property that $C \geq C'$ and $C' \geq C''$ only if $C \geq C''$.

In this way, we have introduced an ordering of the computers for computing infinite sets of natural numbers, and it can be shown that a distributive lattice results. The most important property of this ordering for our present purposes is that there is a set of computers $\geq$ all other computers. In what follows we assume that the computer that is used is a member of this set of fastest computers.

Section XI

We now clarify what we mean by "arbitrarily difficult to compute."

Let $f(n)$ be any computable function that carries natural numbers into natural numbers. Such functions can get big very quickly indeed. For example consider the function $n^{n^{n^{\cdots}}}$ in which there are n^n exponents. There are infinite sets of natural numbers such that, no matter how the computer is programmed, at least $f(n)$ seconds will pass before the computer emits all those elements of the set that are less than or equal to n. Of course, a finite number of exceptions are possible, for any finite part of an infinite set can be computed very quickly by including in the computer's program a table of the first few elements of the set. Note that the difficulty in computing such sets of natural numbers does not lie in the fact that their elements get very big very quickly, for even small elements of such sets require more than astronomical amounts of time to be computed. What is more, there are infinite sets of natural numbers that are arbitrarily difficult to compute and include 90 percent of the natural numbers.

We finally exhibit infinite sets of natural numbers that are arbitrarily difficult to compute, and do not have any infinite subsets essentially easier to compute than the whole set. Consider the following tree of natural numbers (Figure 8 on page 20).[9] The infinite sets of natural numbers that we promised to exhibit are obtained by starting at the root of the tree (that is, at 0) and walking forward, including in the set every natural number that is stepped on.

It is easy to see that no infinite subset of such a set can be computed much more quickly that the whole set. For suppose we are told that n is in such a set. Then we know at once that the greatest integer less than $n/2$ is the previous element of the set. Thus, knowing that 1 000 000 is in the set, we immediately produce all smaller elements in it, by walking backwards through the tree. They are 499 999, 249 999, 124 999, etc. It follows that there is no appreciable difference between generating an infinite subset of such a set, and generating the whole set, for gaps in an incomplete generation can be filled in very quickly.

It is also easy to see that there are sets that can be obtained by walking through this tree and are arbitrarily difficult to compute. These, then, are the sets that we wished to exhibit.

Acknowledgment

The author wishes to express his gratitude to Prof. G. Pollitzer of the University of Buenos Aires, whose constructive criticism much improved the clarity of this presentation.

References

1. M. Davis, Ed., *The Undecidable*. Hewlett, N.Y.: Raven Press, 1965.

2. –, *Computability and Unsolvability*. New York: McGraw-Hill, 1958.

---

[9] This tree is used in Rogers [6], p. 158, in connection with retraceable sets. Retraceable sets are in some ways analogous to those sets that concern us here.

ON THE DIFFICULTY OF COMPUTATIONS

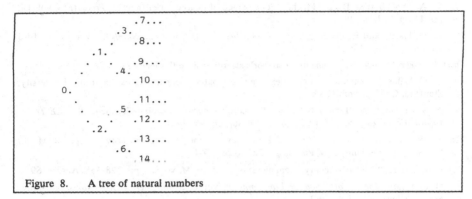

Figure 8. A tree of natural numbers

3. —, "Unsolvable problems: A review," *Proc. Symp. on Mathematical Theory of Automata.* Brooklyn, N.Y.: Polytech. Inst. Brooklyn Press, 1963, pp. 15-22.

4. —, "Applications of recursive function theory to number theory," *Proc. Symp. in Pure Mathematics,* vol. 5. Providence, R.I.: AMS, 1962, pp. 135-138.

5. M. Minsky, *Computation: Finite and Infinite Machines.* Englewood Cliffs, N.J.: Prentice-Hall, 1967.

6. H. Rogers, Jr., *Theory of Recursive Functions and Effective Computability.* New York: McGraw-Hill, 1967.

7. M. A. Arbib, *Theories of Abstract Automata.* Englewood Cliffs, N.J.: Prentice-Hall (to be published).

8. R. J. Solomonoff, "A formal theory of inductive inference," *Inform. and Control,* vol. 7, pp. 1-22, March 1964; pp. 224-254, June 1964.

9. A. N. Kolmogorov, "Three approaches to the definition of the concept 'quantity of information'," *Probl. Peredachi Inform.,* vol. 1, pp. 3-11, 1965.

10. —, *Foundations of the Theory of Probability.* New York: Chelsea, 1950.

11. G. J. Chaitin, "On the length of programs for computing finite binary sequences," *J. ACM,* vol. 13, pp. 547-569, October 1966.

12. —, "On the length of programs for computing finite binary sequences: statistical considerations," *J. ACM,* vol. 16, pp. 145-159, January 1969.

13. —, "On the simplicity and speed of programs for computing infinite sets of natural numbers," *J. ACM,* vol. 16, pp. 407-422, July 1969.

14. P. Martin-Löf, "The definition of random sequences," *Inform. and Control,* vol. 9, pp. 602-619, December 1966.

15. C. E. Shannon and W. Weaver, *The Mathematical Theory of Communication.* Urbana, Ill.: University of Illinois Press, 1949.

16. M. Blum, "A machine-independent theory of the complexity of recursive functions," *J. ACM,* vol. 14, pp. 322-336, April 1967.

17. M. A. Arbib and M. Blum, "Machine dependence of degrees of difficulty," *Proc. AMS,* vol. 16, pp. 442-447, June 1965.

18. G. H. Hardy and E. M. Wright, *An Introduction to the Theory of Numbers.* Oxford: Oxford University Press, 1962.

The following references have come to the author's attention since this lecture was given.

19. D. G. Willis, "Computational complexity and probability constructions," Stanford University, Stanford, Calif., March 1969.

20. A. N. Kolmogorov, "Logical basis for information theory and probability theory," *IEEE Trans. Information Theory,* vol. IT-14, pp. 662-664, September 1968.

21. D. W. Loveland, "A variant of the Kolmogorov concept of complexity," Dept. of Math., Carnegie-Mellon University, Pittsburgh, Pa., Rept. 69-4.

22. P. R. Young, "Toward a theory of enumerations," *J. ACM,* vol. 16, pp. 328-348, April 1969.

23. D. E. Knuth, *The Art of Computer Programming;* vol. 2, *Seminumerical Algorithms.* Reading, Mass.: Addison-Wesley, 1969.

24. *1969 Conf. Rec. of the ACM Symp. on Theory of Computing* (Marina del Rey, Calif.).

INFORMATION-THEORETIC COMPUTATIONAL COMPLEXITY

IEEE Transactions on Information Theory IT-20 (1974), pp. 10-15.

Invited Paper

GREGORY J. CHAITIN[10]

Abstract

This paper attempts to describe, in nontechnical language, some of the concepts and methods of one school of thought regarding computational complexity. It applies the viewpoint of information theory to computers. This will first lead us to a definition of the degree of randomness of individual binary strings, and then to an information-theoretic version of Gödel's theorem on the limitations of the axiomatic method. Finally, we will examine in the light of these ideas the scientific method and von Neumann's views on the basic conceptual problems of biology.

This field's fundamental concept is the complexity of a binary string, that is, a string of bits, of zeros and ones. The complexity of a binary string is the minimum quantity of information needed to define the string. For example, the string of length n consisting entirely of ones is of complexity approximately $\log_2 n$, because only $\log_2 n$ bits of information are required to specify n in binary notation.

However, this is rather vague. Exactly what is meant by the definition of a string? To make this idea precise a computer is used. One says that a string defines another when the first string gives instructions for constructing the second string. In other words, one string defines another when it is a program for a computer to calculate the second string. The fact that a string of n ones is of complexity approximately $\log_2 n$ can now be translated more correctly into the following. There is a program $\log_2 n + c$ bits long that calculates the string of n ones. The program performs a loop for printing ones n times. A fixed number c of bits are needed to program the loop, and $\log_2 n$ bits more for specifying n in binary notation.

Exactly how are the computer and the concept of information combined to define the complexity of a binary string? A computer is considered to take one binary string and perhaps eventually produce another. The first string is the program that has been given to the machine. The second string is the output of this program; it is what this program calculates. Now consider a given string that is to be calculated. How much information must be given to the machine to do this? That is to say, what is the length in bits of the shortest program for calculating the string? This is its complexity.

It can be objected that this is not a precise definition of the complexity of a string, inasmuch as it depends on the computer that one is using. Moreover, a definition should not be based on a machine, but rather on a model that does not have the physical limitations of real computers.

[10]   Manuscript received January 29, 1973; revised July 18, 1973. This paper was presented at the IEEE International Congress of Information Theory, Ashkelon, Israel, June 1973.

The author is at Mario Bravo 249, Buenos Aires, Argentina.

Here we will not define the computer used in the definition of complexity. However, this can indeed be done with all the precision of which mathematics is capable. Since 1936 it has been known how to define an idealized computer with unlimited memory. This was done in a very intuitive way by Turing and also by Post, and there are elegant definitions based on other principles [2]. The theory of recursive functions (or computability theory) has grown up around the questions of what is computable and what is not.

Thus it is not difficult to define a computer mathematically. What remains to be analyzed is which definition should be adopted, inasmuch as some computers are easier to program than others. A decade ago Solomonoff solved this problem [7]. He constructed a definition of a computer whose programs are not much longer than those of any other computer. More exactly, Solomonoff's machine simulates running a program on another computer, when it is given a description of that computer together with its program.

Thus it is clear that the complexity of a string is a mathematical concept, even though here we have not given a precise definition. Furthermore, it is a very natural concept, easy to understand for those who have worked with computers. Recapitulating, the complexity of a binary string is the information needed to define it, that is to say, the number of bits of information that must be given to a computer in order to calculate it, or in other words, the size in bits of the shortest program for calculating it. It is understood that a certain mathematical definition of an idealized computer is being used, but it is not given here, because as a first approximation it is sufficient to think of the length in bits of a program for a typical computer in use today.

Now we would like to consider the most important properties of the complexity of a string. First of all, the complexity of a string of length n is less than $n + c$, because any string of length n can be calculated by putting it directly into a program as a table. This requires n bits, to which must be added c bits of instructions for printing the table. In other words, if nothing betters occurs to us, the string itself can be used as its definition, and this requires only a few more bits than its length.

Thus the complexity of each string of length n is less than $n + c$. Moreover, the complexity of the great majority of strings of length n is approximately n, and very few strings of length n are of complexity much less than n. The reason is simply that there are much fewer programs of length appreciably less than n than strings of length n. More exactly, there are 2^n strings of length n, and less than 2^{n-k} programs of length less than $n - k$. Thus the number of strings of length n and complexity less than $n - k$ decreases exponentially as k increases.

These considerations have revealed the basic fact that the great majority of strings of length n are of complexity very close to n. Therefore, if one generates a binary string of length n by tossing a fair coin n times and noting whether each toss gives head or tail, it is highly probable that the complexity of this string will be very close to n. In 1965 Kolmogorov proposed calling random those strings of length n whose complexity is approximately n [8]. We made the same proposal independently [9]. It can be shown that a string that is random in this sense has the statistical properties that one would expect. For example, zeros and ones appear in such strings with relative frequencies that tend to one-half as the length of the strings increases.

Consequently, the great majority of strings of length n are random, that is, need programs of approximately length n, that is to say, are of complexity approximately n. What happens if one wishes to show that a particular string is random? What if one wishes to prove that the complexity of a certain string is almost equal to its length? What if one wishes to exhibit a specific example of a string of length n and complexity close to n, and assure oneself by means of a proof that there is no shorter program for calculating this string?

It should be pointed out that this question can occur quite naturally to a programmer with a competitive spirit and a mathematical way of thinking. At the beginning of the sixties we attended a course at Columbia University in New York. Each time the professor gave an exercise to be programmed, the students tried to see who could write the shortest program. Even though several times it seemed very difficult to improve upon the best program that had been discovered, we did not fool ourselves. We realized that in order to be sure, for example, that the shortest program for the IBM 650 that prints the prime numbers has, say, 28 instructions, it would be necessary to prove it, not merely to continue for a long time unsuccessfully trying to discover a program with less than 28 instructions. We could never even sketch a first approach to a proof.

It turns out that it was not our fault that we did not find a proof, because we faced a fundamental limitation. One confronts a very basic difficulty when one tries to prove that a string is random, when one attempts to establish a lower bound on its complexity. We will try to suggest why this problem arises by means of a famous paradox, that of Berry [1, p. 153].

Consider the smallest positive integer that cannot be defined by an English phrase with less than 1 000 000 000 characters. Supposedly the shortest definition of this number has 1 000 000 000 or more characters. However, we defined this number by a phrase much less than 1 000 000 000 characters in length when we described it as "the smallest positive integer that cannot be defined by an English phrase with less than 1 000 000 000 characters!"

What relationship is there between this and proving that a string is complex, that its shortest program needs more than n bits? Consider the first string that can be proven to be of complexity greater than 1 000 000 000. Here one more we face a paradox similar to that of Berry, because this description leads to a program with much less than 1 000 000 000 bits that calculates a string supposedly of complexity greater than 1 000 000 000. Why is there a short program for calculating "the first string that can be proven to be of complexity greater than 1 000 000 000?"

The answer depends on the concept of a formal axiom system, whose importance was emphasized by Hilbert [1]. Hilbert proposed that mathematics be made as exact and precise as possible. In order to avoid arguments between mathematicians about the validity of proofs, he set down explicitly the methods of reasoning used in mathematics. In fact, he invented an artificial language with rules of grammar and spelling that have no exceptions. He proposed that this language be used to eliminate the ambiguities and uncertainties inherent in any natural language. The specifications are so precise and exact that checking if a proof written in this artificial language is correct is completely mechanical. We would say today that it is so clear whether a proof is valid or not that this can be checked by a computer.

Hilbert hoped that this way mathematics would attain the greatest possible objectivity and exactness. Hilbert said that there can no longer be any doubt about proofs. The deductive method should be completely clear.

Suppose that proofs are written in the language that Hilbert constructed, and in accordance with his rules concerning the accepted methods of reasoning. We claim that a computer can be programmed to print all the theorems that can be proven. It is an endless program that every now and then writes on the printer a theorem. Furthermore, no theorem is omitted. Each will eventually be printed, if one is very patient and waits long enough.

How is this possible? The program works in the following manner. The language invented by Hilbert has an alphabet with finitely many signs or characters. First the program generates the strings of characters in this alphabet that are one character in length. It checks if one of these strings satisfies the completely mechanical rules for a correct proof and prints all the theorems whose proofs it has found. Then the program generates all the possible proofs that are two characters in length, and examines each of them to determine if it is valid. The program then examines all possible proofs of length three, of length four, and so on. If a theorem can be proven, the program will eventually find a proof for it in this way, and then print it.

Consider again "the first string that can be proven to be of complexity greater than 1 000 000 000." To find this string one generates all theorems until one finds the first theorem that states that a particular string is of complexity greater than 1 000 000 000. Moreover, the program for finding this string is short, because it need only have the number 1 000 000 000 written in binary notation, $\log_2$ 1 000 000 000 bits, and a routine of fixed length c that examines all possible proofs until it finds one that a specific string is of complexity greater than 1 000 000 000.

In fact, we see that there is a program $\log_2 n + c$ bits long that calculates the first string that can be proven to be of complexity greater than n. Here we have Berry's paradox again, because this program of length $\log_2 n + c$ calculates something that supposedly cannot be calculated by a program of length less than or equal to n. Also, $\log_2 n + c$ is much less than n for all sufficiently great values of n, because the logarithm increases very slowly.

What can the meaning of this paradox be? In the case of Berry's original paradox, one cannot arrive at a meaningful conclusion, inasmuch as one is dealing with vague concepts such as an English phrase's defining a positive integer. However our version of the paradox deals with exact concepts that have been defined mathematically. Therefore, it cannot really be a contradiction. It would be absurd for a string not to have a program of length less than or equal to n for calculating it, and at the same time to have such a program. Thus we arrive at the interesting conclusion that such a string cannot exist. For all sufficiently great values of n, one cannot talk about "the first string that can be proven to be of complexity greater than n, " because this string cannot exist. In other words, for all sufficiently great values of n, it cannot be proven that a particular string is of complexity greater than n. If one uses the methods of reasoning accepted by Hilbert, there is an upper bound to the complexity that it is possible to prove that a particular string has.

This is the surprising result that we wished to obtain. Most strings of length n are of complexity approximately n, and a string generated by tossing a coin will almost certainly have this property. Nevertheless, one cannot exhibit individual examples of arbitrarily complex strings using methods of reasoning accepted by Hilbert. The lower bounds on the complexity of specific strings that can be established are limited, and we will never be mathematically certain that a particular string is very complex, even though most strings are random.[11]

In 1931 Gödel questioned Hilbert's ideas in a similar way [1], [2]. Hilbert had proposed specifying once and for all exactly what is accepted as a proof, but Gödel explained that no matter what Hilbert specified so precisely, there would always be true statements about the integers that the methods of reasoning accepted by Hilbert would be incapable of proving. This mathematical result has been considered to be of great philosophical importance. Von Neumann commented that the intellectual shock provoked by the crisis in the foundations of mathematics was equaled only by two other scientific events in this century: the theory of relativity and quantum theory [4].

We have combined ideas from information theory and computability theory in order to define the complexity of a binary string, and have then used this concept to give a definition of a random string and to show that a formal axiom system enables one to prove that a random string is indeed random in only finitely many cases.

Now we would like to examine some other possible applications of this viewpoint. In particular, we would like to suggest that the concept of the complexity of a string and the fundamental

---

[11] This is a particularly perverse example of Kac's comment [13, p. 16] that "as is often the case, it is much easier to prove that an overwhelming majority of objects possess a certain property than to *exhibit* even one such object." The most familiar example of this is Shannon's proof of the coding theorem for a noisy channel; while it is shown that most coding schemes achieve close to the channel capacity, in practice it is difficult to implement a good coding scheme.

INFORMATION-THEORETIC COMPUTATIONAL COMPLEXITY

methodological problems of science are intimately related. We will also suggest that this concept may be of theoretical value in biology.

Solomonoff [7] and the author [9] proposed that the concept of complexity might make it possible to precisely formulate the situation that a scientist faces when he has made observations and wishes to understand them and make predictions. In order to do this the scientist searches for a theory that is in agreement with all his observations. We consider his observations to be represented by a binary string, and a theory to be a program that calculates this string. Scientists consider the simplest theory to be the best one, and that if a theory is too "ad hoc," it is useless. How can we formulate these intuitions about the scientific method in a precise fashion? The simplicity of a theory is inversely proportional to the length of the program that constitutes it. That is to say, the best program for understanding or predicting observations is the shortest one that reproduces what the scientist has observed up to that moment. Also, if the program has the same number of bits as the observations, then it is useless, because it is too "ad hoc." If a string of observations only has theories that are programs with the same length as the string of observations, then the observations are random, and can neither be comprehended nor predicted. They are what they are, and that is all; the scientist cannot have a theory in the proper sense of the concept; he can only show someone else what he observed and say "it was this."

In summary, the value of a scientific theory is that it enables one to compress many observations into a few theoretical hypotheses. There is a theory only when the string of observations is not random, that is to say, when its complexity is appreciably less than its length in bits. In this case the scientist can communicate his observations to a colleague much more economically than by just transmitting the string of observations. He does this by sending his colleague the program that is his theory, and this program must have much fewer bits than the original string of observations.

It is also possible to make a similar analysis of the deductive method, that is to say, of formal axiom systems. This is accomplished by analyzing more carefully the new version of Berry's paradox that was presented. Here we only sketch the three basic results that are obtained in this manner.[12]

1. In a formal system with n bits of axioms it is impossible to prove that a particular binary string is of complexity greater than $n + c$.

2. Contrariwise, there are formal systems with $n + c$ bits of axioms in which it is possible to determine each string of complexity less than n and the complexity of each of these strings, and it is also possible to exhibit each string of complexity greater than or equal to n, but without being able to know by how much the complexity of each of these strings exceeds n.

3. Unfortunately, any formal system in which it is possible to determine each string of complexity less than n has either one grave problem or another. Either it has few bits of axioms and needs incredibly long proofs, or it has short proofs but an incredibly great number of bits of axioms. We say "incredibly" because these quantities increase more quickly than any computable function of n.

It is necessary to clarify the relationship between this and the preceding analysis of the scientific method. There are less than 2^n strings of complexity less than n, but some of them are incredibly long. If one wishes to communicate all of them to someone else, there are two alternatives. The first is to directly show all of them to him. In this case one will have to send him an incredibly long message because some of these strings are incredibly long. The other alternative is to send him a very short message consisting of n bits of axioms from which he can deduce which strings are of complexity less than n. Although the message is very short in this case, he will have to spend an incredibly long time to deduce from these axioms the strings of complexity less than n. This is analogous to the dilemma

---

[12] See the Appendix.

of a scientist who must choose between directly publishing his observations, or publishing a theory that explains them, but requires very extended calculations in order to do this.

Finally, we would like to suggest that the concept of complexity may possibly be of theoretical value in biology.

At the end of his life von Neumann tried to lay the foundation for a mathematics of biological phenomena. His first effort in this direction was his work *Theory of Games and Economic Behavior,* in which he analyzes what is a rational way to behave in situations in which there are conflicting interests [3]. *The Computer and the Brain,* his notes for a lecture series, was published shortly after his death [5]. This book discusses the differences and similarities between the computer and the brain, as a first step to a theory of how the brain functions. A decade later his work *Theory of Self-Reproducing Automata* appeared, in which von Neumann constructs an artificial universe and within it a computer that is capable of reproducing itself [6]. But von Neumann points out that the problem of formulating a mathematical theory of the evolution of life in this abstract setting remains to be solved; and to express mathematically the evolution of the complexity of organisms, one must first define complexity precisely.[13] We submit that "organism" must also be defined, and have tried elsewhere to suggest how this might perhaps be done [10].

We believe that the concept of complexity that has been presented here may be the tool that von Neumann felt is needed. It is by no means accidental that biological phenomena are considered to be extremely complex. Consider how a human being analyzes what he sees, or uses natural languages to communicate. We cannot carry out these tasks by computer because they are as yet too complex for us – the programs would be too long.[14]

Appendix

In this Appendix we try to give a more detailed idea of how the results concerning formal axiom systems that were stated are established.[15]

Two basic mathematical concepts that are employed are the concepts of a recursive function and a partial recursive function. A function is recursive if there is an algorithm for calculating its value when one is given the value of its arguments, in other words, if there is a Turing machine for doing this. If it is possible that this algorithm never terminates and the function is thus undefined for some values of its arguments, then the function is called partial recursive.[16]

In what follows we are concerned with computations involving binary strings. The binary strings are considered to be ordered in the following manner: Λ, 0, 1, 00, 01, 10, 11, 000, 001, 010, ... The natural number n is represented by the nth binary string ($n = 0, 1, 2, ...$). The length of a binary string s is denoted $\lg(s)$. Thus if s is considered to be a natural number, then $\lg(s) = \lceil \log_2 (s + 1) \rceil$. Here $\lceil x \rceil$ is the greatest integer $\leq x$.

Definition 1: A *computer* is a partial recursive function $C(p)$. Its argument p is a binary string. The value of $C(p)$ is the binary string output by the computer C when it is given the program p. If $C(p)$ is undefined, this means that running the program p on C produces an unending computation.

Definition 2: The *complexity* $I_C(s)$ of a binary string s is defined to be the length of the shortest program p that makes the computer C output s, i.e.,

---

[13]   In an important paper [14], Eigen studies these questions from the point of view of thermodynamics and biochemistry.
[14]   Chandrasekaran and Reeker [15] discuss the relevance of complexity to artificial intelligence.
[15]   See [11], [12] for different approaches.
[16]   Full treatments of these concepts can be found in standard texts, e.g., Rogers [16].

$$I_C(s) \;=\; \min{}_{C(p)=s} \lg(p).$$

If no program makes C output s, then $I_C(s)$ is defined to be infinite.

Definition 3: A computer U is *universal* if for any computer C and any binary string s, $I_U(s) \le I_C(s) + c$, where the constant c depends only on C.

It is easy to see that there are universal computers. For example, consider the computer U such that $U(0^i 1 p) = C_i(p)$, where C_i is the ith computer, i.e., a program for U consists of two parts: the left-hand part indicates which computer is to be simulated, and the right-hand part gives the program to be simulated. We now suppose that some particular universal computer U has been chosen as the standard one for measuring complexities, and shall henceforth write $I(s)$ instead of $I_U(s)$.

Definition 4: The *rules of inference* of a class of formal axiom systems is a recursive function $F(a, h)$ (a a binary string, h a natural number) with the property that $F(a, h) \subset F(a, h + 1)$. The value of $F(a, h)$ is the finite (possibly empty) set of theorems that can be proven from the axioms a by means of proofs $\le h$ characters in length. $F(a) = \cup_h F(a, h)$ is the set of theorems that are consequences of the axioms a. The ordered pair $\langle F, a \rangle$, which implies both the choice of rules of inference and axioms, is a particular formal axiom system.

This is a fairly abstract definition, but it retains all those features of formal axiom systems that we need. Note that although one may not be interested in some axioms (e.g., if they are false or incomprehensible), it is stipulated that $F(a, h)$ is always defined.

Theorem 1: a) There is a constant c such that $I(s) \le \lg(s) + c$ for all binary strings s. b) There are less than 2^n binary strings of complexity less than n.

Proof of a): There is a computer C such that $C(p) = p$ for all programs p. Thus for all binary strings s, $I(s) \le I_C(s) + c = \lg(s) + c$.

Proof of b): As there are less than 2^n programs of length less than n, there must be less than this number of binary strings of complexity less than n. Q.E.D.

Thesis: A random binary string s is one having the property that $I(s) \approx \lg(s)$.

Theorem 2: Consider the rules of inference F. Suppose that a proposition of the form "$I(s) \ge n$" is in $F(a)$ only if it is true, i.e., only if $I(s) \ge n$. Then a proposition of the form "$I(s) \ge n$" is in $F(a)$ only if $n \le \lg(a) + c$, where c is a constant that depends only on F.

Proof: Consider that binary string s_k having the shortest proof from the axioms a that it is of complexity $> \lg(a) + 2k$. We claim that $I(s_k) \le \lg(a) + k + c'$, where c' depends only on F. Taking $k = c'$, we conclude that the binary string $s_{c'}$ with the shortest proof from the axioms a that it is of complexity $> \lg(a) + 2c'$ is, in fact, of complexity $\le \lg(a) + 2c'$, which is impossible. It follows that s_k doesn't exist for $k = c'$, that is, no binary string can be proven from the axioms a to be of complexity $> \lg(a) + 2c'$. Thus the theorem is proved with $c = 2c'$.

It remains to verify the claim that $I(s_k) \le \lg(a) + k + c'$. Consider the computer C that does the following when it is given the program $0^k 1 a$. It calculates $F(a, h)$ for $h = 0, 1, 2, \ldots$ until it finds the first theorem in $F(a, h)$ of the form "$I(s) \ge n$" with $n > \lg(a) + 2k$. Finally C outputs the binary string s in the theorem it has found. Thus $C(0^k 1 a)$ is equal to s_k, if s_k exists. It follows that

$$I(s_k) = I(C(0^k 1 a)) \le I_C(C(0^k 1 a)) + c'' \le \lg(0^k 1 a) + c'' = \lg(a) + k + (c'' + 1) = \lg(a) + k + c'.$$

Q.E.D.

Definition 5: A_n is defined to be the kth binary string of length n, where k is the number of programs p of length $< n$ for which $U(p)$ is defined, i.e., A_n has n and this number k coded into it.

Theorem 3: There are rules of inference F^1 such that for all n, $F^1(A_n)$ is the union of the set of all true propositions of the form "$I(s) = k$" with $k < n$ and the set of all true propositions of the form "$I(s) \geq n$."

Proof: From A_n one knows n and for how many programs p of length $< n$ $U(p)$ is defined. One then simulates in parallel, running each program p of length $< n$ on U until one has determined the value of $U(p)$ for each p of length $< n$ for which $U(p)$ is defined. Knowing the value of $U(p)$ for each p of length $< n$ for which $U(p)$ is defined, one easily determines each string of complexity $< n$ and its complexity. What's more, all other strings must be of complexity $\geq n$. This completes our sketch of how all true propositions of the form "$I(s) = k$" with $k < n$ and of the form "$I(s) \geq n$" can be derived from the axiom A_n. Q.E.D.

Recall that we consider the nth binary string to be the natural number n.

Definition 6: The partial function $B(n)$ is defined to be the biggest natural number of complexity $\leq n$, i.e.,

$$B(n) \;=\; \max{}_{I(k) \leq n} k \;=\; \max{}_{\lg(p) \leq n} U(p).$$

Theorem 4: Let f be a partial recursive function that carries natural numbers into natural numbers. Then $B(n) \geq f(n)$ for all sufficiently great values of n.

Proof: Consider the computer C such that $C(p) = f(p)$ for all p.

$$I(f(n)) \;\leq\; I_C(f(n)) + c \;\leq\; \lg(n) + c \;=\; \left[\log_2 (n + 1)\right] + c \;<\; n$$

for all sufficiently great values of n. Thus $B(n) \geq f(n)$ for all sufficiently great values of n. Q.E.D.

Theorem 5: Consider the rules of inference F. Let $F_n = \cup{}_a F(a, B(n))$, where the union is taken over all binary strings a of length $\leq B(n)$, i.e., F_n is the (finite) set of all theorems that can be deduced by means of proofs with not more than $B(n)$ characters from axioms with not more than $B(n)$ bits. Let s_n be the first binary string s not in any proposition of the form "$I(s) = k$" in F_n. Then $I(s_n) \leq n + c$, where the constant c depends only on F.

Proof: We claim that there is a computer C such that if $U(p) = B(n)$, then $C(p) = s_n$. As, by the definition of B, there is a p_0 of length $\leq n$ such that $U(p_0) = B(n)$, it follows that

$$I(s_n) \;\leq\; I_C(s_n) + c \;=\; I_C(C(p_0)) + c \;\leq\; \lg(p_0) + c \;\leq\; n + c,$$

which was to be proved.

It remains to verify the claim that there is a C such that if $U(p) = B(n)$, then $C(p) = s_n$. C works as follows. Given the program p, C first simulates running the program p on U. Once C has determined $U(p)$, it calculates $F(a, U(p))$ for all binary strings a such that $\lg(a) \leq U(p)$, and forms the union of these $2^{U(p)+1} - 1$ different sets of propositions, which is F_n if $U(p) = B(n)$. Finally C outputs the first binary string s not in any proposition of the form "$I(s) = k$" in this set of propositions; s is s_n if $U(p) = B(n)$. Q.E.D.

INFORMATION-THEORETIC COMPUTATIONAL COMPLEXITY

Theorem 6: Consider the rules of inference F. If $F(a, h)$ includes all true propositions of the form "$I(s) = k$" with $k \leq n + c$, then either $\lg(a) > B(n)$ or $h > B(n)$. Here c is a constant that depends only on F.

Proof: This is an immediate consequence of Theorem 5. Q.E.D.

The following theorem gives an upper bound on the size of the proofs in the formal systems $\langle F^1, A_n \rangle$ that were studied in Theorem 3, and also shows that the lower bound on the size of these proofs that is given by Theorem 6 cannot be essentially improved.

Theorem 7: There is a constant c such that for all n $F^1(A_n, B(n + c))$ includes all true propositions of the form "$I(s) = k$" with $k < n$.

Proof: We claim that there is a computer C such that for all n, $C(A_n) = $ the least natural number h such that $F^1(A_n, h)$ includes all true propositions of the form "$I(s) = k$" with $k < n$. Thus the complexity of this value of h is $\leq \lg(A_n) + c = n + c$, and $B(n + c)$ is $\geq$ this value of h, which was to be proved.

It remains to verify the claim. C works as follows when it is given the program A_n. First, it determines each binary string of complexity $< n$ and its complexity, in the manner described in the proof of Theorem 3. Then it calculates $F^1(A_n, h)$ for $h = 0, 1, 2, \ldots$ until all true propositions of the form "$I(s) = k$" with $k < n$ are included in $F^1(A_n, h)$. The final value of h is then output by C. Q.E.D.

References

1. J. van Heijenoort, Ed., *From Frege to Gödel: A Source Book in Mathematical Logic, 1879-1931*. Cambridge, Mass.: Harvard Univ. Press, 1967.

2. M. Davis, Ed., *The Undecidable — Basic Papers on Undecidable Propositions, Unsolvable Problems and Computable Functions*. Hewlett, N.Y.: Raven Press, 1965.

3. J. von Neumann and O. Morgenstern, *Theory of Games and Economic Behavior*. Princeton, N.J.: Princeton Univ. Press, 1944.

4. —, "Method in the physical sciences," in *John von Neumann — Collected Works*. New York: Macmillan, 1963, vol. 6, no. 35.

5. —, *The Computer and the Brain*. New Haven, Conn.: Yale Univ. Press, 1958.

6. —, *Theory of Self-Reproducing Automata*. Urbana, Ill.: Univ. Illinois Press, 1966. (Edited and completed by A. W. Burks.)

7. R. J. Solomonoff, "A formal theory of inductive inference," *Inform. Contr.*, vol. 7, pp. 1-22, Mar. 1964; also, pp. 224-254, June 1964.

8. A. N. Kolmogorov, "Logical basis for information theory and probability theory," *IEEE Trans. Inform. Theory*, vol. IT-14, pp. 662-664, Sept. 1968.

9. G. J. Chaitin, "On the difficulty of computations," *IEEE Trans. Inform. Theory*, vol. IT-16, pp. 5-9, Jan. 1970.

10. —, "To a mathematical definition of 'life'," *ACM SICACT News*, no. 4, pp. 12-18, Jan. 1970.

11. —, "Computational complexity and Gödel's incompleteness theorem," (Abstract) *AMS Notices*, vol. 17, p. 672, June 1970; (Paper) *ACM SIGACT News*, no. 9, pp. 11-12, Apr. 1971.

12. –, "Information-theoretic limitations of formal systems," presented at the Courant Institute Computational Complexity Symp., N.Y., Oct. 1971. A revised version will appear in *J. Ass. Comput. Mach.*

13. M. Kac, *Statistical independence in probability, analysis, and number theory,* Carus Math. Mono., Mathematical Association of America, no. 12, 1959.

14. M. Eigen, "Selforganization of matter and the evolution of biological macromolecules," *Die Naturwissenschaften,* vol. 58, pp. 465-523, Oct. 1971.

15. B. Chandrasekaran and L. H. Reeker, "Artificial intelligence – a case for agnosticism," Ohio State University, Columbus, Ohio, Rep. OSU-CISRC-TR-72-9, Aug. 1972; also, *IEEE Trans. Syst., Man, Cybern.,* vol. SMC-4, pp. 88-94, Jan. 1974.

16. H. Rogers, Jr., *Theory of Recursive Functions and Effective Computability.* New York: McGraw-Hill, 1967.

ALGORITHMIC INFORMATION THEORY

Encyclopedia of Statistical Sciences, Volume 1, Wiley, New York, 1982, pp. 38-41.

The Shannon entropy* concept of classical information theory* [9] is an ensemble notion; it is a measure of the degree of ignorance concerning which possibility holds in an ensemble with a given a priori probability distribution*

$$H(p_1, \ldots, p_n) \equiv - \sum_{k=1}^{n} p_k \log_2 p_k.$$

In algorithmic information theory the primary concept is that of the *information content* of an individual object, which is a measure of how difficult it is to specify or describe how to construct or calculate that object. This notion is also known as *information-theoretic complexity.* For introductory expositions, see refs. 1, 4, and 6. For the necessary background on computability theory and mathematical logic, see refs. 3, 7, and 8. For a more technical survey of algorithmic information theory and a more complete bibliography, see ref. 2. See also ref. 5.

The original formulation of the concept of algorithmic information is independently due to R. J. Solomonoff [22], A. N. Kolmogorov* [19], and G. J. Chaitin [10]. The information content $I(x)$ of a binary string x is defined to be the size in bits (binary digits) of the smallest program for a canonical universal computer U to calculate x. (That the computer U is universal means that for any other computer M there is a prefix μ such that the program μp makes U do exactly the same computation that the program p makes M do.) The *joint information* $I(x, y)$ of two strings is defined to be the size of the smallest program that makes U calculate both of them. And the *conditional* or *relative information* $I(x \mid y)$ of x given y is defined to be the size of the smallest program for U to calculate x from y. The choice of the standard computer U introduces at most an $O(1)$ uncertainty in the numerical value of these concepts. ($O(f)$ is read "order of f" and denotes a function whose absolute value is bounded by a constant times f.)

With the original formulation of these definitions, for most x one has

$$I(x) = |x| + O(1) \tag{1}$$

(here $|x|$ denotes the length or size of the string x, in bits), but unfortunately

$$I(x, y) \le I(x) + I(y) + O(1) \tag{2}$$

holds only if one replaces the $O(1)$ error estimate by $O(\log I(x) I(y))$.

Chaitin [12] and L. A. Levin [20] independently discovered how to reformulate these definitions so that the subadditivity property (2) holds. The change is to require that the set of meaningful computer programs be an instantaneous code, i.e., that no program be a prefix of another. With this modification, (2) now holds, but instead of (1) most x satisfy

$$I(x) = |x| + I(|x|) + O(1)$$
$$= |x| + O(\log |x|).$$

Moreover, in this theory the decomposition of the joint information of two objects into the sum of the information content of the first object added to the relative information of the second one given the first has a different form than in classical information theory. In fact, instead of

$$I(x, y) = I(x) + I(y \mid x) + O(1), \tag{3}$$

one has

$$I(x, y) = I(x) + I(y \mid x, \ I(x)) + O(1). \tag{4}$$

That (3) is false follows from the fact that $I(x, I(x)) = I(x) + O(1)$ and $I(I(x) \mid x)$ is unbounded. This was noted by Chaitin [12] and studied more precisely by Solovay [12, p. 339] and Gac [17].

Two other important concepts of algorithmic information theory are *mutual* or *common information* and *algorithmic independence*. Their importance has been emphasized by Fine [5, p. 141]. The mutual information content of two strings is defined as follows:

$$I(x : y) \equiv I(x) + I(y) - I(x, y).$$

In other words, the mutual information* of two strings is the extent to which it is more economical to calculate them together than to calculate them separately. And x and y are said to be algorithmically independent if their mutual information $I(x : y)$ is essentially zero, i.e., if $I(x, y)$ is approximately equal to $I(x) + I(y)$. Mutual information is symmetrical, i.e., $I(x : y) = I(y : x) + O(1)$. More important, from the decomposition (4) one obtains the following two alternative expressions for mutual information:

$$\begin{aligned} I(x : y) &= I(x) - I(x \mid y, \ I(y)) + O(1) \\ &= I(y) - I(y \mid x, \ I(x)) + O(1). \end{aligned}$$

Thus this notion of mutual information, although is applies to individual objects rather than to ensembles, shares many of the formal properties of the classical version of this concept.

Up until now there have been two principal applications of algorithmic information theory: (a) to provide a new conceptual foundation for probability theory and statistics by making it possible to rigorously define the notion of a *random sequence**, and (b) to provide an information-theoretic approach to metamathematics and the limitative theorems of mathematical logic. A possible application to theoretical mathematical biology is also mentioned below.

A random or patternless binary sequence x_n of length n may be defined to be one of maximal or near-maximal complexity, i.e., one whose complexity $I(x_n)$ is not much less than n. Similarly, an infinite binary sequence x may be defined to be random if its initial segments x_n are all random finite binary sequences. More precisely, x is random if and only if

$$\exists c \forall n [I(x_n) > n - c]. \tag{5}$$

In other words, the infinite sequence x is random if and only if there exists a c such that for all positive integers n, the algorithmic information content of the string consisting of the first n bits of the sequence x, is bounded from below by $n - c$. Similarly, a *random real number* may be defined to be one having the property that the base 2 expansion of its fractional part is a random infinite binary sequence.

These definitions are intended to capture the intuitive notion of a lawless, chaotic, unstructured sequence. Sequences certified as random in this sense would be ideal for use in Monte Carlo* calculations [14], and they would also be ideal as one-time pads for Vernam ciphers or as encryption keys [16]. Unfortunately, as we shall see below, it is a variant of Gödel's famous incompleteness theorem that such certification is impossible. It is a corollary that no pseudo-random number* generator can satisfy these definitions. Indeed, consider a real number x, such as $\sqrt{2}$, π, or e, which has the property that it is possible to compute the successive binary digits of its base 2 expansion. Such x satisfy

$$I(x_n) = I(n) + O(1) = O(\log n)$$

and are therefore maximally nonrandom. Nevertheless, most real numbers are random. In fact, if each bit of an infinite binary sequence is produced by an independent toss of an unbiased coin, then the probability that it will satisfy (5) is 1. We consider next a particularly interesting random real number, Ω, discovered by Chaitin [12, p. 336].

A. M. Turing's theorem that the halting problem is unsolvable is a fundamental result of the theory of algorithms [4]. Turing's theorem states that there is no mechanical procedure for deciding whether or not an arbitrary program p eventually comes to a halt when run on the universal computer U. Let Ω be the probability that the standard computer U eventually halts if each bit of its program p is produced by an independent toss of an unbiased coin. The unsolvability of the halting problem is intimately connected to the fact that the halting probability Ω is a random real number, i.e., its base 2 expansion is a random infinite binary sequence in the very strong sense (5) defined above. From (5) it follows that Ω is normal (a notion due to É. Borel [18]), that Ω is a Kollectiv* with respect to all computable place selection rules (a concept due to R. von Mises and A. Church [15]), and it also follows that Ω satisfies all computable statistical tests of randomness* (this notion being due to P. Martin-Löf [21]). An essay by C. H. Bennett on other remarkable properties of Ω, including its immunity to computable gambling schemes, is contained in ref. 6.

K. Gödel established his famous incompleteness theorem by modifying the paradox of the liar; instead of "This statement is false" he considers "This statement is unprovable." The latter statement is true if and only if it is unprovable; it follows that not all true statements are theorems and thus that any formalization of mathematical logic is incomplete [3, 7, 8]. More relevant to algorithmic information theory is the paradox of "the smallest positive integer that cannot be specified in less than a billion words." The contradiction is that the phrase in quotes has only 14 words, even though at least 1 billion should be necessary. This is a version of the Berry paradox, first published by Russell [7, p. 153]. To obtain a theorem rather than a contradiction, one considers instead "the binary string s which has the shortest proof that its complexity $I(s)$ is greater than 1 billion." The point is that this string s cannot exist. This leads one to the metatheorem that although most bit strings are random and have information content approximately equal to their lengths, it is impossible to prove that a specific string has information content greater than n unless one is using at least n bits of axioms. See ref. 4 for a more complete exposition of this information-theoretic version of Gödel's incompleteness theorem, which was first presented in ref. 11. It can also be shown that n bits of assumptions or postulates are needed to be able to determine the first n bits of the base 2 expansion of the real number Ω.

Finally, it should be pointed out that these concepts are potentially relevant to biology. The algorithmic approach is closer to the intuitive notion of the information content of a biological organism than is the classical ensemble viewpoint, for the role of a computer program and of deoxyribonucleic acid (DNA) are roughly analogous. Reference 13 discusses possible applications of the concept of mutual algorithmic information to theoretical biology; it is suggested that a living organism might be defined as a highly correlated region, one whose parts have high mutual information.

References

General References

1. Chaitin, G. J. (1975). *Sci. Amer.*, **232** (5), 47-52. (An introduction to algorithmic information theory emphasizing the meaning of the basic concepts.)

2. Chaitin, G. J. (1977). *IBM J. Res. Dev.*, **21**, 350-359, 496. (A survey of algorithmic information theory.)

3. Davis, M., ed. (1965). *The Undecidable – Basic Papers on Undecidable Propositions, Unsolvable Problems and Computable Functions*. Raven Press, New York.

4. Davis, M. (1978). In *Mathematics Today: Twelve Informal Essays*. L. A. Steen, ed. Springer-Verlag, New York, pp. 241-267. (An introduction to algorithmic information theory largely devoted to a detailed presentation of the relevant background in computability theory and mathematical logic.)

5. Fine, T. L. (1973). *Theories of Probability: An Examination of Foundations*. Academic Press, New York. (A survey of the remarkably diverse proposals that have been made for formulating probability mathematically. Caution: The material on algorithmic information theory contains some inaccuracies, and it is also somewhat dated as a result of recent rapid progress in this field.)

6. Gardner, M. (1979). *Sci. Amer.*, **241** (5), 20-34. (An introduction to algorithmic information theory emphasizing the fundamental role played by Ω.)

7. Heijenoort, J. van, ed. (1977). *From Frege to Gödel: A Source Book in Mathematical Logic, 1879-1931*. Harvard University Press, Cambridge, Mass. (This book and ref. 3 comprise a stimulating collection of all the classic papers on computability theory and mathematical logic.)

8. Hofstadter, D. R. (1979). *Gödel, Escher, Bach: An Eternal Golden Braid*. Basic Books, New York. (The longest and most lucid introduction to computability theory and mathematical logic.)

9. Shannon, C. E. and Weaver, W. (1949). *The Mathematical Theory of Communication*. University of Illinois Press, Urbana, Ill. (The first and still one of the very best books on classical information theory.)

Additional References

10. Chaitin, G. J. (1966). *J. ACM*, **13**, 547-569; **16**, 145-159 (1969).

11. Chaitin, G. J. (1974). *IEEE Trans. Inf. Theory*, **IT-20**, 10-15.

12. Chaitin, G. J. (1975). *J. ACM*, **22**, 329-340.

13. Chaitin, G. J. (1979). In *The Maximum Entropy Formalism*, R. D. Levine and M. Tribus, eds. MIT Press, Cambridge, Mass., pp. 477-498.

14. Chaitin, G. J. and Schwartz, J. T. (1978). *Commun. Pure Appl. Math.*, **31**, 521-527.

15. Church, A. (1940). *Bull. AMS*, **46**, 130-135.

16. Feistel, H. (1973). *Sci. Amer.*, **228** (5), 15-23.

17. Gac, P. (1974). *Sov. Math. Dokl.*, **15**, 1477-1480.

18. Kac, M. (1959). *Statistical Independence in Probability, Analysis and Number Theory*. Mathematical Association of America, Washington, D.C.

19. Kolmogorov, A. N. (1965). *Problems of Inf. Transmission*, **1**, 1-7.

20. Levin, L. A. (1974). *Problems of Inf. Transmission*, **10**, 206-210.

21. Martin-Löf, P. (1966). *Inf. Control*, **9**, 602-619.

22. Solomonoff, R. J. (1964). *Inf. Control*, **7**, 1-22, 224-254.

(ENTROPY
INFORMATION THEORY
MARTINGALES

MONTE CARLO METHODS
PSEUDO-RANDOM NUMBER GENERATORS
STATISTICAL INDEPENDENCE
TESTS OF RANDOMNESS)

G. J. Chaitin

ALGORITHMIC INFORMATION THEORY

IBM Journal of Research and Development 21 (1977), pp. 350-359, 496.

G. J. Chaitin

Abstract

This paper reviews algorithmic information theory, which is an attempt to apply information-theoretic and probabilistic ideas to recursive function theory. Typical concerns in this approach are, for example, the number of bits of information required to specify an algorithm, or the probability that a program whose bits are chosen by coin flipping produces a given output. During the past few years the definitions of algorithmic information theory have been reformulated. The basic features of the new formalism are presented here and certain results of R. M. Solovay are reported.

Historical Introduction

To our knowledge, the first publication of the ideas of algorithmic information theory was the description of R. J. Solomonoff's ideas given in 1962 by M. L. Minsky in his paper, "Problems of formulation for artificial intelligence" [1]:

> Consider a slightly different form of inductive inference problem. Suppose that we are given a very long "data" sequence of symbols; the problem is to make a prediction about the future of the sequence. This is a problem familiar in discussion concerning "inductive probability." The problem is refreshed a little, perhaps, by introducing the modern notion of universal computer and its associated language of instruction formulas. An instruction sequence will be considered acceptable if it causes the computer to produce a sequence, perhaps infinite, that *begins* with the given finite "data" sequence. Each acceptable instruction sequence thus makes a prediction, and Occam's razor would choose the simplest such sequence and advocate its prediction. (More generally, one could weight the different predictions by weights associated with the simplicities of the instructions.) If the simplicity function is just the length of the instructions, we are then trying to find a minimal description, i.e., an optimally efficient encoding of the data sequence.
>
> Such an induction method could be of interest only if one could show some significant invariance with respect to choice of defining universal machine. There is no such invariance for a fixed pair of data strings. For one could design a machine which would yield the entire first string with a very small input, and the second string only for some very complex input. On the brighter side, one can see that in a sense the induced structure on the space of data strings has some invariance in an "in the large" or "almost everywhere" sense. Given two different universal machines, the induced structures cannot be desperately different. We appeal to the "translation theorem" whereby an arbitrary instruction formula for one machine may be converted into an equivalent instruction formula for the other machine by the addition of a constant prefix text. This text instructs the second machine to simulate the behavior of the first machine in operating on the remainder of the input text. Then for data strings much larger than this translation text (and its inverse) the choice between the two machines cannot greatly affect the induced structure. It would be interesting to see if these intuitive notions could be profitably formalized.

Even if this theory can be worked out, it is likely that it will present overwhelming computational difficulties in application. The recognition problem for minimal descriptions is, in general, unsolvable, and a practical induction machine will have to use heuristic methods. [In this connection it would be interesting to write a program to play R. Abbott's inductive card game [2].]

Algorithmic information theory originated in the independent work of Solomonoff (see [1,3-6]), of A. N. Kolmogorov and P. Martin-Löf (see [7-14]), and of G. J. Chaitin (see [15-26]). Whereas Solomonoff weighted together all the programs for a given result into a probability measure, Kolmogorov and Chaitin concentrated their attention on the size of the smallest program. Recently it has been realized by Chaitin and independently by L. A. Levin that if programs are stipulated to be self-delimiting, these two differing approaches become essentially equivalent. This paper attempts to cast into a unified scheme the recent work in this area by Chaitin [23,24] and by R. M. Solovay [27,28]. The reader may also find it interesting to examine the parallel efforts of Levin (see [29-35]). There has been a substantial amount of other work in this general area, often involving variants of the definitions deemed more suitable for particular applications (see, e.g., [36-47]).

Algorithmic Information Theory of Finite Computations [23]

Definitions

Let us start by considering a class of Turing machines with the following characteristics. Each Turing machine has three tapes: a program tape, a work tape, and an output tape. There is a scanning head on each of the three tapes. The program tape is read-only and each of its squares contains a 0 or a 1. It may be shifted in only one direction. The work tape may be shifted in either direction and may be read and erased, and each of its squares contains a blank, a 0, or a 1. The work tape is initially blank. The output tape may be shifted in only one direction. Its squares are initially blank, and may have a 0, a 1, or a comma written on them, and cannot be rewritten. Each Turing machine of this type has a finite number n of states, and is defined by an $n \times 3$ table, which gives the action to be performed and the next state as a function of the current state and the contents of the square of the work tape that is currently being scanned. The first state in this table is by convention the initial state. There are eleven possible actions: halt, shift work tape left/right, write blank/0/1 on work tape, read square of program tape currently being scanned and copy onto square of work tape currently being scanned and then shift program tape, write 0/1/comma on output tape and then shift output tape, and consult oracle. The oracle is included for the purpose of defining relative concepts. It enables the Turing machine to choose between two possible state transitions, depending on whether or not the binary string currently being scanned on the work tape is in a certain set, which for now we shall take to be the null set.

From each Turing machine M of this type we define a *probability* P, an *entropy* H, and a *complexity* I. $P(s)$ is the probability that M eventually halts with the string s written on its output tape if each square of the program tape is filled with a 0 or a 1 by a separate toss of an unbiased coin. By "string" we shall always mean a finite binary string. From the probability $P(s)$ we obtain the entropy $H(s)$ by taking the negative base-two logarithm, i.e., $H(s)$ is $-\log_2 P(s)$. A string p is said to be a *program* if when it is written on M's program tape and M starts computing scanning the first bit of p, then M eventually halts after reading all of p and without reading any other squares of the tape. A program p is said to be a *minimal program* if no other program makes M produce the same output and has a smaller size. And finally the complexity $I(s)$ is defined to be the least n such that for some contents of its program tape M eventually halts with s written on the output tape after reading precisely n squares of the program tape; i.e., $I(s)$ is the size of a minimal program for s. To summarize,

P is the probability that M calculates s given a random program, H is $-\log_2 P$, and I is the minimum number of bits required to specify an algorithm for M to calculate s.

It is important to note that blanks are not allowed on the program tape, which is imagined to be entirely filled with 0's and 1's. Thus programs are not followed by endmarker blanks. This forces them to be self-delimiting; a program must indicate within itself what size it has. Thus no program can be a prefix of another one, and the programs for M form what is known as a prefix-free set or an instantaneous code. This has two very important effects: It enables a natural probability distribution to be defined on the set of programs, and it makes it possible for programs to be built up from subroutines by concatenation. Both of these desirable features are lost if blanks are used as program endmarkers. This occurs because there is no natural probability distribution on programs with endmarkers; one, of course, makes all programs of the same size equiprobable, but it is also necessary to specify in some arbitrary manner the probability of each particular size. Moreover, if two subroutines with blanks as endmarkers are concatenated, it is necessary to include additional information indicating where the first one ends and the second one begins.

Here is an example of a specific Turing machine M of the above type. M counts the number n of 0's up to the first 1 it encounters on its program tape, then transcribes the next n bits of the program tape onto the output tape, and finally halts. So M outputs s iff it finds length(s) 0's followed by a 1 followed by s on its program tape. Thus $P(s) = \exp_2(-2 \text{ length}(s) - 1)$, $H(s) = 2 \text{ length}(s) + 1$, and $I(s) = 2 \text{ length}(s) + 1$. Here $\exp_2(x)$ is the base-two exponential function 2^x. Clearly this is a very special-purpose computer which embodies a very limited class of algorithms and yields uninteresting functions P, H, and I.

On the other hand it is easy to see that there are "general-purpose" Turing machines that maximize P and minimize H and I; in fact, consider those universal Turing machines which will simulate an arbitrary Turing machine if a suitable prefix indicating the machine to simulate is added to its programs. Such Turing machines yield essentially the same P, H, and I. We therefore pick, somewhat arbitrarily, a particular one of these, U, and the definitive definition of P, H, and I is given in terms of it. The universal Turing machine U works as follows. If U finds i 0's followed by a 1 on its program tape, it simulates the computation that the ith Turing machine of the above type performs upon reading the remainder of the program tape. By the i th Turing machine we mean the one that comes ith in a list of all possible defining tables in which the tables are ordered by size (i.e., number of states) and lexicographically among those of the same size. With this choice of Turing machine, P, H, and I can be dignified with the following titles: $P(s)$ is the *algorithmic probability* of s, $H(s)$ is the *algorithmic entropy* of s, and $I(s)$ is the *algorithmic information* of s. Following Solomonoff [3], $P(s)$ and $H(s)$ may also be called the a priori probability and entropy of s. $I(s)$ may also be termed the descriptive, program-size, or information-theoretic complexity of s. And since P is maximal and H and I are minimal, the above choice of special-purpose Turing machine shows that $P(s) \geq \exp_2(-2 \text{ length}(s) - O(1))$, $H(s) \leq 2 \text{ length}(s) + O(1)$, and $I(s) \leq 2 \text{ length}(s) + O(1)$.

We have defined $P(s)$, $H(s)$, and $I(s)$ for individual strings s. It is also convenient to consider computations which produce finite sequences of strings. These are separated by commas on the output tape. One thus defines the joint probability $P(s_1, \ldots, s_n)$, the joint entropy $H(s_1, \ldots, s_n)$, and the joint complexity $I(s_1, \ldots, s_n)$ of an n-tuple $s_1, \ldots, s_n$. Finally one defines the conditional probability $P(t_1, \ldots, t_m | s_1, \ldots, s_n)$ of the m-tuple $t_1, \ldots, t_m$ given the n-tuple $s_1, \ldots, s_n$ to be the quotient of the joint probability of the n-tuple and the m-tuple divided by the joint probability of the n-tuple. In particular $P(t | s)$ is defined to be $P(s, t)/P(s)$. And of course the conditional entropy is defined to be the negative base-two logarithm of the conditional probability. Thus by definition $H(s, t) = H(s) + H(t | s)$. Finally, in order to extend the above definitions to tuples whose members may either be strings or natural numbers, we identify the natural number n with its binary expansion.

ALGORITHMIC INFORMATION THEORY

Basic Relationships

We now review some basic properties of these concepts. The relation

$$H(s, t) = H(t, s) + O(1)$$

states that the probability of computing the pair s, t is essentially the same as the probability of computing the pair t, s. This is true because there is a prefix that converts any program for one of these pairs into a program for the other one. The inequality

$$H(s) \leq H(s, t) + O(1)$$

states that the probability of computing s is not less than the probability of computing the pair s, t. This is true because a program for s can be obtained from any program for the pair s, t by adding a fixed prefix to it. The inequality

$$H(s, t) \leq H(s) + H(t) + O(1)$$

states that the probability of computing the pair s, t is not less than the product of the probabilities of computing s and t, and follows from the fact that programs are self-delimiting and can be concatenated. The inequality

$$O(1) \leq H(t \mid s) \leq H(t) + O(1)$$

is merely a restatement of the previous two properties. However, in view of the direct relationship between conditional entropy and relative complexity indicated below, this inequality also states that being told something by an oracle cannot make it more difficult to obtain t. The relationship between entropy and complexity is

$$H(s) = I(s) + O(1),$$

i.e., the probability of computing s is essentially the same as $1/\exp_2$(the size of a minimal program for s). This implies that a significant fraction of the probability of computing s is contributed by its minimal programs, and that there are few minimal or near-minimal programs for a given result. The relationship between conditional entropy and relative complexity is

$$H(t \mid s) = I_s(t) + O(1).$$

Here $I_s(t)$ denotes the complexity of t relative to a set having a single element which is a minimal program for s. In other words,

$$I(s, t) = I(s) + I_s(t) + O(1).$$

This relation states that one obtains what is essentially a minimal program for the pair s, t by concatenating the following two subroutines:

- a minimal program for s
- a minimal program for calculating t using an oracle for the set consisting of a minimal program for s.

Algorithmic Randomness

Consider an arbitrary string s of length n. From the fact that

$$H(n) + H(s \mid n) = H(n, s) = H(s) + O(1),$$

it is easy to show that $H(s) \leq n + H(n) + O(1)$, and that less than $\exp_2(n - k + O(1))$ of the s of length n satisfy $H(s) < n + H(n) - k$. It follows that for most s of length n, $H(s)$ is approximately equal to $n + H(n)$. These are the most complex strings of length n, the ones which are most difficult to specify, the ones with highest entropy, and they are said to be the *algorithmically random* strings of length n. Thus a typical string s of length n will have $H(s)$ close to $n + H(n)$, whereas if s has pattern or can be distinguished in some fashion, then it can be compressed or coded into a program that is considerably smaller. That $H(s)$ is usually $n + H(n)$ can be thought of as follows: In order to specify a typical strings s of length n, it is necessary to first specify its size n, which requires $H(n)$ bits, and it is necessary then to specify each of the n bits in s, which requires n more bits and brings the total to $n + H(n)$. In probabilistic terms this can be stated as follows: the sum of the probabilities of all the strings of length n is essentially equal to $P(n)$, and most strings s of length n have probability $P(s)$ essentially equal to $P(n)/2^n$. On the other hand, one of the strings of length n that is least random and that has most pattern is the string consisting entirely of 0's. It is easy to see that this string has entropy $H(n) + O(1)$ and probability essentially equal to $P(n)$, which is another way of saying that almost all the information in it is in its length. Here is an example in the middle: If p is a minimal program of size n, then it is easy to see that $H(p) = n + O(1)$ and $P(p)$ is essentially 2^{-n}. Finally it should be pointed out that since $H(s) = H(n) + H(s \mid n) + O(1)$ if s is of length n, the above definition of randomness is equivalent to saying that the most random strings of length n have $H(s \mid n)$ close to n, while the least random ones have $H(s \mid n)$ close to 0.

Later we shall show that even though most strings are algorithmically random, i.e., have nearly as much entropy as possible, an inherent limitation of formal axiomatic theories is that a lower bound n on the entropy of a specific string can be established only if n is less than the entropy of the axioms of the formal theory. In other words, it is possible to prove that a specific object is of complexity greater than n only if n is less than the complexity of the axioms being employed in the demonstration. These statements may be considered to be an information-theoretic version of Gödel's famous incompleteness theorem.

Now let us turn from finite random strings to infinite ones, or equivalently, by invoking the correspondence between a real number and its dyadic expansion, to random reals. Consider an infinite string X obtained by flipping an unbiased coin, or equivalently a real x uniformly distributed in the unit interval. From the preceding considerations and the Borel-Cantelli lemma it is easy to see that with probability one there is a c such that $H(X_n) > n - c$ for all n, where X_n denotes the first n bits of X, that is, the first n bits of the dyadic expansion of x. We take this property to be our definition of an algorithmically random infinite string X or real x.

Algorithmic randomness is a clear-cut property for infinite strings, but in the case of finite strings it is a matter of degree. If a cutoff were to be chosen, however, it would be well to place it at about the point at which $H(s)$ is equal to length(s). Then an infinite random string could be defined to be one for which all initial segments are finite random strings, within a certain tolerance.

Now consider the real number Ω defined as the halting probability of the universal Turing machine U that we used to define P, H, and I; i.e., Ω is the probability that U eventually halts if each square of its program tape is filled with a 0 or a 1 by a separate toss of an unbiased coin. Then it is not difficult to see that Ω is in fact an algorithmically random real, because if one were given the first n bits of the dyadic expansion of Ω, then one could use this to tell whether each program for U of size less than n ever halts or not. In other words, when written in binary the probability of halting Ω is a random or incompressible infinite string. Thus the basic theorem of recursive function theory that the halting problem is unsolvable corresponds in algorithmic information theory to the theorem that the probability of halting is algorithmically random if the program is chosen by coin flipping.

This concludes our review of the most basic facts regarding the probability, entropy, and complexity of finite objects, namely strings and tuples of strings. Before presenting some of Solovay's remarkable results regarding these concepts, and in particular regarding Ω, we would like to review the most important facts which are known regarding the probability, entropy, and complexity of infinite objects, namely recursively enumerable sets of strings.

Algorithmic Information Theory of Infinite Computations [24]

In order to define the probability, entropy, and complexity of r.e. (recursively enumerable) sets of strings it is necessary to consider unending computations performed on our standard universal Turing machine U. A computation is said to produce an r.e. set of strings A if all the members of A and only members of A are eventually written on the output tape, each followed by a comma. It is important that U not be required to halt if A is finite. The members of the set A may be written in arbitrary order, and duplications are ignored. A technical point: If there are only finitely many strings written on the output tape, and the last one is infinite or is not followed by a comma, then it is considered to be an "unfinished" string and is also ignored. Note that since computations may be endless, it is now possible for a semi-infinite portion of the program tape to be read.

The definitions of the probability $P(A)$, the entropy $H(A)$, and the complexity $I(A)$ of an r.e. set of strings A may now be given. $P(A)$ is the probability that U produces the output set A if each square of its program tape is filled with a 0 or a 1 by a separate toss of an unbiased coin. $H(A)$ is the negative base-two logarithm of $P(A)$. And $I(A)$ is the size in bits of a minimal program that produces the output set A, i.e., $I(A)$ is the least n such that there is a program tape contents that makes U undertake a computation in the course of which it reads precisely n squares of the program tape and produces the set of strings A. In order to define the joint and conditional probability and entropy we need a mechanism for encoding two r.e. sets A and B into a single set A join B. To obtain A join B one prefixes each string in A with a 0 and each string in B with a 1 and takes the union of the two resulting sets. Enumerating A join B is equivalent to simultaneously enumerating A and B. So the joint probability $P(A, B)$ is $P(A$ join $B)$, the joint entropy $H(A, B)$ is $H(A$ join $B)$, and the joint complexity $I(A, B)$ is $I(A$ join $B)$. These definitions can obviously be extended to more than two r.e. sets, but it is unnecessary to do so here. Lastly, the conditional probability $P(B \mid A)$ of B given A is the quotient of $P(A, B)$ divided by $P(A)$, and the conditional entropy $H(B \mid A)$ is the negative base-two logarithm of $P(B \mid A)$. Thus by definition $H(A, B) = H(A) + H(B \mid A)$.

As before, one obtains the following basic inequalities:

1. $H(A, B) \;=\; H(B, A) + O(1)$,

2. $H(A) \;\leq\; H(A, B) + O(1)$,

3. $H(A, B) \;\leq\; H(A) + H(B) + O(1)$,

4. $O(1) \;\leq\; H(B \mid A) \;\leq\; H(B) + O(1)$,

5. $I(A, B) \;\leq\; I(A) + I(B) + O(1)$.

In order to demonstrate the third and the fifth of these relations one imagines two unending computations to be occurring simultaneously. Then one interleaves the bits of the two programs in the order in which they are read. Putting a fixed size prefix in front of this, one obtains a single program for performing both computations simultaneously whose size is $O(1)$ plus the sum of the sizes of the original programs.

So far things look much as they did for individual strings. But the relationship between entropy and complexity turns out to be more complicated for r.e. sets than it was in the case of individual strings. Obviously the entropy $H(A)$ is always less than or equal to the complexity $I(A)$, because of the probability contributed by each minimal program for A:

- $H(A) \le I(A)$.

But how about bounds on $I(A)$ in terms of $H(A)$? First of all, it is easy to see that if A is a singleton set whose only member is the string s, then $H(A) = H(s) + O(1)$ and $I(A) = I(s) + O(1)$. Thus the theory of the algorithmic information of individual strings is contained in the theory of the algorithmic information of r.e. sets as the special case of sets having a single element:

- For singleton A, $I(A) = H(A) + O(1)$.

There is also a close but not an exact relationship between H and I in the case of sets consisting of initial segments of the set of natural numbers (recall we identify the natural number n with its binary representation). Let us use the adjective "initial" for any set consisting of all natural numbers less than a given one:

- For initial A, $I(A) = H(A) + O(\log H(A))$.

Moreover, it is possible to show that there are infinitely many initial sets A for which $I(A) > H(A) + O(\log H(A))$. This is the greatest known discrepancy between I and H for r.e. sets. It is demonstrated by showing that occasionally the number of initial sets A with $H(A) < n$ is appreciably greater than the number of initial sets A with $I(A) < n$. On the other hand, with the aid of a crucial game-theoretic lemma of D. A. Martin, Solovay [28] has shown that

- $I(A) \le 3 H(A) + O(\log H(A))$.

These are the best results currently known regarding the relationship between the entropy and the complexity of an r.e. set; clearly much remains to be done. Furthermore, what is the relationship between the conditional entropy and the relative complexity of r.e. sets? And how many minimal or near-minimal programs for an r.e. set are there?

We would now like to mention some other results concerning these concepts. Solovay has shown that:

There are $\exp_2(n - H(n) + O(1))$ singleton sets A with $H(A) < n$,
There are $\exp_2(n - H(n) + O(1))$ singleton sets A with $I(A) < n$.

We have extended Solovay's result as follows:

There are $\exp_2(n - H'(n) + O(1))$ finite sets A with $H(A) < n$,
There are $\exp_2(n - H(L_n) + O(\log H(L_n)))$ sets A with $I(A) < n$,
There are $\exp_2(n - H'(L_n) + O(\log H'(L_n)))$ sets A with $H(A) < n$.

Here L_n is the set of natural numbers less than n, and H' is the entropy relative to the halting problem; if U is provided with an oracle for the halting problem instead of one for the null set, then the probability, entropy, and complexity measures one obtains are P', H', and I' instead of P, H, and I. Two final results:

- $I'(A$, the complement of $A) \le H(A) + O(1)$;
- the probability that the complement of an r.e. set has cardinality n is essentially equal to the probability that a set r.e. in the halting problem has cardinality n.

More Advanced Results [27]

The previous sections outline the basic features of the new formalism for algorithmic information theory obtained by stipulating that programs be self-delimiting instead of having endmarker blanks. Error terms in the basic relations which were logarithmic in the previous approach [9] are now of the order of unity.

In the previous approach the complexity of n is usually $\log_2 n + O(1)$, there is an information-theoretic characterization of recursive infinite strings [25,26], and much is known about complexity oscillations in random infinite strings [14]. The corresponding properties in the new approach have been elucidated by Solovay in an unpublished paper [27]. We present some of his results here. For related work by Solovay, see the publications [28,48,49].

Recursive Bounds on H(n)

Following [23, p. 337], let us consider recursive upper and lower bounds on $H(n)$. Let f be an unbounded recursive function, and consider the series $\Sigma \exp_2(-f(n))$ summed over all natural numbers n. If this infinite series converges, then $H(n) < f(n) + O(1)$ for all n. And if it diverges, then the inequalities $H(n) > f(n)$ and $H(n) < f(n)$ each hold for infinitely many n. Thus, for example, for any $\varepsilon > 0$,

$$H(n) \; < \; \log n + \log\log n + (1 + \varepsilon)\log\log\log n + O(1)$$

for all n, and

$$H(n) \; > \; \log n + \log\log n + \log\log\log n$$

for infinitely many n, where all logarithms are base two. See [50] for the results on convergence used to prove this.

Solovay has obtained the following results regarding recursive upper bounds on H, i.e., recursive h such that $H(n) < h(n)$ for all n. First he shows that there is a recursive upper bound on H which is almost correct infinitely often, i.e., $|H(n) - h(n)| < c$ for infinitely many values of n. In fact, the lim sup of the fraction of values of i less than n such that $|H(i) - h(i)| < c$ is greater than 0. However, he also shows that the values of n for which $|H(n) - h(n)| < c$ must in a certain sense be quite sparse. In fact, he establishes that if h is any recursive upper bound on H then there cannot exist a tolerance c and a recursive function f such that there are always at least n different natural numbers i less than $f(n)$ at which $h(i)$ is within c of $H(i)$. It follows that the lim inf of the fraction of values of i less than n such that $|H(i) - h(i)| < c$ is zero.

The basic idea behind his construction of h is to choose f so that $\Sigma \exp_2(-f(n))$ converges "as slowly" as possible. As a by-product he obtains a recursive convergent series of rational numbers Σa_n such that if Σb_n is any recursive convergent series of rational numbers, then lim sup a_n/b_n is greater than zero.

Nonrecursive Infinite Strings with Simple Initial Segments

At the high-order end of the complexity scale for infinite strings are the random strings, and the recursive strings are at the low order end. Is anything else there? More formally, let X be an infinite binary string, and let X_n be the first n bits of X. If X is recursive, then we have $H(X_n) = H(n) + O(1)$. What about the converse, i.e., what can be said about X given only that $H(X_n) = H(n) + O(1)$? Obviously

$$H(X_n) \; = \; H(n, X_n) + O(1) \; = \; H(n) + H(X_n \mid n) + O(1).$$

So $H(X_n) = H(n) + O(1)$ iff $H(X_n \mid n) = O(1)$. Then using a relativized version of the proof in [37, pp. 525-526], one can show that X is recursive in the halting problem. Moreover, by using a priority argument Solovay is actually able to construct a nonrecursive X that satisfies $H(X_n) = H(n) + O(1)$.

Equivalent Definitions of an Algorithmically Random Real

Pick a recursive enumeration O_0, O_1, O_2, ... of all open intervals with rational endpoints. A sequence of open sets U_0, U_1, U_2, ... is said to be *simultaneously r.e.* if there is a recursive function h such that U_n is the union of those O_i whose index i is of the form $h(n,j)$, for some natural number j. Consider a real number x in the unit interval. We say that x has the Solovay randomness property if the following holds. Let U_0, U_1, U_2, ... be any simultaneously r.e. sequence of open sets such that the sum of the usual Lebesgue measure of the U_n converges. Then x is in only finitely many of the U_n. We say that x has the Chaitin randomness property if there is a c such that $H(X_n) > n - c$ for all n, where X_n is the string consisting of the first n bits of the dyadic expansion of x. Solovay has shown that these randomness properties are equivalent to each other, and that they are also equivalent to Martin-Löf's definition [10] of randomness.

The Entropy of Initial Segments of Algorithmically Random and of Ω-like Reals

Consider a random real x. By the definition of randomness, $H(X_n) > n + O(1)$. On the other hand, for any infinite string X, random or not, we have $H(X_n) \leq n + H(n) + O(1)$. Solovay shows that the above bounds are each sometimes sharp. More precisely, consider a random X and a recursive function f such that $\Sigma \exp_2(-f(n))$ diverges (e.g., $f(n) = $ integer part of $\log_2 n$). Then there are infinitely many natural numbers n such that $H(X_n) \geq n + f(n)$. And consider an unbounded monotone increasing recursive function g (e.g., $g(n) = $ integer part of $\log\log n$). There infinitely many natural numbers n such that it is simultaneously the case that $H(X_n) \leq n + g(n)$ and $H(n) \geq f(n)$.

Solovay has obtained much more precise results along these lines about Ω and a class of reals which he calls "Ω-like." A real number is said to be an *r.e. real* if the set of all rational numbers less than it is an r.e. subset of the rational numbers. Roughly speaking, an r.e. real x is Ω-like if for any r.e. real y one can get in an effective manner a good approximation to y from any good approximation to x, and the quality of the approximation to y is at most $O(1)$ binary digits worse than the quality of the approximation to x. The formal definition of Ω-like is as follows. The real x is said to *dominate* the real y if there is a partial recursive function f and a constant c with the property that if q is any rational number that is less than x, then $f(q)$ is defined and is a rational number that is less than y and satisfies the inequality $c |x - q| \geq |y - f(q)|$. And a real number is said to be Ω-like if it is an r.e. real that dominates all r.e. reals. Solovay proves that Ω is in fact Ω-like, and that if x and y are Ω-like, then $H(X_n) = H(Y_n) + O(1)$, where X_n and Y_n are the first n bits in the dyadic expansions of x and y. It is an immediate corollary that if x is Ω-like then $H(X_n) = H(\Omega_n) + O(1)$, and that all Ω-like reals are algorithmically random. Moreover Solovay shows that the algorithmic probability $P(s)$ of any string s is always an Ω-like real.

In order to state Solovay's results contrasting the behavior of $H(\Omega_n)$ with that of $H(X_n)$ for a typical real number x, it is necessary to define two extremely slowly growing monotone functions α and α'. $\alpha(n) = \min H(j)$ $(j \geq n)$, and α' is defined in the same manner as α except that H is replaced by H', the algorithmic entropy relative to the halting problem. It can be shown (see [29, pp. 90-91]) that α goes to infinity, but more slowly than any monotone partial recursive function does. More precisely, if f is an unbounded nondecreasing partial recursive function, then $\alpha(n)$ is less than $f(n)$ for almost all n for which $f(n)$ is defined. Similarly α' goes to infinity, but more slowly than any monotone partial function recursive in α does. More precisely, if f is an unbounded nondecreasing partial function recursive in the halting problem, then $\alpha'(n)$ is less than $f(n)$ for almost all n for which $f(n)$ is defined. In particular, $\alpha'(n)$ is less than $\alpha(\alpha(n))$ for almost all n.

We can now state Solovay's results. Consider a real number x uniformly distributed in the unit interval. With probability one there is a c such that $H(X_n) > n + H(n) - c$ holds for infinitely many

n. And with probability one, $H(X_n) > n + \alpha(n) + O(\log \alpha(n))$. Whereas if x is Ω-like, then the following occurs:

$$H(X_n) < n + H(n) - \alpha(n) + O(\log \alpha(n)),$$

and for infinitely many n we have

$$H(X_n) < n + \alpha'(n) + O(\log \alpha'(n)).$$

This shows that the complexity of initial segments of the dyadic expansions of Ω-like reals is atypical. It is an open question whether $H(\Omega_n) - n$ tends to infinity; Solovay suspects that it does.

Algorithmic Information Theory and Metamathematics

There is something paradoxical about being able to prove that a specific finite string is random; this is perhaps captured in the following antinomies from the writings of M. Gardner [51] and B. Russell [52]. In reading them one should interpret "dull," "uninteresting," and "indefinable" to mean "random," and "interesting" and "definable" to mean "nonrandom."

> [Natural] numbers can of course be interesting in a variety of ways. The number 30 was interesting to George Moore when he wrote his famous tribute to "the woman of 30," the age at which he believed a married woman was most fascinating. To a number theorist 30 is more likely to be exciting because it is the largest integer such that all smaller integers with which it has no common divisor are prime numbers... The question arises: Are there any uninteresting numbers? We can prove that there are none by the following simple steps. If there are dull numbers, we can then divide all numbers into two sets – interesting and dull. In the set of dull numbers there will be only one number that is smallest. Since it is the smallest uninteresting number it becomes, *ipso facto*, an interesting number. [Hence there are no dull numbers!] [51]

> Among transfinite ordinals some can be defined, while others cannot; for the total number of possible definitions is $\aleph_0$, while the number of transfinite ordinals exceeds $\aleph_0$. Hence there must be indefinable ordinals, and among these there must be a least. But this is defined as "the least indefinable ordinal," which is a contradiction. [52]

Here is our incompleteness theorem for formal axiomatic theories whose arithmetic consequences are true. The setup is as follows: The axioms are a finite string, the rules of inference are an algorithm for enumerating the theorems given the axioms, and we fix the rules of inference and vary the axioms. Within such a formal theory a specific string cannot be proven to be of entropy more than $O(1)$ greater than the entropy of the axioms of the theory. Conversely, there are formal theories whose axioms have entropy $n + O(1)$ in which it is possible to establish all true propositions of the form "H(specific string) $\geq n$."

Proof: Consider the enumeration of the theorems of the formal axiomatic theory in order of the size of their proofs. For each natural number k, let s^* be the string in the theorem of the form "$H(s) \geq n$" with n greater than H(axioms) $+ k$ which appears first in the enumeration. On the one hand, if all theorems are true, then $H(s^*) > H$(axioms) $+ k$. On the other hand, the above prescription for calculating s^* shows that $H(s^*) \leq H$(axioms) $+ H(k) + O(1)$. It follows that $k < H(k) + O(1)$. However, this inequality is false for all $k \geq k^*$, where k^* depends only on the rules of inference. The apparent contradiction is avoided only if s^* does not exist for $k = k^*$, i.e., only if it is impossible to prove in the formal theory that a specific string has H greater than H(axioms) $+ k^*$.

Proof of Converse: The set T of all true propositions of the form "$H(s) < k$" is r.e. Choose a fixed enumeration of T without repetitions, and for each natural number n let s^* be the string in the last proposition of the form "$H(s) < n$" in the enumeration. It is not difficult to see that $H(s^*, n) = n + O(1)$. Let p be a minimal program for the pair s^*, n. Then p is the desired axiom, for $H(p) = n + O(1)$ and to obtain all true propositions of the form "$H(s) \geq n$" from p one enumerates T until all s with $H(s) < n$ have been discovered. All other s have $H(s) \geq n$.

We developed this information-theoretic approach to metamathematics before being in possession of the notion of self-delimiting programs (see [20-22] and also [53]); the technical details are somewhat different when programs have blanks as endmarkers. The conclusion to be drawn from all this is that even though most strings are random, we will never be able to explicitly exhibit a string of reasonable size which demonstrably possess this property. A less pessimistic conclusion to be drawn is that it is reasonable to measure the power of formal axiomatic theories in information-theoretic terms. The fact that in some sense one gets out of a formal theory no more than one puts in should not be taken too seriously: a formal theory is at its best when a great many apparently independent theorems are shown to be closely interrelated by reducing them to a handful of axioms. In this sense a formal axiomatic theory is valuable for the same reason as a scientific theory; in both cases information is being compressed, and one is also concerned with the trade-off between the degree of compression and the length of proofs of interesting theorems or the time required to compute predictions.

Algorithmic Information Theory and Biology

Above we have pointed out a number of open problems. In our opinion, however, the most important challenge is to see if the ideas of algorithmic information theory can contribute in some form or manner to theoretical mathematical biology in the style of von Neumann [54], in which genetic information is considered to be an extremely large and complicated program for constructing organisms. We alluded briefly to this in a previous paper [21], and discussed it at greater length in a publication [19] of somewhat limited access.

Von Neumann wished to isolate the basic conceptual problems of biology from the detailed physics and biochemistry of life as we know it. The gist of his message is that it should be possible to formulate mathematically and to answer in a quite general setting such fundamental questions as "How is self-reproduction possible?", "What is an organism?", "What is its degree of organization?", and "How probable is evolution?". He achieved this for the first question; he showed that exact self-reproduction of universal Turing machines is possible in a particular deterministic model universe.

There is such an enormous difference between dead and organized living matter that it must be possible to give a quantitative structural characterization of this difference, i.e., of degree of organization. One possibility [19] is to characterize an organism as a highly interdependent region, one for which the complexity of the whole is much less than the sum of the complexities of its parts. C. H. Bennett [55] has suggested another approach based on the notion of "logical depth." A structure is deep "if it is superficially random but subtly redundant, in other words, if almost all its algorithmic probability is contributed by slow-running programs. A string's logical depth should reflect the amount of computational work required to expose its buried redundancy." It is Bennett's thesis that "a priori the most probable explanation of 'organized information' such as the sequence of bases in a naturally occurring DNA molecule is that it is the product of an extremely long evolutionary process." For related work by Bennett, see [56].

This, then, is the fundamental problem of theoretical biology that we hope the ideas of algorithmic information theory may help to solve: to set up a nondeterministic model universe, to formally define what it means for a region of space-time in that universe to be an organism and what is its degree of organization, and to rigorously demonstrate that, starting from simple initial conditions, organisms will appear and evolve in degree of organization in a reasonable amount of time and with high probability.

Acknowledgments

The quotation by M. L. Minsky in the first section is reprinted with the kind permission of the publisher American Mathematical Society from *Mathematical Problems in the Biological Sciences, Proceedings.of Symposia in Applied Mathematics XIV*, pp. 42-43, copyright © 1962. We are grateful to R. M. Solovay for permitting us to include several of his unpublished results in the section entitled "More advanced results." The quotation by M. Gardner in the section on algorithmic information theory and metamathematics is reprinted with his kind permission, and the quotation by B. Russell in that section is reprinted with permission of the Johns Hopkins University Press. We are grateful to C. H. Bennett for permitting us to present his notion of logical depth in print for the first time in the section on algorithmic information theory and biology.

References

1. M. L. Minsky, "Problems of Formulation for Artificial Intelligence," *Mathematical Problems in the Biological Sciences, Proceedings of Symposia in Applied Mathematics XIV*, R. E. Bellman, ed., American Mathematical Society, Providence, RI, p. 35.

2. M. Gardner, "An Inductive Card Game," *Sci. Amer.* **200**, No. 6, 160 (1959).

3. R. J. Solomonoff, "A Formal Theory of Inductive Inference," *Info. Control* **7**, 1, 224 (1964).

4. D. G. Willis, "Computational Complexity and Probability Constructions," *J. ACM* **17**, 241 (1970).

5. T. M. Cover, "Universal Gambling Schemes and the Complexity Measures of Kolmogorov and Chaitin," *Statistics Department Report 12,* Stanford University, CA, October, 1974.

6. R. J. Solomonoff, "Complexity Based Induction Systems: Comparisons and Convergence Theorems," *Report RR-329,* Rockford Research, Cambridge, MA, August, 1976.

7. A. N. Kolmogorov, "On Tables of Random Numbers," *Sankhya* **A25**, 369 (1963).

8. A. N. Kolmogorov, "Three Approaches to the Quantitative Definition of Information," *Prob. Info. Transmission* **1**, No. 1, 1 (1965).

9. A. N. Kolmogorov, "Logical Basis for Information Theory and Probability Theory," *IEEE Trans. Info. Theor.* **IT-14**, 662 (1968).

10. P. Martin-Löf, "The Definition of Random Sequences," *Info. Control* **9**, 602 (1966).

11. P. Martin-Löf, "Algorithms and Randomness," *Intl. Stat. Rev.* **37**, 265 (1969).

12. P. Martin-Löf, "The Literature on von Mises' Kollektivs Revisited," *Theoria* **35**, Part 1, 12 (1969).

13. P. Martin-Löf, "On the Notion of Randomness," *Intuitionism and Proof Theory*, A. Kino, J. Myhill, and R. E. Vesley, eds., North-Holland Publishing Co., Amsterdam, 1970, p. 73.

14. P. Martin-Löf, "Complexity Oscillations in Infinite Binary Sequences," *Z. Wahrscheinlichk. verwand. Geb.* **19**, 225 (1971).

15. G. J. Chaitin, "On the Length of Programs for Computing Finite Binary Sequences," *J. ACM* **13**, 547 (1966).

16. G. J. Chaitin, "On the Length of Programs for Computing Finite Binary Sequences: Statistical Considerations," *J. ACM* **16**, 145 (1969).

17. G. J. Chaitin, "On the Simplicity and Speed of Programs for Computing Infinite Sets of Natural Numbers," *J. ACM* **16**, 407 (1969).

18. G. J. Chaitin, "On the Difficulty of Computations," *IEEE Trans. Info. Theor.* **IT-16**, 5 (1970).

19. G. J. Chaitin, "To a Mathematical Definition of 'Life'," *ACM SICACT News* **4**, 12 (1970).

20. G. J. Chaitin, "Information-theoretic Limitations of Formal Systems," *J. ACM* **21**, 403 (1974).

21. G. J. Chaitin, "Information-theoretic Computational Complexity," *IEEE Trans. Info. Theor.* **IT-20**, 10 (1974).

22. G. J. Chaitin, "Randomness and Mathematical Proof," *Sci. Amer.* **232**, No. 5, 47 (1975). (Also published in the Japanese and Italian editions of *Sci. Amer.*)

23. G. J. Chaitin, "A Theory of Program Size Formally Identical to Information Theory," *J. ACM* **22**, 329 (1975).

24. G. J. Chaitin, "Algorithmic Entropy of Sets," *Comput. & Math. Appls.* **2**, 233 (1976).

25. G. J. Chaitin, "Information-theoretic Characterizations of Recursive Infinite Strings," *Theoret. Comput. Sci.* **2**, 45 (1976).

26. G. J. Chaitin, "Program Size, Oracles, and the Jump Operation," *Osaka J. Math.*, to be published in Vol. 14, No. 1, 1977.

27. R. M. Solovay, "Draft of a paper... on Chaitin's work... done for the most part during the period of Sept.-Dec. 1974," unpublished manuscript, IBM Thomas J. Watson Research Center, Yorktown Heights, NY, May, 1975.

28. R. M. Solovay, "On Random R. E. Sets," *Proceedings of the Third Latin American Symposium on Mathematical Logic,* Campinas, Brazil, July, 1976. To be published.

29. A. K. Zvonkin and L. A. Levin, "The Complexity of Finite Objects and the Development of the Concepts of Information and Randomness by Means of the Theory of Algorithms," *Russ. Math. Surv.* **25**, No. 6, 83 (1970).

30. L. A. Levin, "On the Notion of a Random Sequence," *Soviet Math. Dokl.* **14**, 1413 (1973).

31. P. Gac, "On the Symmetry of Algorithmic Information," *Soviet Math. Dokl.* **15**, 1477 (1974). "Corrections," *Soviet Math. Dokl.* **15**, No. 6, v (1974).

32. L. A. Levin, "Laws of Information Conservation (Nongrowth) and Aspects of the Foundation of Probability Theory," *Prob. Info. Transmission* **10**, 206 (1974).

33. L. A. Levin, "Uniform Tests of Randomness," *Soviet Math. Dokl.* **17**, 337 (1976).

34. L. A. Levin, "Various Measures of Complexity for Finite Objects (Axiomatic Description)," *Soviet Math. Dokl.* **17**, 522 (1976).

35. L. A. Levin, "On the Principle of Conservation of Information in Intuitionistic Mathematics," *Soviet Math. Dokl.* **17**, 601 (1976).

36. D. E. Knuth, *Seminumerical Algorithms. The Art of Computer Programming,* Volume 2, Addison-Wesley Publishing Co., Inc., Reading, MA, 1969. See Ch. 2, "Random Numbers," p. 1.

37. D. W. Loveland, "A Variant of the Kolmogorov Concept of Complexity," *Info. Control* **15**, 510 (1969).

38. T. L. Fine, *Theories of Probability – An Examination of Foundations,* Academic Press, Inc., New York, 1973. See Ch. V, "Computational Complexity, Random Sequences, and Probability," p. 118.

39. J. T. Schwartz, *On Programming: An Interim Report on the SETL Project. Installment I: Generalities,* Lecture Notes, Courant Institute of Mathematical Sciences, New York University, 1973. See Item 1, "On the Sources of Difficulty in Programming," p. 1, and Item 2, "A Second General Reflection on Programming," p. 12.

40. T. Kamae, "On Kolmogorov's Complexity and Information," *Osaka J. Math.* **10,** 305 (1973).

41. C. P. Schnorr, "Process Complexity and Effective Random Tests," *J. Comput. Syst. Sci.* **7,** 376 (1973).

42. M. E. Hellman, "The Information Theoretic Approach to Cryptography," Information Systems Laboratory, Center for Systems Research, Stanford University, April, 1974.

43. W. L. Gewirtz, "Investigations in the Theory of Descriptive Complexity," *Courant Computer Science Report 5,* Courant Institute of Mathematical Sciences, New York University, October, 1974.

44. R. P. Daley, "Minimal-program Complexity of Pseudo-recursive and Pseudo-random Sequences," *Math. Syst. Theor.* **9,** 83 (1975).

45. R. P. Daley, "Noncomplex Sequences: Characterizations and Examples," *J. Symbol. Logic* **41,** 626 (1976).

46. J. Gruska, "Descriptional Complexity (of Languages) – A Short Survey," *Mathematical Foundations of Computer Science 1976,* A. Mazurkiewicz, ed., *Lecture Notes in Computer Science 45,* Springer-Verlag, Berlin, 1976, p. 65.

47. J. Ziv, "Coding Theorems for Individual Sequences," undated manuscript, Bell Laboratories, Murray Hill, NJ.

48. R. M. Solovay, "A Model of Set-theory in which Every Set of Reals is Lebesgue Measurable," *Ann. Math.* **92,** 1 (1970).

49. R. Solovay and V. Strassen, "A Fast Monte-Carlo Test for Primality," *SIAM J. Comput.* **6,** 84 (1977).

50. G. H. Hardy, *A Course of Pure Mathematics,* Tenth edition, Cambridge University Press, London, 1952. See Section 218, "Logarithmic Tests of Convergence for Series and Integrals," p. 417.

51. M. Gardner, "A Collection of Tantalizing Fallacies of Mathematics," *Sci. Amer.* **198,** No. 1, 92 (1958).

52. B. Russell, "Mathematical Logic as Based on the Theory of Types," *From Frege to Gödel: A Source Book in Mathematical Logic, 1879-1931,* J. van Heijenoort, ed., Harvard University Press, Cambridge, MA, 1967, p. 153; reprinted from *Amer. J. Math.* **30,** 222 (1908).

53. M. Levin, "Mathematical Logic for Computer Scientists," *MIT Project MAC TR-131,* June, 1974, pp. 145, 153.

54. J. von Neumann, *Theory of Self-reproducing Automata,* University of Illinois Press, Urbana, 1966; edited and completed by A. W. Burks.

55. C. H. Bennett, "On the Thermodynamics of Computation," undated manuscript, IBM Thomas J. Watson Research Center, Yorktown Heights, NY.

56. C. H. Bennett, "Logical Reversibility of Computation," *IBM J. Res. Develop.* **17,** 525 (1973).

Received February 2, 1977; revised March 9, 1977

The author is located at the IBM Thomas J. Watson Research Center, Yorktown Heights, New York 10598.

ALGORITHMIC INFORMATION THEORY

Part II—Applications to Metamathematics

GÖDEL'S THEOREM AND INFORMATION

International Journal of Theoretical Physics 22 (1982), pp. 941-954.

Gregory J. Chaitin
IBM Research, P.O. Box 218,
Yorktown Heights, New York 10598

Received April 14, 1982

Abstract

Gödel's theorem may be demonstrated using arguments having an information-theoretic flavor. In such an approach it is possible to argue that if a theorem contains more information than a given set of axioms, then it is impossible for the theorem to be derived from the axioms. In contrast with the traditional proof based on the paradox of the liar, this new viewpoint suggests that the incompleteness phenomenon discovered by Gödel is natural and widespread rather than pathological and unusual.

1. Introduction

To set the stage, let us listen to Hermann Weyl (1946), as quoted by Eric Temple Bell (1951):

> We are less certain than ever about the ultimate foundations of (logic and) mathematics. Like everybody and everything in the world today, we have our "crisis." We have had it for nearly fifty years. Outwardly it does not seem to hamper our daily work, and yet I for one confess that it has had a considerable practical influence on my mathematical life: it directed my interests to fields I considered relatively "safe," and has been a constant drain on the enthusiasm and determination with which I pursued my research work. This experience is probably shared by other mathematicians who are not indifferent to what their scientific endeavors mean in the context of man's whole caring and knowing, suffering and creative existence in the world.

And these are the words of John von Neumann (1963):

> ...there have been within the experience of people now living at least three serious crises... There have been two such crises in physics – namely, the conceptual soul-searching connected with the discovery of relativity and the conceptual difficulties connected with discoveries in quantum theory... The third crisis was in mathematics. It was a very serious conceptual crisis, dealing with rigor and the proper way to carry out a correct mathematical proof. In view of the earlier notions of the absolute rigor of mathematics, it is surprising that such a thing could have happened, and even more surprising that it could have happened in these latter days when miracles are not supposed to take place. Yet it did happen.

At the time of its discovery, Kurt Gödel's incompleteness theorem was a great shock and caused much uncertainty and depression among mathematicians sensitive to foundational issues, since it seemed to pull the rug out from under mathematical certainty, objectivity, and rigor. Also, its proof was considered to be extremely difficult and recondite. With the passage of time the situation has been reversed. A great many different proofs of Gödel's theorem are now known, and the result is now considered easy to prove and almost obvious: It is equivalent to the unsolvability of the halting problem, or alternatively to the assertion that there is an r.e. (recursively enumerable) set that is not recursive. And it has had no lasting impact on the daily lives of mathematicians or on their working habits; no one loses sleep over it any more.

Gödel's original proof constructed a paradoxical assertion that is true but not provable within the usual formalizations of number theory. In contrast I would like to measure the power of a set of axioms and rules of inference. I would like to able to say that if one has ten pounds of axioms and a twenty-pound theorem, then that theorem cannot be derived from those axioms. And I will argue that this approach to Gödel's theorem does suggest a change in the daily habits of mathematicians, and that Gödel's theorem cannot be shrugged away.

To be more specific, I will apply the viewpoint of thermodynamics and statistical mechanics to Gödel's theorem, and will use such concepts as probability, randomness, entropy, and information to study the incompleteness phenomenon and to attempt to evaluate how widespread it is. On the basis of this analysis, I will suggest that mathematics is perhaps more akin to physics than mathematicians have been willing to admit, and that perhaps a more flexible attitude with respect to adopting new axioms and methods of reasoning is the proper response to Gödel's theorem. Probabilistic proofs of primality via sampling (Chaitin and Schwartz, 1978) also suggest that the sources of mathematical truth are wider than usually thought. Perhaps number theory should be pursued more openly in the spirit of experimental science (Pólya, 1959)!

I am indebted to John McCarthy and especially to Jacob Schwartz for making me realize that Gödel's theorem is not an obstacle to a practical AI (artificial intelligence) system based on formal logic. Such an AI would take the form of an intelligent proof checker. Gottfried Wilhelm Liebnitz and David Hilbert's dream that disputes could be settled with the words "Gentlemen, let us compute!" and that mathematics could be formalized, should still be a topic for active research. Even though mathematicians and logicians have erroneously dropped this train of thought dissuaded by Gödel's theorem, great advances have in fact been made "covertly," under the banner of computer science, LISP, and AI (Cole et al., 1981; Dewar et al., 1981; Levin, 1974; Wilf, 1982).

To speak in metaphors from Douglas Hofstadter (1979), we shall now stroll through an art gallery of proofs of Gödel's theorem, to the tune of Moussorgsky's pictures at an exhibition! Let us start with some traditional proofs (Davis, 1978; Hofstadter, 1979; Levin, 1974; Post, 1965).

2. Traditional Proofs of Gödel's Theorem

Gödel's original proof of the incompleteness theorem is based on the paradox of the liar: "This statement is false." He obtains a theorem instead of a paradox by changing this to: "This statement is unprovable." If this assertion is unprovable, then it is true, and the formalization of number theory in question is incomplete. If this assertion is provable, then it is false, and the formalization of number theory is inconsistent. The original proof was quite intricate, much like a long program in machine language. The famous technique of Gödel numbering statements was but one of the many ingenious ideas brought to bear by Gödel to construct a number-theoretic assertion which says of itself that it is unprovable.

Gödel's original proof applies to a particular formalization of number theory, and was to be followed by a paper showing that the same methods applied to a much broader class of formal axiomatic systems. The modern approach in fact applies to all formal axiomatic systems, a concept which could not even be defined when Gödel wrote his original paper, owing to the lack of a mathematical definition of effective procedure or computer algorithm. After Alan Turing succeeded in defining effective procedure by inventing a simple idealized computer now called the Turing machine (also done independently by Emil Post), it became possible to proceed in a more general fashion.

Hilbert's key requirement for a formal mathematical system was that there be an objective criterion for deciding if a proof written in the language of the system is valid or not. In other words, there must be an algorithm, a computer program, a Turing machine, for checking proofs. And the compact modern definition of formal axiomatic system as a recursively enumerable set of assertions is an immediate consequence if one uses the so-called British Museum algorithm. One applies the proof checker in turn to all possible proofs, and prints all the theorems, which of course would actually take astronomical amounts of time. By the way, in practice LISP is a very convenient programming language in which to write a simple proof checker (Levin, 1974).

Turing showed that the halting problem is unsolvable, that is, that there is no effective procedure or algorithm for deciding whether or not a program ever halts. Armed with the general definition of a formal axiomatic system as an r.e. set of assertions in a formal language, one can immediately deduce a version of Gödel's incompleteness theorem from Turing's theorem. I will sketch three different proofs of the unsolvability of the halting problem in a moment; first let me derive Gödel's theorem from it. The reasoning is simply that if it were always possible to prove whether or not particular programs halt, since the set of theorems is r.e., one could use this to solve the halting problem for any particular program by enumerating all theorems until the matter is settled. But this contradicts the unsolvability of the halting problem.

Here come three proofs that the halting problem is unsolvable. One proof considers that function $F(N)$ defined to be either one more than the value of the Nth computable function applied to the natural number N, or zero if this value is undefined because the Nth computer program does not halt on input N. F cannot be a computable function, for if program N calculated it, then one would have $F(N) = F(N) + 1$, which is impossible. But the only way that F can fail to be computable is because one cannot decide if the Nth program ever halts when given input N.

The proof I have just given is of course a variant of the diagonal method which Georg Cantor used to show that the real numbers are more numerous than the natural numbers (Courant and Robbins, 1941). Something much closer to Cantor's original technique can also be used to prove Turing's theorem. The argument runs along the lines of Bertrand Russell's paradox (Russell, 1967) of the set of all things that are not members of themselves. Consider programs for enumerating sets of natural numbers, and number these computer programs. Define a set of natural numbers consisting of the numbers of all programs which do not include their own number in their output set. This set of natural numbers cannot be recursively enumerable, for if it were listed by computer program N, one arrives at Russell's paradox of the barber in a small town who shaves all those and only those who do not shave themselves, and can neither shave himself nor avoid doing so. But the only way that this set can fail to be recursively enumerable is if it is impossible to decide whether or not a program ever outputs a specific natural number, and this is a variant of the halting problem.

For yet another proof of the unsolvability of the halting problem, consider programs which take no input and which either produce a single natural number as output or loop forever without ever producing an output. Think of these programs as being written in binary notation, instead of as natural numbers as before. I now define a so-called Busy Beaver function: BB of N is the largest natural number output by any program less than N bits in size. The original Busy Beaver function measured program size in terms of the number of states in a Turing machine instead of using the more correct information-theoretic measure, bits. It is easy to see that BB of N grows more quickly than any computable function, and is therefore not computable, which as before implies that the halting problem is unsolvable.

In a beautiful and easy to understand paper Post (1965) gave versions of Gödel's theorem based on his concepts of simple and creative r.e. sets. And he formulated the modern abstract form of Gödel's theorem, which is like a Japanese haiku: there is an r.e. set of natural numbers that is not recursive. This set has the property that there are programs for printing all the members of the set in some order, but not in ascending order. One can eventually realize that a natural number is a member of the set, but there is no algorithm for deciding if a given number is in the set or not. The set is r.e. but its complement is not. In fact, the set of (numbers of) halting programs is such a set. Now consider a particular formal axiomatic system in which one can talk about natural numbers and computer programs and such, and let X be any r.e. set whose complement is not r.e. It follows immediately that not all true assertions of the form "the natural number N is not a member of the set X" are theorems in the formal axiomatic system. In fact, if X is what Post called a simple r.e. set, then only finitely many of these assertions can be theorems.

These traditional proofs of Gödel's incompleteness theorem show that formal axiomatic systems are incomplete, but they do not suggest ways to measure the power of formal axiomatic systems, to rank their degree of completeness or incompleteness. Actually, Post's concept of a simple set contains the germ of the information-theoretic versions of Gödel's theorem that I will give later, but this is only visible in retrospect. One could somehow choose a particular simple r.e. set X and rank formal axiomatic systems according to how many different theorems of the form "N is not in X" are provable. Here are three other quantitative versions of Gödel's incompleteness theorem which do sort of fall within the scope of traditional methods.

Consider a particular formal axiomatic system in which it is possible to talk about total recursive functions (computable functions which have a natural number as value for each natural number input) and their running time computational complexity. It is possible to construct a total recursive function which grows more quickly than any function which is provably total recursive in the formal axiomatic system. It is also possible to construct a total recursive function which takes longer to compute than any provably total recursive function. That is to say, a computer program which produces a natural number output and then halts whenever it is given a natural number input, but this cannot be proved in the formal axiomatic system, because the program takes too long to produce its output.

It is also fun to use constructive transfinite ordinal numbers (Hofstadter, 1979) to measure the power of formal axiomatic systems. A constructive ordinal is one which can be obtained as the limit from below of a computable sequence of smaller constructive ordinals. One measures the power of a formal axiomatic system by the first constructive ordinal which cannot be proved to be a constructive ordinal within the system. This is like the paradox of the first unmentionable or indefinable ordinal number (Russell, 1967)!

Before turning to information-theoretic incompleteness theorems, I must first explain the basic concepts of algorithmic information theory (Chaitin, 1975b, 1977, 1982).

3. Algorithmic Information Theory

Algorithmic information theory focuses on individual objects rather than on the ensembles and probability distributions considered in Claude Shannon and Norbert Wiener's information theory. How many bits does it take to describe how to compute an individual object? In other words, what is the size in bits of the smallest program for calculating it? It is easy to see that since general-purpose computers (universal Turing machines) can simulate each other, the choice of computer as yardstick is not very important and really only corresponds to the choice of origin in a coordinate system.

The fundamental concepts of this new information theory are: algorithmic information content, joint information, relative information, mutual information, algorithmic randomness, and algorithmic independence. These are defined roughly as follows.

The algorithmic information content $I(X)$ of an individual object X is defined to be the size of the smallest program to calculate X. Programs must be self-delimiting so that subroutines can be combined by concatenating them. The joint information $I(X, Y)$ of two objects X and Y is defined to be the size of the smallest program to calculate X and Y simultaneously. The relative or conditional information content $I(X \mid Y)$ of X given Y is defined to be the size of the smallest program to calculate X from a minimal program for Y.

Note that the relative information content of an object is never greater than its absolute information content, for being given additional information can only help. Also, since subroutines can be concatenated, it follows that joint information is subadditive. That is to say, the joint information content is bounded from above by the sum of the individual information contents of the objects in question. The extent to which the joint information content is less than this sum leads to the next fundamental concept, mutual information.

The mutual information content $I(X : Y)$ measures the commonality of X and Y: it is defined as the extent to which knowing X helps one to calculate Y, which is essentially the same as the extent to which knowing Y helps one to calculate X, which is also the same as the extent to which it is cheaper to calculate them together than separately. That is to say,

$$I(X : Y) = I(X) - I(X \mid Y) = I(Y) - I(Y \mid X)$$
$$= I(X) + I(Y) - I(X, Y).$$

Note that this implies that

$$I(X, Y) = I(X) + I(Y \mid X) = I(Y) + I(X \mid Y).$$

I can now define two very fundamental and philosophically significant notions: algorithmic randomness and algorithmic independence. These concepts are, I believe, quite close to the intuitive notions that go by the same name, namely, that an object is chaotic, typical, unnoteworthy, without structure, pattern, or distinguishing features, and is irreducible information, and that two objects have nothing in common and are unrelated.

Consider, for example, the set of all N-bit long strings. Most such strings S have $I(S)$ approximately equal to N plus $I(N)$, which is N plus the algorithmic information contained in the base-two numeral for N, which is equal to N plus order of log N. No N-bit long S has information content greater than this. A few have less information content; these are strings with a regular structure or pattern. Those S of a given size having greatest information content are said to be random or patternless or algorithmically incompressible. The cutoff between random and nonrandom is somewhere around $I(S)$ equal to N if the string S is N bits long.

Similarly, an infinite binary sequence such as the base-two expansion of π is random if and only if all its initial segments are random, that is, if and only if there is a constant C such that no initial segment has information content less than C bits below its length. Of course, π is the extreme opposite of a random string: it takes only $I(N)$ which is order of log N bits to calculate π's first N bits. But the probability that an infinite sequence obtained by independent tosses of a fair coin is algorithmically random is unity.

Two strings are algorithmically independent if their mutual information is essentially zero, more precisely, if their mutual information is as small as possible. Consider, for example, two arbitrary strings X and Y each N bits in size. Usually, X and Y will be random to each other, excepting the fact that they have the same length, so that $I(X : Y)$ is approximately equal to $I(N)$. In other words, knowing one of them is no help in calculating the other, excepting that it tells one the other string's size.

To illustrate these ideas, let me give an information-theoretic proof that there are infinitely many prime numbers (Chaitin, 1979). Suppose on the contrary that there are only finitely many primes, in fact, K of them. Consider an algorithmically random natural number N. On the one hand, we know that $I(N)$ is equal to $\log_2 N$ + order of loglog N, since the base-two numeral for N is an algorithmically random ($\log_2 N$) -bit string. On the other hand, N can be calculated from the exponents in its prime factorization, and vice versa. Thus $I(N)$ is equal to the joint information of the K exponents in its prime factorization. By subadditivity, this joint information is bounded from above by the sum of the information contents of the K individual exponents. Each exponent is of order log N. The information content of each exponent is thus of order loglog N. Hence $I(N)$ is simultaneously equal to $\log_2 N + O(\text{loglog } N)$ and less than or equal to $K\ O(\text{loglog } N)$, which is impossible.

The concepts of algorithmic information theory are made to order for obtaining quantitative incompleteness theorems, and I will now give a number of information-theoretic proofs of Gödel's theorem (Chaitin, 1974a, 1974b, 1975a, 1977, 1982; Chaitin and Schwartz, 1978; Gardner, 1979).

4. Information-Theoretic Proofs of Gödel's Theorem

I propose that we consider a formal axiomatic system to be a computer program for listing the set of theorems, and measure its size in bits. In other words, the measure of the size of a formal axiomatic system that I will use is quite crude. It is merely the amount of space it takes to specify a proof-checking algorithm and how to apply it to all possible proofs, which is roughly the amount of space it takes to be very precise about the alphabet, vocabulary, grammar, axioms, and rules of inference. This is roughly proportional to the number of pages it takes to present the formal axiomatic system in a textbook.

Here is the first information-theoretic incompleteness theorem. Consider an N-bit formal axiomatic system. There is a computer program of size N which does not halt, but one cannot prove this within the formal axiomatic system. On the other hand, N bits of axioms can permit one to deduce precisely which programs of size less than N halt and which ones do not. Here are two different N-bit axioms which do this. If God tells one how many different programs of size less than N halt, this can be expressed as an N-bit base-two numeral, and from it one could eventually deduce which of these programs halt and which do not. An alternative divine revelation would be knowing that program of size less than N which takes longest to halt. (In the current context, programs have all input contained within them.)

Another way to thwart an N-bit formal axiomatic system is to merely toss an unbiased coin slightly more than N times. It is almost certain that the resulting binary string will be algorithmically random, but it is not possible to prove this within the formal axiomatic system. If one believes the postulate of quantum mechanics that God plays dice with the universe (Albert Einstein did not), then physics provides a means to expose the limitations of formal axiomatic systems. In fact, within an N-bit formal axiomatic system it is not even possible to prove that a particular object has algorithmic information content greater than N, even though almost all (all but finitely many) objects have this property.

The proof of this closely resembles G. G. Berry's paradox of "the first natural number which cannot be named in less than a billion words," published by Russell at the turn of the century (Russell, 1967). The version of Berry's paradox that will do the trick is "that object having the shortest proof that its algorithmic information content is greater than a billion bits." More precisely, "that object having the shortest proof within the following formal axiomatic system that its information content is greater than the information content of the formal axiomatic system: ...," where the dots are to be filled in with a complete description of the formal axiomatic system in question.

By the way, the fact that in a given formal axiomatic system one can only prove that finitely many specific strings are random, is closely related to Post's notion of a simple r.e. set. Indeed, the set of nonrandom or compressible strings is a simple r.e. set. So Berry and Post had the germ of my incompleteness theorem!

In order to proceed, I must define a fascinating algorithmically random real number between zero and one, which I like to call Ω (Chaitin, 1975b; Gardner, 1979). Ω is a suitable subject for worship by mystical cultists, for as Charles Bennett (Gardner, 1979) has argued persuasively, in a sense Ω contains all constructive mathematical truth, and expresses it as concisely and compactly as possible. Knowing the numerical value of Ω with N bits of precision, that is to say, knowing the first N bits of Ω's base-two expansion, is another N-bit axiom that permits one to deduce precisely which programs of size less than N halt and which ones do not.

Ω is defined as the halting probability of whichever standard general-purpose computer has been chosen, if each bit of its program is produced by an independent toss of a fair coin. To Turing's theorem in recursive function theory that the halting problem is unsolvable, there corresponds in algorithmic information theory the theorem that the base-two expansion of Ω is algorithmically random. Therefore it takes N bits of axioms to be able to prove what the first N bits of Ω are, and these bits seem completely accidental like the products of a random physical process. One can therefore measure the power of a formal axiomatic system by how much of the numerical value of Ω it is possible to deduce from its axioms. This is sort of like measuring the power of a formal axiomatic system in terms of the size in bits of the shortest program whose halting problem is undecidable within the formal axiomatic system.

It is possible to dress this incompleteness theorem involving Ω so that no direct mention is made of halting probabilities, in fact, in rather straight-forward number-theoretic terms making no mention of computer programs at all. Ω can be represented as the limit of a monotone increasing computable sequence of rational numbers. Its Nth bit is therefore the limit as T tends to infinity of a computable function of N and T. Thus the Nth bit of Ω can be expressed in the form $\exists\, X\, \forall\, Y\, [$computable predicate of X, Y, and $N]$. Complete chaos is only two quantifiers away from computability! Ω can also be expressed via a polynomial P in, say, one hundred variables, with integer coefficients and exponents (Davis et al., 1976): the Nth bit of Ω is a 1 if and only if there are infinitely many natural numbers K such that the equation $P(N, K, X_1, \ldots, X_{98}) = 0$ has a solution in natural numbers.

Of course, Ω has the very serious problem that it takes much too long to deduce theorems from it, and this is also the case with the other two axioms we considered. So the ideal, perfect mathematical axiom is in fact useless! One does not really want the most compact axiom for deducing a given set of assertions. Just as there is a trade-off between program size and running time, there is a trade-off between the number of bits of axioms one assumes and the size of proofs. Of course, random or irreducible truths cannot be compressed into axioms shorter than themselves. If, however, a set of assertions is not algorithmically independent, then it takes fewer bits of axioms to deduce them all than the sum of the number of bits of axioms it takes to deduce them separately, and this is desirable as long as the proofs do not get too long. This suggests a pragmatic attitude toward mathematical truth, somewhat more like that of physicists.

Ours has indeed been a long stroll through a gallery of incompleteness theorems. What is the conclusion or moral? It is time to make a final statement about the meaning of Gödel's theorem.

5. The Meaning of Gödel's Theorem

Information theory suggests that the Gödel phenomenon is natural and widespread, not pathological and unusual. Strangely enough, it does this via counting arguments, and without exhibiting individual assertions which are true but unprovable! Of course, it would help to have more proofs that particular interesting and natural true assertions are not demonstrable within fashionable formal axiomatic systems.

The real question is this: Is Gödel's theorem a mandate for revolution, anarchy, and license?! Can one give up after trying for two months to prove a theorem, and add it as a new axiom? This sounds ridiculous, but it is sort of what number theorists have done with Bernhard Riemann's ζ conjecture (Pólya, 1959). Of course, two months is not enough. New axioms should be chosen with care, because of their usefulness and large amounts of evidence suggesting that they are correct, in the same careful manner, say, in practice in the physics community.

Gödel himself has espoused this view with remarkable vigor and clarity, in his discussion of whether Cantor's continuum hypothesis should be added to set theory as a new axiom (Gödel, 1964):

...even disregarding the intrinsic necessity of some new axiom, and even in case it has no intrinsic necessity at all, a probable decision about its truth is possible also in another way, namely, inductively by studying its "success." Success here means fruitfulness in consequences, in particular in "verifiable" consequences, i.e., consequences demonstrable without the new axiom, whose proofs with the help of the new axiom, however, are considerably simpler and easier to discover, and make it possible to contract into one proof many different proofs. The axioms for the system of real numbers, rejected by intuitionists, have in this sense been verified to some extent, owing to the fact that analytical number theory frequently allows one to prove number-theoretical theorems which, in a more cumbersome way, can subsequently be verified by elementary methods. A much higher degree of verification than that, however, is conceivable. There might exist axioms so abundant in their verifiable consequences, shedding so much light upon a whole field, and yielding such powerful methods for solving problems (and even solving them constructively, as far as that is possible) that, no matter whether or not they are intrinsically necessary, they would have to be accepted at least in the same sense as any well-established physical theory.

Later in the same discussion Gödel refers to these ideas again:

It was pointed out earlier... that, besides mathematical intuition, there exists another (though only probable) criterion of the truth of mathematical axioms, namely their fruitfulness in mathematics and, one may add, possibly also in physics... The simplest case of an application of the criterion under discussion arises when some... axiom has number-theoretical consequences verifiable by computation up to any given integer.

Gödel also expresses himself in no uncertain terms in a discussion of Russell's mathematical logic (Gödel, 1964):

The analogy between mathematics and a natural science is enlarged upon by Russell also in another respect... axioms need not be evident in themselves, but rather their justification lies (exactly as in physics) in the fact that they make it possible for these "sense perceptions" to be deduced... I think that... this view has been largely justified by subsequent developments, and it is to be expected that it will be still more so in the future. It has turned out that solution of certain arithmetical problems requires the use of assumptions essentially transcending arithmetic... Furthermore it seems likely that for deciding certain questions of abstract set theory and even for certain related questions of the theory of real numbers new axioms based on some hitherto unknown idea will be necessary. Perhaps also the apparently insurmountable difficulties which some other mathematical problems have been presenting for many years are due to the fact that the necessary axioms have not yet been found. Of course, under these circumstances mathematics may lose a great deal of its "absolute certainty;" but, under the influence of the modern criticism of the foundations, this has already happened to a large extent...

I end as I began, with a quotation from Weyl (1949): "A truly realistic mathematics should be conceived, in line with physics, as a branch of the theoretical construction of the one real world, and should adopt the same sober and cautious attitude toward hypothetic extensions of its foundations as is exhibited by physics."

6. Directions for Future Research

a. Prove that a famous mathematical conjecture is unsolvable in the usual formalizations of number theory. Problem: if Pierre Fermat's "last theorem" is undecidable then it is true, so this is hard to do.

b. Formalize all of college mathematics in a practical way. One wants to produce textbooks that can be run through a practical formal proof checker and that are not too much larger than the usual ones. LISP (Levin, 1974) and SETL (Dewar et al., 1981) might be good for this.

c. Is algorithmic information theory relevant to physics, in particular, to thermodynamics and statistical mechanics? Explore the thermodynamics of computation (Bennett, 1982) and determine the ultimate physical limitations of computers.

d. Is there a physical phenomenon that computes something noncomputable? Contrariwise, does Turing's thesis that anything computable can be computed by a Turing machine constrain the physical universe we are in?

e. Develop measures of self-organization and formal proofs that life must evolve (Chaitin, 1979; Eigen and Winkler, 1981; von Neumann, 1966).

f. Develop formal definitions of intelligence and measures of its various components; apply information theory and complexity theory to AI.

References

Let me give a few pointers to the literature. The following are my previous publications on Gödel's theorem: Chaitin, 1974a, 1974b, 1975a, 1977, 1982; Chaitin and Schwartz, 1978. Related publications by other authors include Davis, 1978; Gardner, 1979; Hofstadter, 1979; Levin, 1974; Post, 1965. For discussions of the epistemology of mathematics and science, see Einstein, 1944, 1954; Feynman, 1965; Gödel, 1964; Pólya, 1959; von Neumann, 1956, 1963; Taub, 1961; Weyl, 1946, 1949.

- Bell, E. T. (1951). *Mathematics, Queen and Servant of Science,* McGraw-Hill, New York.

- Bennett, C. H. (1982). The thermodynamics of computation – a review, *International Journal of Theoretical Physics,* **21**, 905-940.

- Chaitin, G. J. (1974a). Information-theoretic computational complexity, *IEEE Transactions on Information Theory,* **IT-20**, 10-15.

- Chaitin, G. J. (1974b). Information-theoretic limitations of formal systems, *Journal of the ACM,* **21**, 403-424.

- Chaitin, G. J. (1975a). Randomness and mathematical proof, *Scientific American,* **232** (5) (May 1975), 47-52. (Also published in the French, Japanese, and Italian editions of *Scientific American.*)

- Chaitin, G. J. (1975b). A theory of program size formally identical to information theory, *Journal of the ACM,* **22**, 329-340.

- Chaitin, G. J. (1977). Algorithmic information theory, *IBM Journal of Research and Development,* **21**, 350-359, 496.

- Chaitin, G. J., and Schwartz, J. T. (1978). A note on Monte Carlo primality tests and algorithmic information theory, *Communications on Pure and Applied Mathematics,* **31**, 521-527.

- Chaitin, G. J. (1979). Toward a mathematical definition of "life," in *The Maximum Entropy Formalism*, R. D. Levine and M. Tribus (eds.), MIT Press, Cambridge, Massachusetts, pp. 477-498.

- Chaitin, G. J. (1982). Algorithmic information theory, *Encyclopedia of Statistical Sciences*, Vol. 1, Wiley, New York, pp. 38-41.

- Cole, C. A., Wolfram, S., et al. (1981). *SMP: a symbolic manipulation program*, California Institute of Technology, Pasadena, California.

- Courant, R., and Robbins, H. (1941). *What is Mathematics?*, Oxford University Press, London.

- Davis, M., Matijasevic, Y., and Robinson, J. (1976). Hilbert's tenth problem. Diophantine equations: positive aspects of a negative solution, in *Mathematical Developments Arising from Hilbert Problems, Proceedings of Symposia in Pure Mathematics*, Vol. XXVII, American Mathematical Society, Providence, Rhode Island, pp. 323-378.

- Davis, M. (1978). What is a computation?, in *Mathematics Today: Twelve Informal Essays*, L. A. Steen (ed.), Springer-Verlag, New York, pp. 241-267.

- Dewar, R. B. K., Schonberg, E., and Schwartz, J. T. (1981). *Higher Level Programming: Introduction to the Use of the Set-Theoretic Programming Language SETL*, Courant Institute of Mathematical Sciences, New York University, New York.

- Eigen, M., and Winkler, R. (1981). *Laws of the Game*, Knopf, New York.

- Einstein, A. (1944). Remarks on Bertrand Russell's theory of knowledge, in *The Philosophy of Bertrand Russell*, P. A. Schilpp (ed.), Northwestern University, Evanston, Illinois, pp. 277-291.

- Einstein, A. (1954). *Ideas and Opinions*, Crown, New York, pp. 18-24.

- Feynman, A. (1965). *The Character of Physical Law*, MIT Press, Cambridge, Massachusetts.

- Gardner, M. (1979). The random number Ω bids fair to hold the mysteries of the universe, Mathematical Games Dept., *Scientific American*, **241** (5) (November 1979), 20-34.

- Gödel, K. (1964). Russell's mathematical logic, and What is Cantor's continuum problem?, in *Philosophy of Mathematics*, P. Benacerraf and H. Putnam (eds.), Prentice-Hall, Englewood Cliffs, New Jersey, pp. 211-232, 258-273.

- Hofstadter, D. R. (1979). *Gödel, Escher, Bach: an Eternal Golden Braid*, Basic Books, New York.

- Levin, M. (1974). *Mathematical Logic for Computer Scientists*, MIT Project MAC report MAC TR-131, Cambridge, Massachusetts.

- Pólya, G. (1959). Heuristic reasoning in the theory of numbers, *American Mathematical Monthly*, **66**, 375-384.

- Post, E. (1965). Recursively enumerable sets of positive integers and their decision problems, in *The Undecidable: Basic Papers on Undecidable Propositions, Unsolvable Problems and Computable Functions*, M. Davis (ed.), Raven Press, Hewlett, New York, pp. 305-337.

- Russell, B. (1967). Mathematical logic as based on the theory of types, in *From Frege to Gödel: A Source Book in Mathematical Logic, 1879-1931*, J. van Heijenoort (ed.), Harvard University Press, Cambridge, Massachusetts, pp. 150-182.

- Taub, A. H. (ed.) (1961). *J. von Neumann – Collected Works,* Vol. I, Pergamon Press, New York, pp. 1-9.

- von Neumann, J. (1956). The mathematician, in *The World of Mathematics,* Vol. 4, J. R. Newman (ed.), Simon and Schuster, New York, pp. 2053-2063.

- von Neumann, J. (1963). The role of mathematics in the sciences and in society, and Method in the physical sciences, in *J. von Neumann – Collected Works,* Vol. VI, A. H. Taub (ed.), McMillan, New York, pp. 477-498.

- von Neumann, J. (1966). *Theory of Self-Reproducing Automata,* A. W. Burks (ed.), University of Illinois Press, Urbana, Illinois.

- Weyl, H. (1946). Mathematics and logic, *American Mathematical Monthly,* **53**, 1-13.

- Weyl, H. (1949). *Philosophy of Mathematics and Natural Science,* Princeton University Press, Princeton, New Jersey.

- Wilf, H. S. (1982). The disk with the college education, *American Mathematical Monthly,* **89**, 4-8.

Reprinted by permission from *Proceedings of 1985 Solvay Conference, Brussels* (in press).

RANDOMNESS AND GÖDEL'S THEOREM

IBM Research Report RC 11582 (December 1985).

G. J. Chaitin, IBM Research Division

Abstract

Complexity, non-predictability and randomness not only occur in quantum mechanics and non-linear dynamics, they also occur in pure mathematics and shed new light on the limitations of the axiomatic method. In particular, we discuss a Diophantine equation exhibiting randomness, and how it yields a proof of Gödel's incompleteness theorem.

Our view of the physical world has certainly changed radically during the past hundred years, as unpredictability, randomness and complexity have replaced the comfortable world of classical physics. Amazingly enough, the same thing has occurred in the world of pure mathematics, in fact, in number theory, a branch of mathematics that is concerned with the properties of the positive integers. How can an uncertainty principle apply to number theory, which has been called the queen of mathematics, and is a discipline that goes back to the ancient Greeks and is concerned with such things as the primes and their properties?

Following Davis (1982), consider an equation of the form

$$P(x, n, y_1, \dots, y_m) = 0,$$

where P is a polynomial with integer coefficients, and x, n, m, $y_1, \dots, y_m$ are positive integers. Here n is to be regarded as a parameter, and for each value of n we are interested in the set D_n of those values of x for which there exist y_1 to y_m such that $P = 0$. Thus a particular polynomial P with integer coefficients in $m + 2$ variables serves to define a set D_n of values of x as a function of the choice of the parameter n.

The study of equations of this sort goes back to the ancient Greeks, and the particular type of equation we have described is called a polynomial Diophantine equation.

One of the most remarkable mathematical results of this century has been the discovery that there is a "universal" polynomial P such that by varying the parameter n, the corresponding set D_n of solutions that is obtained can be any set of positive integers that can be generated by a computer program. In particular, there is a value of n such that the set of prime numbers is obtained. This immediately yields a prime-generating polynomial

$$x \left[1 - \left(P(x, n, y_1, \dots, y_m) \right)^2 \right],$$

whose set of positive values, as the values of x and y_1 to y_m vary over all the positive integers, is precisely equal to the primes. This is a remarkable result that surely would have amazed Fermat and Euler, and it is obtained as a trivial corollary to a much more general theorem!

The proof that there is such a universal P may be regarded as the culmination of Gödel's original proof of his famous incompleteness theorem. In thinking about P, it is helpful to regard the parameter n as the Gödel number of a computer program, and to regard the set of solutions x as the output of this computer program, and to think of the auxiliary variables y_1 to y_m as a kind of multidimensional time variable. In other words,

$$P(x, n, y_1, \ldots, y_m) = 0$$

if and only if the nth computer program outputs the positive integer x at time $(y_1, \ldots, y_m)$.

Let us prove Gödel's incompleteness theorem by making use of this universal polynomial P and Cantor's famous diagonal method, which Cantor originally used to prove that the real numbers are more numerous than the integers. Recall that D_n denotes the set of positive integers x for which there exist positive integers y_1 to y_m such that $P = 0$. I.e.,

$$D_n = \left\{ x \mid (\exists y_1, \ldots, y_m)[P(x, n, y_1, \ldots, y_m) = 0] \right\}.$$

Consider the "diagonal" set

$$V = \left\{ n \mid n \notin D_n \right\}$$

of all those positive integers n that are not contained in the corresponding set D_n. It is easy to see that V cannot be generated by a computer program, because V differs from the set generated by the nth computer program regarding the membership of n. It follows that there can be no algorithm for deciding, given n, whether or not the equation

$$P(n, n, y_1, \ldots, y_m) = 0$$

has a solution. And if there cannot be an algorithm for deciding if this equation has a solution, no fixed system of axioms and rules of inference can permit one to prove whether or not it has a solution. For if there were a formal axiomatic theory for proving whether or not there is a solution, given any particular value of n one could in principle use this formal theory to decide if there is a solution, by searching through all possible proofs within the formal theory in size order, until a proof is found one way or another. It follows that no single set of axioms and rules of inference suffice to enable one to prove whether or not a polynomial Diophantine equation has a solution. This is a version of Gödel's incompleteness theorem.

What does this have to do with randomness, uncertainty and unpredictability? The point is that the solvability or unsolvability of the equation

$$P(n, n, y_1, \ldots, y_m) = 0$$

in positive integers is in a sense mathematically uncertain and jumps around unpredictably as the parameter n varies. In fact, it is possible to construct another polynomial P' with integer coefficients for which the situation is much more dramatic.

Instead of asking whether $P' = 0$ can be solved, consider the question of whether or not there are infinitely many solutions. Let D_n' be the set of positive integers x such that

$$P'(x, n, y_1, \ldots, y_m) = 0$$

has a solution. P' has the remarkable property that the truth or falsity of the assertion that the set D_n' is infinite, is completely random. Indeed, this infinite sequence of true/false values is indistinguishable from the result of successive independent tosses of an unbiased coin. In other words, the truth or falsity of each of these assertions is an independent mathematical fact with probability one-half! These independent facts cannot be compressed into a smaller amount of information, i.e., they are irreducible mathematical information. In order to be able to prove whether or not D_n' is infinite for the first k values of the parameter n, one needs at least k bits of axioms and rules of inference, i.e., the formal theory must be based on at least k independent choices between equally likely alternative assumptions. In other words, a system of axioms and rules of inference, considered as a computer program for generating theorems, must be at least k bits in size if it enables one to prove whether or not D_n' is infinite for $n = 1, 2, 3, \ldots, k$.

This is a dramatic extension of Gödel's theorem. Number theory, the queen of mathematics, is infected with uncertainty and randomness! Simple properties of Diophantine equations escape the power of any particular formal axiomatic theory! To mathematicians, accustomed as they often are to believe that mathematics offers absolute certainty, this may appear to be a serious blow. Mathematicians often deride the non-rigorous reasoning used by physicists, but perhaps they have something to learn from them. Physicists know that new experiments, new domains of experience, often require fundamentally new physical principles. They have a more pragmatic attitude to truth than mathematicians do. Perhaps mathematicians should acquire some of this flexibility from their colleagues in the physical sciences!

Appendix

Let me say a few words about where P' comes from. P' is closely related to the fascinating random real number which I like to call Ω. Ω is defined to be the halting probability of a universal Turing machine when its program is chosen by coin tossing, more precisely, when a program n bits in size has probability 2^{-n} [see Gardner (1979)]. One could in principle try running larger and larger programs for longer and longer amounts of time on the universal Turing machine. Thus if a program ever halts, one would eventually discover this; if the program is n bits in size, this would contribute 2^{-n} more to the total halting probability Ω. Hence Ω can be obtained as the limit from below of a computable sequence $r_1 \leq r_2 \leq r_3 \leq \cdots$ of rational numbers:

$$\Omega = \lim_{k \to \infty} r_k;$$

this sequence converges very slowly, in fact, in a certain sense, as slowly as possible. The polynomial P' is constructed from the sequence r_k by using the theorem that "a set of tuples of positive integers is Diophantine if and only if it is recursively enumerable" [see Davis (1982)]: the equation

$$P'(k, \ n, \ y_1, \ldots, \ y_m) = 0$$

has a solution if and only if the nth bit of the base-two expansion of r_k is a "1". Thus D_n', the set of x such that

$$P'(x, \ n, \ y_1, \ldots, \ y_m) = 0$$

has a solution, is infinite if and only if the nth bit of the base-two expansion of Ω is a "1". Knowing whether or not D_n' is infinite for $n = 1, 2, 3, \ldots, k$ is therefore equivalent to knowing the first k bits of Ω.

Bibliography

1. G. J. Chaitin (1975), "Randomness and mathematical proof," *Scientific American* 232 (5), pp. 47-52.

2. M. Davis (1978), "What is a computation?", *Mathematics Today: Twelve Informal Essays,* L. A. Steen, Springer-Verlag, New York, pp. 241-267.

3. D. R. Hofstadter (1979), *Gödel, Escher, Bach: an Eternal Golden Braid,* Basic Books, New York.

4. M. Gardner (1979), "The random number Ω bids fair to hold the mysteries of the universe," Mathematical Games Dept., *Scientific American* 241 (5), pp. 20-34.

5. G. J. Chaitin (1982), "Gödel's theorem and information," *International Journal of Theoretical Physics* 22, pp. 941-954.
6. M. Davis (1982), "Hilbert's Tenth Problem is Unsolvable," *Computability & Unsolvability*, Dover, New York, pp. 199-235.

AN ALGEBRAIC EQUATION FOR THE HALTING PROBABILITY

The Universal Turing Machine—A Half-Century Survey, Rolf Herken (ed.), Verlag Kammerer & Unverzagt, in press. `

Gregory J. Chaitin

Abstract

We outline our construction of a single equation involving only addition, multiplication, and exponentiation of non-negative integer constants and variables with the following remarkable property. One of the variables is considered to be a parameter. Take the parameter to be 0,1,2,... obtaining an infinite series of equations from the original one. Consider the question of whether each of the derived equations has finitely or infinitely many non-negative integer solutions. The original equation is constructed in such a manner that the answers to these questions about the derived equations are independent mathematical facts that cannot be compressed into any finite set of axioms. To produce this equation, we start with a universal Turing machine in the form of the LISP universal function EVAL written as a register machine program about 300 lines long. Then we "compile" this register machine program into a universal exponential Diophantine equation. The resulting equation is about 200 pages long and has about 17,000 variables. Finally, we substitute for the program variable in the universal Diophantine equation the Gödel number of a LISP program for Ω, the halting probability of a universal Turing machine if n-bit programs have measure 2^{-n}. Full details appear in a book.[17]

More than half a century has passed since the famous papers of Gödel (1931) and Turing (1936-7) that shed so much light on the foundations of mathematics, and that simultaneously promulgated mathematical formalisms for specifying algorithms, in one case via primitive recursive function definitions, and in the other case via Turing machines. The development of computer hardware and software technology during this period has been phenomenal, and as a result we now know much better how to do the high-level functional programming of Gödel, and how to do the low-level machine language programming found in Turing's paper. And we can actually run our programs on machines and debug them, which Gödel and Turing could not do.

I believe that the best way to actually program a universal Turing machine is John McCarthy's universal function EVAL. In 1960 McCarthy proposed LISP as a new mathematical foundation for the theory of computation (McCarthy 1960). But by a quirk of fate LISP has largely been ignored by theoreticians and has instead become the standard programming language for work on artificial intelligence. I believe that pure LISP is in precisely the same role in computational mathematics that set theory is in theoretical mathematics, in that it provides a beautifully elegant and extremely powerful formalism which enables concepts such as that of numbers and functions to be defined from a handful of more primitive notions.

Simultaneously there have been profound theoretical advances. Gödel and Turing's fundamental undecidable proposition, the question of whether an algorithm ever halts, is equivalent to the question

---

AN ALGEBRAIC EQUATION FOR THE HALTING PROBABILITY

of whether it ever produces any output. In another paper (Chaitin 1987a) I have shown that much more devastating undecidable propositions arise if one asks whether an algorithm produces an infinite amount of output or not.

Gödel expended much effort to express his undecidable proposition as an arithmetical fact. Here too there has been considerable progress. In my opinion the most beautiful proof is the recent one of Jones and Matijasevic (1984), based on three simple ideas:

1. the observation that $11^0 = 1$, $11^1 = 11$, $11^2 = 121$, $11^3 = 1331$, $11^4 = 14641$ reproduces Pascal's triangle, makes it possible to express binomial coefficients as the digits of powers of 11 written in high enough bases;

2. an appreciation of E. Lucas's hundred-year-old remarkable theorem that the binomial coefficient $\binom{n}{k}$ is odd if and only if each bit in the base-two numeral for k implies the corresponding bit in the base-two numeral for n;

3. the idea of using register machines rather than Turing machines, and of encoding computational histories via variables which are vectors giving the contents of a register as a function of time.

Their work gives a simple straight-forward proof, using almost no number theory, that there is an exponential Diophantine equation with one parameter p which has a solution if and only if the pth computer program (i.e., the program with Gödel number p) ever halts. Similarly, one can use their method to arithmetize my undecidable proposition. The result is an exponential Diophantine equation with the parameter n and the property that it has infinitely many solutions if and only if the nth bit of Ω is a 1. Here Ω is the halting probability of a universal Turing machine if an n-bit program has measure 2^{-n} (Chaitin 1985, 1986). Ω is an algorithmically random real number in the sense that the first N bits of the base-two expansion of Ω cannot be compressed into a program shorter than N bits, from which it follows that the successive bits of Ω cannot be distinguished from the result of independent tosses of a fair coin. It can also be shown that an N-bit program cannot calculate the positions and values of more than N scattered bits of Ω, not just the first N bits (Chaitin 1987a). This implies that there are exponential Diophantine equations with one parameter n which have the property that no formal axiomatic theory can enable one to settle whether the number of solutions of the equation is finite or infinite for more than a finite number of values of the parameter n.

What is gained by asking if there are infinitely many solutions rather than whether or not a solution exists? The question of whether or not an exponential Diophantine equation has a solution is in general undecidable, but the answers to such questions are not independent. Indeed, if one considers such an equation with one parameter k, and asks whether or not there is a solution for $k = 0, 1, 2, \ldots, N - 1$, the N answers to these N questions really only constitute $\log_2 N$ bits of information. The reason for this is that we can in principle determine which equations have a solution if we know how many of them are solvable, for the set of solutions and of solvable equations is r.e. On the other hand, if we ask whether the number of solutions is finite or infinite, then the answers can be independent, if the equation is constructed properly.

In view of the philosophical impact of exhibiting an algebraic equation with the property that the number of solutions jumps from finite to infinite at random as a parameter is varied, I have taken the trouble of explicitly carrying out the construction outlined by Jones and Matijasevic. That is to say, I have encoded the halting probability Ω into an exponential Diophantine equation. To be able to actually do this, one has to start with a program for calculating Ω, and the only language I can think of in which actually writing such a program would not be an excruciating task is pure LISP. It is in fact necessary to go beyond the ideas of McCarthy in three fundamental ways:

1. First of all, we simplify LISP by only allowing atoms to be one character long. (This is similar to McCarthy's "linear LISP.")

2. Secondly, EVAL must not lose control by going into an infinite loop. In other words, we need a safe EVAL that can execute garbage for a limited amount of time, and always results in an error message or a valid value of an expression. This is similar to the notion in modern operating systems that the supervisor should be able to give a user task a time slice of CPU, and that the supervisor should not abort if the user task has an abnormal error termination.

3. Lastly, in order to program such a safe time-limited EVAL, it greatly simplifies matters if we stipulate "permissive" LISP semantics with the property that the only way a syntactically valid LISP expression can fail to have a value is if it loops forever. Thus, for example, the head (CAR) and tail (CDR) of an atom is defined to be the atom itself, and the value of an unbound variable is the variable.

Proceeding in this spirit, we have defined a class of abstract computers which, as in Jones and Matijasevic's treatment, are register machines. However, our machine's finite set of registers each contain a LISP *S*-expression in the form of a character string with balanced left and right parentheses to delimit the list structure. And we use a small set of machine instructions, instructions for testing, moving, erasing, and setting one character at a time. In order to be able to use subroutines more effectively, we have also added an instruction for jumping to a subroutine after putting into a register the return address, and an indirect branch instruction for returning to the address contained in a register. The complete register machine program for a safe time-limited LISP universal function (interpreter) EVAL is about 300 instructions long. To test this LISP interpreter written for an abstract machine, we have written in 370 machine language a register machine simulator. We have also re-written this LISP interpreter directly in 370 machine language, representing LISP *S*-expressions by binary trees of pointers rather than as character strings, in the standard manner used in practical LISP implementations. We have then run a large suite of tests through the very slow interpreter on the simulated register machine, and also through the extremely fast 370 machine language interpreter, in order to make sure that identical results are produced by both implementations of the LISP interpreter.

Our version of pure LISP also has the property that in it we can write a short program to calculate Ω in the limit from below. The program for calculating Ω is only a few pages long, and by running it (on the 370 directly, not on the register machine!), we have obtained a lower bound of $^{127}/_{128}$-ths for the particular definition of Ω we have chosen, which depends on our choice of a self-delimiting universal computer.

The final step was to write a compiler that compiles a register machine program into an exponential Diophantine equation. This compiler consists of about 700 lines of code in a very nice and easy to use programming language invented by Mike Cowlishaw called REXX (Cowlishaw 1985). REXX is a pattern-matching string processing language which is implemented by means of a very efficient interpreter. It takes the compiler only a few minutes to convert the 300-line LISP interpreter into a 200-page 17,000-variable universal exponential Diophantine equation. The resulting equation is a little large, but the ideas used to produce it are simple and few, and the equation results from the straight-forward application of these ideas.

I have published the details of this adventure (but not the full equation!) as a book (Chaitin 1987b). My hope is that this book will convince mathematicians that randomness not only occurs in non-linear dynamics and quantum mechanics, but that it even happens in rather elementary branches of number theory.

References

- Chaitin, G. J.

1985   Randomness and Gödel's theorem. *Report RC 11582.* Yorktown Heights, NY: IBM Watson Research Center (1985).

1986   Information-theoretic computational complexity *and* Gödel's theorem and information. In: *New Directions in the Philosophy of Mathematics,* ed. T. Tymoczko. Boston: Birkhäuser (1986).

1987a   Incompleteness theorems for random reals. *Adv. Appl. Math.* **8** (1987) 119-146.

1987b   *Algorithmic Information Theory.* Cambridge, England: Cambridge University Press (1987).

●   Cowlishaw, M. F.

1985   *The REXX Language.* Englewood Cliffs, NJ: Prentice-Hall (1985).

●   Gödel, K.

1931   On formally undecidable propositions of *Principia mathematica* and related systems I. In: *Kurt Gödel: Collected Works, Volume I: Publications 1929-1936,* ed. S. Feferman. New York: Oxford University Press (1986).

●   Jones, J. P., and Y. V. Matijasevic

1984   Register machine proof of the theorem on exponential Diophantine representation of enumerable sets. *J. Symb. Log.* **49** (1984) 818-829.

●   McCarthy, J.

1960   Recursive functions of symbolic expressions and their computation by machine, Part I. *ACM Comm.* **3** (1960) 184-195.

●   Turing, A. M.

1936-7   On computable numbers, with an application to the Entscheidungsproblem. *P. Lond. Math. Soc. (2)* **42** (1936) 230-265; with a correction, *Ibid. (2)* **43** (1936-7) 544-546; reprinted in: *The Undecidable,* ed. M. Davis. Hewlett, NY: Raven Press (1965).

Open Problems in Communication and Computation. Edited by Thomas M. Cover and B. Gopinath.
Copyright 1988, Springer-Verlag. ISBN 96621-8

COMPUTING THE BUSY BEAVER FUNCTION

IBM Research Report RC 10722 (September 1984).

Gregory J. Chaitin
IBM Research Division, P. O. Box 218
Yorktown Heights, NY 10598, U.S.A.

Abstract

Efforts to calculate values of the noncomputable Busy Beaver function are discussed in the light of algorithmic information theory.

I would like to talk about some impossible problems that arise when one combines information theory with recursive function or computability theory. That is to say, I'd like to look at some unsolvable problems which arise when one examines computation unlimited by any practical bound on running time, from the point of view of information theory. The result is what I like to call "algorithmic information theory" [5].

In the Computer Recreations department of a recent issue of *Scientific American* [7], A. K. Dewdney discusses efforts to calculate the Busy Beaver function Σ. This is a very interesting endeavor for a number of reasons.

First of all, the Busy Beaver function is of interest to information theorists, because it measures the capability of computer programs as a function of their size, as a function of the amount of information which they contain. $\Sigma(n)$ is defined to be the largest number which can be computed by an n-state Turing machine; to information theorists it is clear that the correct measure is bits, not states. Thus it is more correct to define $\Sigma(n)$ as the largest natural number whose program-size complexity or algorithmic information content is less than or equal to n. Of course, the use of states has made it easier and a definite and fun problem to calculate values of Σ(number of states); to deal with Σ(number of bits) one would need a model of a binary computer as simple and compelling as the Turing machine model, and no obvious natural choice is at hand.

Perhaps the most fascinating aspect of Dewdney's discussion is that it describes successful attempts to calculate the initial values $\Sigma(1)$, $\Sigma(2)$, $\Sigma(3)$, ... of an uncomputable function Σ. Not only is Σ uncomputable, but it grows faster than any computable function can. In fact, it is not difficult to see that $\Sigma(n)$ is greater than the computable function $f(n)$ as soon as n is greater than (the program-size complexity or algorithmic information content of f) + $O(1)$. Indeed, to compute $f(n) + 1$ it is sufficient to know (a minimum-size program for f), and the value of the integer (n − the program-size complexity of f). Thus the program-size complexity of $f(n) + 1$ is $\leq$ (the program-size complexity of f) + $O(\log | n$ − the program-size complexity of $f |$), which is $< n$ if n is greater than $O(1)$ + the program-size complexity of f. Hence $f(n) + 1$ is included in $\Sigma(n)$, that is, $\Sigma(n) \geq f(n) + 1$, if n is greater than $O(1)$ + the program-size complexity of f.

Yet another reason for interest in the Busy Beaver function is that, when properly defined in terms of bits, it immediately provides an information-theoretic proof of an extremely fundamental fact of recursive function theory, namely Turing's theorem that the halting problem is unsolvable [2]. Turing's original proof involves the notion of a computable real number, and the observation that it cannot be decided whether or not the nth computer program ever outputs an nth digit, because otherwise one could carry out Cantor's diagonal construction and calculate a paradoxical real number whose nth digit is chosen to differ from the nth digit output by the nth program, and which therefore cannot actually be a computable real number after all. To use the noncomputability of Σ to demonstrate the unsolvability of the halting problem, it suffices to note that in principle, if one were very patient, one could calculate $\Sigma(n)$ by checking each program of size less than or equal to n to determine whether or not it halts, and then running each of the programs which halt to determine what their output is, and then taking the largest output. Contrariwise, if Σ were computable, then it would provide a solution to the halting problem, for an n-bit program either halts in time less than $\Sigma(n + O(1))$, or else it never halts.

The Busy Beaver function is also of considerable metamathematical interest; in principle it would be extremely useful to know larger values of $\Sigma(n)$. For example, this would enable one to settle the Goldbach conjecture and the Riemann hypothesis, and in fact any conjecture such as Fermat's which can be refuted by a numerical counterexample. Let P be a computable predicate of a natural number, so that for any specific natural number n it is possible to compute in a mechanical fashion whether or not $P(n)$, P of n, is true or false, that is, to determine whether or not the natural number n has property P. How could one use the Busy Beaver function to decide if the conjecture that P is true for all natural numbers is correct? An experimental approach is to use a fast computer to check whether or not P is true, say for the first billion natural numbers. To convert this empirical approach into a proof, it would suffice to have a bound on how far it is necessary to test P before settling the conjecture in the affirmative if no counterexample has been found, and of course rejecting it if one was discovered. Σ provides this bound, for if P has program-size complexity or algorithmic information content k, then it suffices to examine the first $\Sigma(k + O(1))$ natural numbers to decide whether or not P is always true. Note that the program-size complexity or algorithmic information content of a famous conjecture P is usually quite small; it is hard to get excited about a conjecture that takes a hundred pages to state.

For all these reasons, it is really quite fascinating to contemplate the successful efforts which have been made to calculate some of the initial values of $\Sigma(n)$. In a sense these efforts simultaneously penetrate to "mathematical bedrock" and are "storming the heavens," to use images of E. T. Bell. They amount to a systematic effort to settle all finitely refutable mathematical conjectures, that is, to determine all constructive mathematical truth. And these efforts fly in the face of fundamental information-theoretic limitations on the axiomatic method [1,2,6], which amount to an information-theoretic version of Gödel's famous incompleteness theorem [3].

Here is the Busy Beaver version of Gödel's incompleteness theorem: n bits of axioms and rules of inference cannot enable one to prove what is the value of $\Sigma(k)$ for any k greater than $n + O(1)$. The proof of this fact is along the lines of the Berry paradox. Contrariwise, there is an n-bit axiom which does enable one to demonstrate what is the value of $\Sigma(k)$ for any k less than $n - O(1)$. To get such an axiom, one either asks God for the number of programs less than n bits in size which halt, or one asks God for a specific n-bit program which halts and has the maximum possible running time or the maximum possible output before halting. Equivalently, the divine revelation is a conjecture $\forall k\ P(k)$ (with P of program-size complexity or algorithmic information content $\leq n$) which is false and for which (the smallest counterexample i with $\neg P(i)$) is as large as possible. Such an axiom would pack quite a wallop, but only in principle, because it would take about $\Sigma(n)$ steps to deduce from it whether or not a specific program halts and whether or not a specific mathematical conjecture is true for all natural numbers.

These considerations involving the Busy Beaver function are closely related to another fascinating noncomputable object, the halting probability of a universal Turing machine on random input, which

I like to call Ω, and which is the subject of an essay by my colleague Charles Bennett that was published in the Mathematical Games department of *Scientific American* some years ago [4].

References

1. G. J. Chaitin, "Randomness and mathematical proof," *Scientific American* **232**, No. 5 (May 1975), 47-52.

2. M. Davis, "What is a computation?" in *Mathematics Today: Twelve Informal Essays*, L. A. Steen (ed.), Springer-Verlag, New York, 1978, 241-267.

3. D. R. Hofstadter, *Gödel, Escher, Bach: an Eternal Golden Braid*, Basic Books, New York, 1979.

4. M. Gardner, "The random number Ω bids fair to hold the mysteries of the universe," Mathematical Games Dept., *Scientific American* **241**, No. 5 (Nov. 1979), 20-34.

5. G. J. Chaitin, "Algorithmic information theory," in *Encyclopedia of Statistical Sciences*, Volume 1, Wiley, New York, 1982, 38-41.

6. G. J. Chaitin, "Gödel's theorem and information," *International Journal of Theoretical Physics* **22** (1982), 941-954.

7. A. K. Dewdney, "A computer trap for the busy beaver, the hardest-working Turing machine," Computer Recreations Dept., *Scientific American* **251**, No. 2 (Aug. 1984), 19-23.

Part III—Applications to Biology

TO A MATHEMATICAL DEFINITION OF "LIFE"

ACM SICACT News 4 (January 1970), pp. 12-18.

G. J. Chaitin

Abstract

"Life" and its "evolution" are fundamental concepts that have not yet been formulated in precise mathematical terms, although some efforts in this direction have been made. We suggest a possible point of departure for a mathematical definition of "life." This definition is based on the computer and is closely related to recent analyses of "inductive inference" and "randomness." A living being is a unity; it is simpler to view a living organism as a whole than as the sum of its parts. If we want to compute a complete description of the region of space-time that is a living being, the program will be smaller in size if the calculation is done all together, than if it is done by independently calculating descriptions of parts of the region and then putting them together.

1. The problem

"Life" and its "evolution" from the lifeless are fundamental concepts of science. According to Darwin and his followers, we can expect living organisms to evolve under very general conditions. Yet this theory has never been formulated in precise mathematical terms. Supposing Darwin is right, it should be possible to formulate a general definition of "life" and to prove that under certain conditions we can expect it to "evolve." If mathematics can be made out of Darwin, then we will have added something basic to mathematics; while if it cannot, then Darwin must be wrong, and life remains a miracle which has not been explained by science.

The point is that the view that life has spontaneously evolved, and the very concept of life itself, are very general concepts, which it should be possible to study without getting involved in, for example, the details of quantum chemistry. We can idealize the laws of physics and simplify them and make them complete, and then study the resulting universe. It is necessary to do two things in order to study the evolution of life within our model universe. First of all, we must define "life," we must characterize a living organism in a precise fashion. At the same time it should become clear what the complexity of an organism is, and how to distinguish primitive forms of life from advanced forms. Then we must study our universe in the light of the definition. Will an evolutionary process occur? What is the expected time for a certain level of complexity to be reached? Or can we show that life will probably not evolve?

2. Previous Work

Von Neumann devoted much attention to the analysis of fundamental biological questions from a mathematical point of view.[18] He considered a universe consisting of an infinite plane divided into squares. Time is quantized, and at any moment each square is in one of 29 states. The state of a square at any time depends only on its previous state and the previous states of its four neighboring squares. The universe is homogeneous; the state transitions of all squares are governed by the same law. It is a deterministic universe. Von Neumann showed that a self-reproducing general-purpose computer can exist in his model universe.

A large amount of work on these questions has been done since von Neumann's initial investigations, and a complete bibliography would be quite lengthy. We may mention Moore (1962), Arbib (1966,1967), and Codd (1968).

The point of departure of all this work has been the identification of "life" with "self-reproduction," and this identification has both helped and hindered. It has helped, because it has not allowed fundamental conceptual difficulties to tie up work, but has instead permitted much that is very interesting to be accomplished. But it has hindered because, in the end, these fundamental difficulties must be faced. At present the problem has evidenced itself as a question of "good taste." As von Neumann remarks,[19] good taste is required in building one's universe. If its elementary parts are assumed to be very powerful, self-reproduction is immediate. Arbib (1966) is an intermediate case.

What is the relation between self-reproduction and life? A man may be sterile, but no one would doubt he is alive. Children are not identical to their parents. Self-reproduction is not exact; if it were, evolution would be impossible. What's more, a crystal reproduces itself, yet we would not consider it to have much life. As von Neumann comments,[20] the matter is the other way around. We can deduce self-reproduction as a property which must be possessed by many living beings, if we ask ourselves what kinds of living beings are likely to be around. Obviously, a species that did not reproduce would die out. Thus, if we ask what kinds of living organisms are likely to evolve, we can draw conclusions concerning self-reproduction.

3. Simplicity and Complexity

"Complexity" is a concept whose importance and vagueness von Neumann emphasized many times in his lectures.[21] Due to the work of Solomonoff, Kolmogorov, Chaitin, Martin-Löf, Willis, and Loveland, we now understand this concept a great deal better than it was understood while von Neumann worked. Obviously, to understand the evolution of the complexity of living beings from primitive, simple life to today's very complex organisms, we need to make precise a measure of complexity. But it also seems that perhaps a precise concept of complexity will enable us to define "living organism" in an exact and general fashion. Before suggesting the manner in which this may perhaps be done, we shall review the recent developments which have converted "simplicity" and "complexity" into precise concepts.

We start by summarizing Solomonoff's work.[22] Solomonoff proposes the following model of the predicament of the scientist. A scientist is continually observing increasingly larger initial segments of an infinite sequence of 0's and 1's. This is his experimental data. He tries to find computer programs which compute infinite binary sequences which begin with the observed sequence. These are his theories. In order to predict his future observations, he could use any of the theories. But there will always be one theory that predicts that all succeeding observations will be 1's, as well as others that take more account of the previous observations. Which of the infinitely many theories should he use to make the prediction? According to Solomonoff, the principle that the simplest theory is the best should guide him.[23] What is the simplicity of a theory in the present context? It is the size of the

---

[18] See in particular his fifth lecture delivered at the University of Illinois in December of 1949, "Re-evaluation of the problem of complicated automata – Problems of hierarchy and evolution," and his unfinished *The Theory of Automata: Construction, Reproduction, Homogeneity.* Both are posthumously published in von Neumann (1966).

[19] See pages 76-77 of von Neumann (1966).

[20] See page 78 of von Neumann (1966).

[21] See especially pages 78-80 of von Neumann (1966).

TO A MATHEMATICAL DEFINITION OF "LIFE"

computer program. Larger computer programs embody more complex theories, and smaller programs embody simpler theories.

Willis has further studied the above proposal, and also has introduced the idea of a hierarchy of finite approximations to it. To my knowledge, however, the success which predictions made on this basis will have has not been made completely clear.

We must discuss a more technical aspect of Solomonoff's work. He realized that the simplicity of theories, and thus also the predictions, will depend on the computer which one is using. Let us consider only computers whose programs are finite binary sequences, and measure the size of a binary sequence by its length. Let us denote by $C(T)$ the complexity of a theory T. By definition, $C(T)$ is the size of the smallest program which makes our computer C compute T. Solomonoff showed that there are "optimal" binary computers C that have the property that for any other binary computer C', $C(T) \leq C'(T) + d$, for all T. Here d is a constant that depends on C and C', not on T. Thus, these are the most efficient binary computers, for their programs are shortest. Any two of these optimal binary computers C_1 and C_2 result in almost the same complexity measure, for from $C_1(T) \leq C_2(T) + d_{12}$ and $C_2(T) \leq C_1(T) + d_{21}$, it follows that the difference between $C_1(T)$ and $C_2(T)$ is bounded. The optimal binary computers are transparent theoretically, they are enormously convenient from the technical point of view. What's more, their optimality makes them a very natural choice.[24] Kolmogorov and Chaitin later independently hit upon the same kind of computer in their search for a suitable computer upon which to base a definition of "randomness."

However, the naturalness and technical convenience of the Solomonoff approach should not blind us to the fact that it is by no means the only possible one. Chaitin first based his definition of randomness on Turing machines, taking as the complexity measure the number of states in the machine, and he later used bounded-transfer Turing machines. Although these computers are quite different, they lead to similar definitions of randomness. Later it became clear that using the usual 3-tape-symbol Turing machine and taking its size to be the number of states leads to a complexity measure $C_3(T)$ which is asymptotically just a Solomonoff measure $C(T)$ with its scale changed: $C(T)$ is asymptotic to $2 C_3(T) \log_2 C_3(T)$. It appears that people interested in computers may still study other complexity measures, but to apply these concepts of simplicity/complexity it is at present most convenient to use Solomonoff measures.

We now turn to Kolmogorov's and Chaitin's proposed definition of randomness or patternlessness. Let us consider once more the scientist confronted by experimental data, a long binary sequence. This time he in not interested in predicting future observations, but only in determining if there is a pattern in his observations, if there is a simple theory that explains them. If he found a way of compressing his observations into a short computer program which makes the computer calculate them, he would say that the sequence follows a law, that it has pattern. But if there is no short program, then the sequence has no pattern – it is random. That is to say, the complexity $C(S)$ of a finite binary sequence S is the size of the smallest program which makes the computer calculate it. Those binary sequences S of a given length n for which $C(S)$ is greatest are the most complex binary sequences of length n, the random or patternless ones. This is a general formulation of the definition. If we use one of Solomonoff's optimal binary computers, this definition becomes even clearer. Most binary sequences of any given length n require programs of about length n. These are the patternless or random sequences. Those binary sequences which can be compressed into programs appreciably

---

22   The earliest generally available appearance in print of Solomonoff's ideas of which we are aware is Minsky's summary of them on pages 41-43 of Minsky (1962). A more recent reference is Solomonoff (1964).

23   Solomonoff actually proposes weighing together all the theories into the prediction, giving the simplest theories the largest weight.

24   Solomonoff's approach to the size of programs has been extended in Chaitin (1969a) to the speed of programs.

shorter than themselves are the sequences which have pattern. Chaitin and Martin-Löf have studied the statistical properties of these sequences, and Loveland has compared several variants of the definition.

This completes our summary of the new rigorous meaning which has been given to simplicity/complexity – the complexity of something is the size of the smallest program which computes it or a complete description of it. Simpler things require smaller programs. We have emphasized here the relation between these concepts and the philosophy of the scientific method. In the theory of computing the word "complexity" is usually applied to the speed of programs or the amount of auxiliary storage they need for scratch-work. These are completely different meanings of complexity. When one speaks of a simple scientific theory, one refers to the fact that few arbitrary choices have been made in specifying the theoretical structure, not to the rapidity with which predictions can be made.

4. What is life?

Let us once again consider a scientist in a hypothetical situation. He wishes to understand a universe very different from his own which he has been observing. As he observes it, he comes eventually to distinguish certain objects. These are highly interdependent regions of the universe he is observing, so much so, that he comes to view them as wholes. Unlike a gas, which consists of independent particles that do not interact, these regions of the universe are unities, and for this reason he has distinguished them as single entities.

We believe that the most fundamental property of living organisms is the enormous interdependence between their components. A living being is a unity; it is much simpler to view it as a whole than as the sum of parts. That is to say, if we want to compute a complete description of a region of space-time that is a living being, the program will be smaller in size if the calculation is done all together, than if it is done by independently calculating descriptions of parts of the region and then putting them together. What is the complexity of a living being, how can we distinguish primitive life from complex forms? The interdependence in a primitive unicellular organism is far less than that in a human being.

A living being is indeed a unity. All the atoms in it cooperate and work together. If Mr. Smith is afraid of missing the train to his office, all his incredibly many molecules, all his organs, all his cells, will be cooperating so that he finishes breakfast quickly and runs to the train station. If you cut the leg of an animal, all of it will cooperate to escape from you, or to attack you and scare you away, in order to protect its leg. Later the wound will heal. How different from what happens if you cut the leg of a table. The whole table will neither come to the defense of its leg, nor will it help it to heal. In the more intelligent living creatures, there is also a very great deal of interdependence between an animal's past experience and its present behavior; that is to say, it learns, its behavior changes with time depending on its experiences. Such enormous interdependence must be a monstrously rare occurrence in a universe, unless it has evolved gradually.

In summary, the case is the whole versus the sum of its parts. If both are equally complex, the parts are independent (do not interact). If the whole is very much simpler than the sum of its parts, we have the interdependence that characterizes a living being.[25] Note finally that we have introduced something new into the study of the size of programs (= complexity). Before we compared the sizes of programs that calculate different things. Now we are interested in comparing the sizes of programs

TO A MATHEMATICAL DEFINITION OF "LIFE"

that calculate the same things in different ways. That is to say, the method by which a calculation is done is now of importance to us; in the previous section it was not.

5. Numerical Examples

In this paper, unfortunately, we can only suggest a possible point of departure for a mathematical definition of life. A great amount of work must be done; it is not even clear what is the formal mathematical counterpart to the informal definition of the previous section. A possibility is sketched here.

Consider a computer C_1 which accepts programs P which are binary sequences consisting of a number of subsequences B, C, $P_1, \ldots, P_k$, A.

B, the leftmost subsequence, is a program for breaking the remainder of P into C, $P_1, \ldots, P_k$, and A. B is self-describing; it starts with a binary sequence which results from writing the length of B in base-two notation, doubling each of its bits, and then placing a pair of unequal bits at the right end. Also, B is not allowed to see whether any of the remaining bits of P are 0's or 1's, only to separate them into groups.[26]

C is the description of a computer C_2. For example, C_2 could be one of Solomonoff's optimal binary computers, or a computer which emits the program without processing it.

$P_1, \ldots, P_k$ are programs which are processed by k different copies of the computer C_2. $R_1, \ldots, R_k$ are the resulting outputs. These outputs would be regions of space-time, a space-time which, like von Neumann's, has been cut up into little cubes with a finite number of states.

A is a program for adding together $R_1, \ldots, R_k$ to produce R, a single region of space-time. A merely juxtaposes the intermediate results $R_1, \ldots, R_k$ (perhaps with some overlapping); it is not allowed to change any of the intermediate results. In the examples below, we shall only compute regions R which are one-dimensional strings of 0's and 1's, so that A need only indicate that R is the concatenation of $R_1, \ldots, R_k$, in that order.

R is the output of the computer C_1 produced by processing the program P.

We now define a family of complexity measures $C(d, R)$, the complexity of a region R of space-time when it is viewed as the sum of independent regions of diameter not greater than d. $C(d, R)$ is the length of the smallest program P which makes the computer C_1 output R, among all those P such that the intermediate results R_1 to R_k are all less than or equal to d in diameter. $C(d, R)$ where d equals the diameter of R is to within a bounded difference just the usual Solomonoff complexity measure. But as d decreases, we may be forced to forget any patterns in R that are more than d in diameter, and the complexity $C(d, R)$ increases.

We present below a table with four examples. In each of the four cases, R is a 1-dimensional region, a binary sequence of length n. R_1 is a random binary sequence of length n ("gas"). R_2 consists of n repetitions of 1 ("crystal"). The left half of R_3 is a random binary sequence of length $n/2$. The right half of R_3 is produced by rotating the left half about R_3's midpoint ("bilateral symmetry"). R_4 consists of two identical copies of a random binary sequence of length $n/2$ ("twins").

---

[25]   The whole cannot be more complex than the sum of its parts, because one of the ways of looking at it is as the sum of its parts, and this bounds its complexity.

[26]   The awkwardness of this part of the definition is apparently its chief defect.

C(d,R) = approx. ?	R = R1 "gas"	R = R2 "crystal"	R = R3 "bilateral symmetry"	R = R4 "twins"
d = n	n	log2 n note 1	n/2	n/2
d = n/k (k>1 fixed, n large)	n	k log2 n notes 1,2	n-(n/2k) note 2	n note 2
d = 1	n	n	n	n

Note 1. This supposes that n is represented in base-two notation by a random binary sequence. These values are too high in those rare cases where this is not true.

Note 2. These are conjectured values. We can only show that $C(d, R)$ is approximately less than or equal to these values.

Bibliography

- Arbib, M. A. (1962). "Simple self-reproducing automata," *Information and Control.*

- Arbib, M. A. (1967). "Automata theory and development: Part 1," *Journal of Theoretical Biology.*

- Arbib, M. A. "Self-reproducing automata – some implications for theoretical biology."

- Biological Science Curriculum Study. (1968). *Biological Science: Molecules to Man,* Houghton Mifflin Co.

- Chaitin, G. J. (1966). "On the length of programs for computing finite binary sequences," *Journal of the Association for Computing Machinery.*

- Chaitin, G. J. (1969a). "On the length of programs for computing finite binary sequences: Statistical considerations," *ibid.*

- Chaitin, G. J. (1969b). "On the simplicity and speed of programs for computing infinite sets of natural numbers," *ibid.*

- Chaitin, G. J. (1970). "On the difficulty of computations," *IEEE Transactions on Information Theory.*

- Codd, E. F. (1968). *Cellular Automata.* Academic Press.

- Kolmogorov, A. N. (1965). "Three approaches to the definition of the concept 'amount of information'," *Problemy Peredachi Informatsii.*

- Kolmogorov, A. N. (1968). "Logical basis for information theory and probability theory," *IEEE Transactions on Information Theory.*

- Loveland, D. W. "A variant of the Kolmogorov concept of complexity," report 69-4, Math. Dept., Carnegie-Mellon University.

- Loveland, D. W. (1969). "On minimal program complexity measures," *Conference Record of the ACM Symposium on Theory of Computing,* May 1969.

- Martin-Löf, P. (1966). "The definition of random sequences," *Information and Control.*

- Minsky, M. L. (1962). "Problems of formulation for artificial intelligence," *Mathematical Problems in the Biological Science,* American Math. Society.

- Moore, E. F. (1962). "Machine models of self-reproduction," *ibid.*

- von Neumann, J. (1966). *Theory of Self-Reproducing Automata.* (Edited by A. W. Burks.) University of Illinois Press.

- Solomonoff, R. J. (1964). "A formal theory of inductive inference," *Information and Control.*

- Willis, D. G. (1969). "Computational complexity and probability constructions," Stanford University.

TOWARD A MATHEMATICAL DEFINITION OF "LIFE"

The Maximum Entropy Formalism, R. D. Levine and M. Tribus (eds.), MIT Press, 1979, pp. 477-498.

Gregory J. Chaitin

Abstract

In discussions of the nature of life, the terms "complexity," "organism," and "information content," are sometimes used in ways remarkably analogous to the approach of algorithmic information theory, a mathematical discipline which studies the amount of information necessary for computations. We submit that this is not a coincidence and that it is useful in discussions of the nature of life to be able to refer to analogous precisely defined concepts whose properties can be rigorously studied. We propose and discuss a measure of degree of organization and structure of geometrical patterns which is based on the algorithmic version of Shannon's concept of mutual information. This paper is intended as a contribution to von Neumann's program of formulating mathematically the fundamental concepts of biology in a very general setting, i.e. in highly simplified model universes.

1. Introduction

Here are two quotations from works dealing with the origins of life and exobiology:

> These vague remarks can be made more precise by introducing the idea of information. Roughly speaking, the information content of a structure is the minimum number of instructions needed to specify the structure. Once can see intuitively that many instructions are needed to specify a complex structure. On the other hand, a simple repeating structure can be specified in rather few instructions. [1]

> The traditional concept of life, therefore, may be too narrow for our purpose... We should try to break away from the four properties of growth, feeding, reaction, and reproduction... Perhaps there is a clue in the way we speak of living *organisms*. They are *highly organized*, and perhaps this is indeed their essence... What, then, is organization? What sets it apart from other similarly vague concepts? Organization is perhaps viewed best as "complex interrelatedness"... A book is complex; it only resembles an organism in that passages in one paragraph or chapter refer to others elsewhere. A dictionary or thesaurus shows more organization, for every entry refers to others. A telephone directory shows less, for although it is equally elaborate, there is little cross-reference between its entries... [2]

If one compares the first quotation with any introductory article on algorithmic information theory (e.g. [3-4]), and compares the second quotation with a preliminary version of this paper [5], one is struck by the similarities. As these quotations show, there has been a great deal of thought about how to define "life," "complexity," "organism," and "information content of organism." The attempted contribution of this paper is that we propose a rigorous quantitative definition of these concepts and are able to prove theorems about them. We do not claim that our proposals are in any sense definitive, but, following von Neumann [6-7], we submit that a precise mathematical definition must be given.

Some preliminary considerations: We shall find it useful to distinguish between the notion of degree of interrelatedness, interdependence, structure, or organization, and that of information content. Two extreme examples are an ideal gas and a perfect crystal. The complete microstate at a given time of the first one is very difficult to describe fully, and for the second one this is trivial to do, but neither is organized. In other words, white noise is the most informative message possible, and a constant pitch tone is least informative, but neither is organized. Neither a gas nor a crystal should count as organized (see Theorems 1 and 2 in Section 5), nor should a whale or elephant be considered more organized than a person simply because it requires more information to specify the precise details of the current position of each molecule in its much larger bulk. Also note that following von Neumann [7] we deal with a discrete model universe, a cellular automata space, each of whose cells has only a finite number of states. Thus we impose a certain level of granularity in our idealized description of the real world.

We shall now propose a rigorous theoretical measure of degree of organization or structure. We use ideas from the new algorithmic formulation of information theory, in which one considers individual objects and the amount of information in bits needed to compute, construct, describe, generate or produce them, as opposed to the classical formulation of information theory in which one considers an ensemble of possibilities and the uncertainty as to which of them is actually the case. In that theory the uncertainty or "entropy" of a distribution is defined to be

$$-\sum_{i<k} p_i \log p_i,$$

and is a measure of one's ignorance of which of the k possibilities actually holds given that the *a priori* probability of the ith alternative is p_i. (Throughout this paper "log" denotes the base-two logarithm.) In contrast, in the newer formulation of information theory one can speak of the information content of an individual book, organism, or picture, without having to imbed it in an ensemble of all possible such objects and postulate a probability distribution on them.

We believe that the concepts of algorithmic information theory are extremely basic and fundamental. Witness the light they have shed on the scientific method [8], the meaning of randomness and the Monte Carlo method [9], the limitations of the deductive method [3-4], and now, hopefully, on theoretical biology. An information-theoretic proof of Euclid's theorem that there are infinitely many prime numbers should also be mentioned (see Appendix 2).

The fundamental notion of algorithmic information theory is $H(X)$, the algorithmic information content (or, more briefly, "complexity") of the object X. $H(X)$ is defined to be the smallest possible number of bits in a program for a general-purpose computer to print out X. In other words, $H(X)$ is the amount of information necessary to describe X sufficiently precisely for it to be constructed. Two objects X and Y are said to be (algorithmically) independent if the best way to describe them both is simply to describe each of them separately. That is to say, X and Y are independent if $H(X,Y)$ is approximately equal to $H(X) + H(Y)$, i.e. if the joint information content of X and Y is just the sum of the individual information contents of X and Y. If, however, X and Y are related and have something in common, one can take advantage of this to describe X and Y together using much fewer bits than the total number that would be needed to describe them separately, and so $H(X,Y)$ is much less than $H(X) + H(Y)$. The quantity $H(X:Y)$ which is defined as follows

$$H(X:Y) = H(X) + H(Y) - H(X,Y)$$

is called the mutual information of X and Y and measures the degree of interdependence between X and Y. This concept was defined, in an ensemble rather than an algorithmic setting, in Shannon's original paper [10] on information theory, noisy channels, and coding.

We now explain our definition of the degree of organization or structure in a geometrical pattern. The d-diameter complexity $H_d(X)$ of an object X is defined to be the minimum number of bits needed to describe X as the "sum" of separate parts each of diameter not greater than d. Let us be more precise. Given d and X, consider all possible ways of partitioning X into nonoverlapping pieces each of diameter $\leq d$. Then $H_d(X)$ is the sum of the number of bits needed to describe each of the pieces separately, plus the number of bits needed to specify how to reassemble them into X. Each piece must have a separate description which makes no cross-references to any of the others. And one is interested in those partitions of X and reassembly techniques α which minimize this sum. That is to say,

$$H_d(X) = \min \left[H(\alpha) + \sum_{i<k} H(X_i) \right],$$

the minimization being taken over all partitions of X into nonoverlapping pieces

$$X_0, X_1, X_2, \ldots, X_{k-1}$$

all of diameter $\leq d$.

Thus $H_d(X)$ is the minimum number of bits needed to describe X as if it were the sum of independent pieces of size $\leq d$. For d larger than the diameter of X, $H_d(X)$ will be the same as $H(X)$. If X is unstructured and unorganized, then as d decreases $H_d(X)$ will stay close to $H(X)$. However if X has structure, then $H_d(X)$ will rapidly increase as d decreases and one can no longer take advantage of patterns of size $> d$ in describing X. Hence $H_d(X)$ as a function of d is a kind of "spectrum" or "Fourier transform" of X. $H_d(X)$ will increase as d decreases past the diameter of significant patterns in X, and if X is organized hierarchically this will happen at each level in the hierarchy.

Thus the faster the difference increases between $H_d(X)$ and $H(X)$ as d decreases, the more interrelated, structured, and organized X is. Note however that X may be a "scene" containing many independent structures or organisms. In that case their degrees of organization are summed together in the measure

$$H_d(X) - H(X).$$

Thus the organisms can be defined as the minimal parts of the scene for which the amount of organization of the whole can be expressed as the sum of the organization of the parts, i.e. pieces for which the organization decomposes additively. Alternatively, one can use the notion of the mutual information of two pieces to obtain a theoretical prescription of how to separate a scene into independent patterns and distinguish a pattern from an unstructured background in which it is imbedded (see Section 6).

Let us enumerate what we view as the main points in favor of this definition of organization: It is general, i.e. following von Neumann the details of the physics and chemistry of this universe are not involved; it measures organized structure rather than unstructured details; and it passes the spontaneous generation or "Pasteur" test, i.e. there is a very low probability of creating organization by chance without a long evolutionary process (this may be viewed as a way of restating Theorem 1 in Section 5). The second point is worth elaborating: The information content of an organism includes much irrelevant detail, and a bigger animal is necessarily more complex in this sense. *But if it were possible to calculate the mutual information of two arbitrary cells in a body at a given moment, we surmise that this would give a measure of the genetic information in a cell. This is because the irrelevant details in each of them, such as the exact position and velocity of each molecule, are uncorrelated and would cancel each other out.*

In addition to providing a definition of information content and of degree of organization, this approach also provides a definition of "organism" in the sense that a theoretical prescription is given for dissecting a scene into organisms and determining their boundaries, so that the measure of degree of organization can then be applied separately to each organism. However a strong note of caution is in order: We agree with [1] that a definition of "life" is valid as long as anything that satisfies the

TOWARD A MATHEMATICAL DEFINITION OF "LIFE"

definition and is likely to appear in the universe under consideration, either is alive or is a by-product of living beings or their activities. There certainly are structures satisfying our definition that are not alive (see Theorems 3 to 6 in Section 5); however, we believe that they would only be likely to arise as by-products of the activities of living beings.

In the succeeding sections we shall do the following: give a more formal presentation of the basic concepts of algorithmic information theory; discuss the notions of the independence and mutual information of groups of more than two objects; formally define H_d; evaluate $H_d(R)$ for some typical one-dimensional geometrical patterns R which we dub "gas," "crystal," "twins," "bilateral symmetry," and "hierarchy;" consider briefly the problem of decomposing scenes containing several independent patterns, and of determining the boundary of a pattern which is imbedded in an unstructured background; discuss briefly the two and higher dimension cases; and mention some alternative definitions of mutual information which have been proposed.

The next step in this program of research would be to proceed from static snapshots to time-varying situations, in other words, to set up a discrete universe with probabilistic state transitions and to show that there is a certain probability that a certain level of organization will be reached by a certain time. More generally, one would like to determine the probability distribution of the maximum degree of organization of any organism at time $t + \Delta$ as a function of it at time t. Let us propose an initial proof strategy for setting up a nontrivial example of the evolution of organisms: construct a series of intermediate evolutionary forms [11], argue that increased complexity gives organisms a selective advantage, and show that no primitive organism is so successful or lethal that it diverts or blocks this gradual evolutionary pathway. What would be the intellectual flavor of the theory we desire? It would be a quantitative formulation of Darwin's theory of evolution in a very general model universe setting. It would be the opposite of ergodic theory. Instead of showing that things mix and become uniform, it would show that variety and organization will probably increase.

Some final comments: Software is fast approaching biological levels of complexity, and hardware, thanks to very large scale integration, is not far behind. Because of this, we believe that the computer is now becoming a valid metaphor for the entire organism, not just for the brain [12]. Perhaps the most interesting example of this is the evolutionary phenomenon suffered by extremely large programs such as operating systems. It becomes very difficult to make changes in such programs, and the only alternative is to add new features rather than modify existing ones. The genetic program has been "patched up" much more and over a much longer period of time than even the largest operating systems, and Nature has accomplished this in much the same manner as systems programmers have, by carrying along all the previous code as new code is added [11]. The experimental proof of this is that ontogeny recapitulates phylogeny, i.e. each embryo to a certain extent recapitulates in the course of its development the evolutionary sequence that led to it. In this connection we should also mention the thesis developed in [13] that the information contained in the human brain is now comparable with the amount of information in the genes, and that intelligence plus education may be characterized as a way of getting around the limited modifiability and channel capacity of heredity. In other words, Nature, like computer designers, has decided that it is much more flexible to build general-purpose computers than to use heredity to "hardwire" each behavior pattern instinctively into a special-purpose computer.

2. Algorithmic Information Theory

We first summarize some of the basic concepts of algorithmic information theory in its most recent formulation [14-16].

This new approach leads to a formalism that is very close to that of classical probability theory and information theory, and is based on the notion that the tape containing the Turing machine's program is infinite and entirely filled with 0's and 1's. This forces programs to be self-delimiting; i.e. they must contain within themselves information about their size, since the computer cannot rely on a blank at the end of the program to indicate where it ends.

Consider a universal Turing machine U whose programs are in binary and are self-delimiting. By "self-delimiting" we mean, as was just explained, that they do not have blanks appended as endmarkers. By "universal" we mean that for any other Turing machine M whose programs p are in binary and are self-delimiting, there is a prefix μ such that $U(\mu p)$ always carries out the same computation as $M(p)$.

$H(X)$, the algorithmic information content of the finite object X, is defined to be the size in bits of the smallest self-delimiting program for U to compute X. This includes the proviso that U halt after printing X. There is absolutely no restriction on the running time or storage space used by this program. For example, X can be a natural number or a bit string or a tuple of natural numbers or bit strings. Note that variations in the definition of U give rise to at most $O(1)$ differences in the resulting H, by the definition of universality.

The self-delimiting requirement is adopted so that one gets the following basic subadditivity property of H:

$$H(\langle X,Y\rangle) \leq H(X) + H(Y) + O(1).$$

This inequality holds because one can concatenate programs. It expresses the notion of "adding information," or, in computer jargon, "using subroutines."

Another important consequence of this requirement is that a natural probability measure P, which we shall refer to as the algorithmic probability, can be associated with the result of any computation. $P(X)$ is the probability that X is obtained as output if the standard universal computer U is started running on a program tape filled with 0's and 1's by separate tosses of a fair coin. The algorithmic probability P and the algorithmic information content H are related as follows [14]:

$$H(X) = -\log P(X) + O(1). \tag{1}$$

Consider a binary string s. Define the function L as follows:

$$L(n) = \max \{H(s) : \text{length}(s) = n\}.$$

It can be shown [14] that $L(n) = n + H(n) + O(1)$, and that an overwhelming majority of the s of length n have $H(s)$ very close to $L(n)$. Such s have maximum information content and are highly random, patternless, incompressible, and typical. They are said to be "algorithmically random." The greater the difference between $H(s)$ and $L(\text{length}(s))$, the less random s is. It is convenient to say that "s is k-random" if $H(s) \geq L(n) - k$, where $n = \text{length}(s)$. There are at most

$$2^{n-k+O(1)}$$

n-bit strings which aren't k-random. As for natural numbers, most n have $H(n)$ very close to $L(\text{floor}(\log n))$ Here floor(x) is the greatest integer $\leq x$. Strangely enough, though most strings are random it is impossible to prove that specific strings have this property. For an explanation of this paradox and further references, see the section on metamathematics in [15], and also see [9].

We now make a few observations that will be needed later. First of all, $H(n)$ is a smooth function of n:

$$|H(n) - H(m)| = O(\log |n - m|). \tag{2}$$

(Note that this is not strictly true if $|n - m|$ is equal to 0 or 1, unless one considers the log of 0 and 1 to be 1; this convention is therefore adopted throughout this paper.) For a proof, see [16]. The following upper bound on $H(n)$ is an immediate corollary of this smoothness property: $H(n) = O(\log n)$. Hence if s is an n-bit string, then $H(s) \leq n + O(\log n)$. Finally, note that changes in the value of the argument of the function L produce nearly equal changes in the value of L. Thus, for any ε there is a δ such that $L(n) \geq L(m) + \varepsilon$ if $n \geq m + \delta$. This is because of the fact that $L(n) = n + H(n) + O(1)$ and the smoothness property (2) of H.

An important concept of algorithmic information theory that hasn't been mentioned yet is the conditional probability $P(Y|X)$, which by definition is $P(\langle X,Y \rangle)/P(X)$. To the conditional probability there corresponds the relative information content $H(Y|X^*)$, which is defined to be the size in bits of the smallest programs for the standard universal computer U to output Y if it is given X^*, a canonical minimum-size program for calculating X. X^* is defined to be the first $H(X)$-bit program to compute X that one encounters in a fixed recursive enumeration of the graph of U (i.e. the set of all ordered pairs of the form $\langle p, U(p) \rangle$). Note that there are partial recursive functions which map X^* to $\langle X, H(X) \rangle$ and back again, and so X^* may be regarded as an abbreviation for the ordered pair whose first element is the string X and whose second element is the natural number that is the complexity of X. We should also note the immediate corollary of (1) that minimum-size or nearly minimum-size programs are essentially unique: For any ε there is a δ such that for all X the cardinality of {the set of all programs for U to calculate X that are within ε bits of the minimum size $H(X)$} is less than δ. It is possible to prove the following theorem relating the conditional probability and the relative information content [14]:

$$H(Y^*|X) = -\log P(Y|X) + O(1). \qquad (3)$$

From (1) and (3) and the definition $P(\langle X,Y \rangle) = P(X)P(Y|X)$, one obtains this very basic decomposition:

$$H(\langle X,Y \rangle) = H(X) + H(Y|X^*) + O(1). \qquad (4)$$

3. Independence and Mutual Information

It is an immediate corollary of (4) that the following four quantities are all within $O(1)$ of each other:

$$\begin{cases} H(X) - H(X|Y^*), \\ H(Y) - H(Y|X^*), \\ H(X) + H(Y) - H(\langle X,Y \rangle), \\ H(Y) + H(X) - H(\langle Y,X \rangle). \end{cases}$$

These four quantities are known as the mutual information $H(X:Y)$ of X and Y; they measure the extent to which X and Y are interdependent. For if $P(\langle X,Y \rangle) \approx P(X)P(Y)$, then $H(X:Y) = O(1)$; and if Y is a recursive function of X, then $H(Y|X^*) = O(1)$ and $H(X:Y) = H(Y) + O(1)$. In fact,

$$H(X:Y) = -\log \left[\frac{P(X)P(Y)}{P(\langle X,Y \rangle)} \right] + O(1),$$

which shows quite clearly that $H(X:Y)$ is a symmetric measure of the independence of X and Y. Note that in algorithmic information theory, what is of importance is an approximate notion of independence and a measure of its degree (mutual information), rather than the exact notion. This is because the algorithmic probability may vary within a certain percentage depending on the choice of universal computer U. Conversely, information measures in algorithmic information theory should not vary by more than $O(1)$ depending on the choice of U.

To motivate the definition of the d-diameter complexity, we now discuss how to generalize the notion of independence and mutual information from a pair to an n-tuple of objects. In what follows classical and algorithmic probabilities are distinguished by using curly brackets for the first one and parentheses for the second. In probability theory the mutual independence of a set of n events $\{A_k : k < n\}$ is defined by the following 2^n equations:

$$\prod_{k \in S} P\{A_k\} = P\{\bigcap_{k \in S} A_k\}$$

for all $S \subset n$. Here the set-theoretic convention due to von Neumann is used that identifies the natural number n with the set $\{k : k < n\}$. In algorithmic probability theory the analogous condition would be to require that

$$\prod_{k \in S} P(A_k) \approx P(\bigodot_{k \in S} A_k) \tag{5}$$

for all $S \subset n$. Here ΘA_k denotes the tuple forming operation for a variable length tuple, i.e.

$$\bigodot_{k<n} A_k = \langle A_0, A_1, A_2, \ldots, A_{n-1} \rangle.$$

It is a remarkable fact that these 2^n conditions (5) are equivalent to the single requirement that

$$\prod_{k<n} P(A_k) \approx P(\bigodot_{k<n} A_k). \tag{6}$$

To demonstrate this it is necessary to make use of special properties of algorithmic probability that are not shared by general probability measures. In the case of a general probability space,

$$P\{A \cap B\} \geq P\{A\} + P\{B\} - 1$$

is the best lower bound on $P\{A \cap B\}$ that can in general be formulated in terms of $P\{A\}$ and $P\{B\}$. For example, it is possible for $P\{A\}$ and $P\{B\}$ to both be $1/2$, while $P\{A \cap B\} = 0$. In algorithmic information theory the situation is quite different. In fact one has:

$$P(\langle A,B \rangle) \geq c_2 P(A)P(B),$$

and this generalizes to any fixed number of objects:

$$P(\bigodot_{k<n} A_k) \geq c_n \prod_{k<n} P(A_k).$$

Thus if the joint algorithmic probability of a subset of the n-tuple of objects were significantly greater than the product of their individual algorithmic probabilities, then this would also hold for the entire n-tuple of objects. More precisely, for any $S \subset n$ one has

$$P(\bigodot_{k<n} A_k) \geq \hat{c}_n P(\bigodot_{k \in S} A_k) P(\bigodot_{k \in n-S} A_k) \geq \tilde{c}_n P(\bigodot_{k \in S} A_k) \prod_{k \in n-S} P(A_k).$$

Then if one assumes that

$$P(\bigodot_{k \in S} A_k) >> \prod_{k \in S} P(A_k)$$

(here $>>$ denotes "much greater than"), it follows that

TOWARD A MATHEMATICAL DEFINITION OF "LIFE"

$$P(\underset{k<n}{\ominus} A_k) \; >> \; \prod_{k<n} P(A_k)$$

We conclude that in algorithmic probability theory (5) and (6) are equivalent and thus (6) is a necessary and sufficient condition for an n-tuple to be mutually independent. Therefore the following measure of mutual information for n-tuples accurately characterizes the degree of interdependence of n objects:

$$\left[\sum_{k<n} H(A_k)\right] - H(\underset{k<n}{\ominus} A_k).$$

This measure of mutual information subsumes all others in the following precise sense:

$$\left[\sum_{k<n} H(A_k)\right] - H(\underset{k<n}{\ominus} A_k) \;=\; \max \; \{\left[\sum_{k\in S} H(A_k)\right] - H(\underset{k\in S}{\ominus} A_k)\} + O(1),$$

where the maximum is taken over all $S \subseteq n$.

4. Formal Definition of Hd

We can now present the definition of the d-diameter complexity $H_d(R)$. We assume a geometry: graph paper of some finite number of dimensions that is divided into unit cubes. Each cube is black or white, opaque or transparent, in other words, contains a 1 or a 0. Instead of requiring an output tape which is multidimensional, our universal Turing machine U outputs tuples giving the coordinates and the contents (0 or 1) of each unit cube in a geometrical object that it wishes to print. Of course geometrical objects are considered to be the same if they are translation equivalent. We choose for this geometry the city-block metric

$$D(X,Y) \;=\; \max \; |x_i - y_i| \,,$$

which is more convenient for our purposes than the usual metric. By a region we mean a set of unit cubes with the property that from any cube in it to any other one there is a path that only goes through other cubes in the region. To this we add the constraint which in the 3-dimensional case is that the connecting path must only pass through the interior and faces of cubes in the region, not through their edges or vertices. The diameter of an arbitrary region R is denoted by $|R|$, and is defined to be the minimum diameter $2r$ of a "sphere"

$$\{X : D(X,X_0) \leq r\}$$

which contains R. $H_d(R)$, the size in bits of the smallest programs which calculate R as the "sum" of independent regions of diameter $\leq d$, is defined as follows:

$$H_d(R) \;=\; \min \; \left[\alpha + \sum_{i<k} H(R_i)\right],$$

where

$$\alpha \;=\; H(R \mid \underset{i<k}{\ominus} R_i) + H(k),$$

the minimization being taken over all k and partitions of R into k-tuples $\ominus R_i$ of nonoverlapping regions with the property that $|R_i| < d$ for all $i < k$.

The discussion in Section 3 of independence and mutual information shows that $H_d(R)$ is a natural measure to consider. Excepting the α term, $H_d(R) - H(R)$ is simply the minimum attainable mutual information over any partition of R into nonoverlapping pieces all of size not greater than d. We shall see in Section 5 that in practice the min is attained with a small number of pieces and the α term is not very significant.

A few words about α, the number of bits of information needed to know how to assemble the pieces: The $H(k)$ term is included in α, as illustrated in Lemma 1 below, because it is the number of bits needed to tell U how many descriptions of pieces are to be read. The $H(R \mid \Theta R_i)$ term is included in α because it is the number of bits needed to tell U how to compute R given the k-tuple of its pieces. This is perhaps the most straight-forward formulation, and the one that is closest in spirit to Section 5 [5]. However, less information may suffice, e.g.

$$H(R \mid \langle k^*, \underset{i<k}{\Theta} (R_i^*) \rangle) + H(k)$$

bits. In fact, one could define α to be the minimum number of bits in a string which yields a program to compute the entire region when it is concatenated with minimum-size programs for all the pieces of the region; i.e. one could take

$$\alpha = \min \{ |p| : U(pR_0^*R_1^*R_2^* \ldots R_{k-1}^*) = R \}.$$

Here are two basic properties of H_d: If $d \geq |R|$, then $H_d(R) = H(R) + O(1)$; $H_d(R)$ increases monotonically as d decreases. $H_d(R) = H(R) + O(1)$ if $d \geq |R|$ because we have included the α term in the definition of $H_d(R)$. $H_d(R)$ increases as d decreases because one can no longer take advantage of patterns of diameter greater than d to describe R. The curve showing $H_d(R)$ as a function of d may be considered a kind of "Fourier spectrum" of R. Interesting things will happen to the curve at d which are the sizes of significant patterns in R.

Lemma 1: ("Subadditivity for n-tuples")

$$H(\underset{k<n}{\Theta} A_k) \leq c_n + \sum_{k<n} H(A_k).$$

Proof:

$$H(\underset{k<n}{\Theta} A_k)$$
$$= H(\langle n, \underset{k<n}{\Theta} A_k \rangle) + O(1)$$
$$= H(n) + H(\underset{k<n}{\Theta} A_k \mid n^*) + O(1)$$
$$\leq \hat{c} + H(n) + \sum_{k<n} H(A_k).$$

Hence one can take

$$c_n = \hat{c} + H(n).$$

5. Evaluation of Hd for Typical One-Dimensional Geometrical Patterns

Before turning to the examples, we present a lemma needed for estimating $H_d(R)$. The idea is simply that sufficiently large pieces of a random string are also random. It is required that the pieces be sufficiently large for the following reason: It is not difficult to see that for any j, there is an n so large that random strings of size greater than n must contain all 2^j possible subsequences of length j. In fact, for n sufficiently large the relative frequency of occurrence of all 2^j possible subsequences must approach the limit 2^{-j}.

Lemma 2: ("Random parts of random strings")
Consider an n-bit string s to be a loop. For any natural numbers i and j between 1 and n, consider the sequence u of contiguous bits from s starting at the i th and continuing around the loop to the jth. Then if s is k-random, its subsequence u is $(k + O(\log n))$-random.

Proof: The number of bits in u is $j - i + 1$ if j is $\geq i$, and is $n + j - i + 1$ if j is $< i$. Let v be the remainder of the loop s after u has been excised. Then we have $H(u) + H(v) + H(i) + O(1) \geq H(s)$. Thus $H(u) + n - |u| + O(\log n) \geq H(s)$, or $H(u) \geq H(s) - n + |u| + O(\log n)$. Thus if s is k-random, i.e. $H(s) \geq L(n) - k = n + H(n) - k + O(1)$, then u is x-random, where x is determined as follows: $H(u) \geq n + H(n) - k - n + |u| + O(\log n) = |u| + H(|u|) - k + O(\log n)$. That is to say, if s is k-random, then its subsequence u is $(k + O(\log n))$-random.

Lemma 3: ("Random prefixes of random strings")
Consider an n-bit string s. For any natural number j between 1 and n, consider the sequence u consisting of the first j bits of s. Then if s is k-random, its j-bit prefix u is $(O(\log j) + k)$-random.

Proof: Let the $(n - j)$-bit string v be the remainder of s after u is excised. Then we have $H(u) + H(v) + O(1) \geq H(s)$, and therefore $H(u) \geq H(s) - L(n - j) + O(1) = L(n) - k - L(n - j) + O(1)$ since s is k-random. Note that $L(n) - L(n - j) = j + H(n) - H(n - j) + O(1) = j + O(\log j)$, by the smoothness property (2) of H. Hence $H(u) \geq j + O(\log j) - k$. Thus if u is x-random (x as small as possible), we have $L(j) - x = j + O(\log j) - x \geq j + O(\log j) - k$. Hence $x \leq O(\log j) + k$.

Remark: Conversely, any random n-bit string can be extended by concatenating k bits to it in such a manner that the result is a random $(n + k)$-bit string. We shall not use this converse result, but it is included here for the sake of completeness.

Lemma 4: ("Random extensions of random strings")
Assume the string s is x-random. Consider a natural number k. Then there is a k-bit string e such that se is y-random, as long as k, x, and y satisfy a condition of the following form:

$$y \geq x + O(\log x) + O(\log k).$$

Proof: Assume on the contrary that the x-random string s has no y-random k-bit extension and $y \geq x + O(\log x) + O(\log k)$, i.e. $x < y + O(\log y) + O(\log k)$. From this assumption we shall derive a contradiction by using the fact that most strings of any particular size are y-random, i.e. the fraction of them that are y-random is at least

$$1 - 2^{-y + O(1)}.$$

It follows that the fraction of $|s|$-bit strings which have no y-random k-bit extension is less than

$$2^{-y+O(1)}.$$

Since by hypothesis no k-bit extension of s is y-random, we can uniquely determine s if we are given y and k and the ordinal number of the position of s in {the set of all $|s|$-bit strings which have no y-random k-bit extension} expressed as an $(|s| - y + O(1))$-bit string. Hence $H(s)$ is less than $L(|s| - y + O(1)) + H(y) + H(k) + O(1)$. In as much as $L(n) = n + H(n) + O(1)$ and $|H(n) - H(m)| = O(\log |n - m|)$, it follows that $H(s)$ is less than $L(|s|) - [y + O(\log y) + O(\log k)]$. Since s is by assumption x-random, i.e. $H(s) \geq L(|s|) - x$, we obtain a lower bound on x of the form $y + O(\log y) + O(\log k)$, which contradicts our original assumption that $x < y + O(\log y) + O(\log k)$.

Theorem 1: ("Gas")

Suppose that the region R is an $O(\log n)$-random n-bit string. Consider $d = n/k$, where n is large, and k is fixed and greater than zero. Then

$$H(R) = n + O(\log n), \quad \text{and} \quad H_d(R) = H(R) + O(\log H(R)).$$

Proof that: $H_d(R) \leq H(R) + O(\log H(R))$

Let β be concatenation of tuples of strings, i.e.

$$\beta\left(\bigodot_{i \leq k} R_i\right) = R_0 R_1 R_2 \ldots R_k.$$

Note that

$$H\left(\beta\left(\bigodot_{i \leq k} R_i\right) \,\middle|\, \bigodot_{i \leq k} R_i\right) = O(1).$$

Divide R into k successive strings of size floor($|R|/k$), with one (possibly null) string of size less than k left over at the end. Taking this choice of partition ΘR_i in the definition of $H_d(R)$, and using the fact that $H(s) \leq |s| + O(\log |s|)$, we see that

$$H_d(R) \leq O(1) + H(k+1) + \sum_{i \leq k} \{|R_i| + O(\log |R_i|)\}$$
$$\leq O(1) + n + (k+2)O(\log n) = n + O(\log n).$$

Proof that: $H_d(R) \geq H(R) + O(\log H(R))$

This follows immediately from the fact that $H_{|R|}(R) = H(R) + O(1)$ and $H_d(R)$ increases monotonically as d decreases.

Theorem 2: ("Crystal")

Suppose that the region R is an n-bit string consisting entirely of 1's, and that the base-two numeral for n is $O(\log\log n)$-random. Consider $d = n/k$, where n is large, and k is fixed and greater than zero. Then

$$H(R) = \log n + O(\log\log n), \quad \text{and} \quad H_d(R) = H(R) + O(\log H(R)).$$

Proof that: $H_d(R) \leq H(R) + O(\log H(R))$

If one considers using the concatenation function β for assembly as was done in the proof of Theorem 1, and notes that $H(1^n) = H(n) + O(1)$, one sees that it is sufficient to partition the natural number

TOWARD A MATHEMATICAL DEFINITION OF "LIFE"

n into $O(k)$ summands none of which is greater than n/k in such a manner that $H(n) + O(\log\log n)$ upper bounds the sum of the complexities of the summands. Division into equal size pieces will not do, because $H(\text{floor}(n/k)) = H(n) + O(1)$, and one only gets an upper bound of $kH(n) + O(1)$. It is necessary to proceed as follows: Let m be the greatest natural number such that $2^m \le n/k$. And let p be the smallest natural number such that $2^p > n$. By converting n to base-two notation, one can express n as the sum of $\le p$ distinct non-negative powers of two. Divide all these powers of two into two groups: those that are less than 2^m and those that are greater than or equal to 2^m. Let f be the sum of all the powers in the first group. f is $< 2^m \le n/k$. Let s be the sum of all the powers in the second group. s is a multiple of 2^m; in fact, it is of the form $t2^m$ with $t = O(k)$. Thus $n = f + s = f + t2^m$, where $f \le n/k$, $2^m \le n/k$, and $t = O(k)$. The complexity of 2^m is $H(m) + O(1) = O(\log m) = O(\log\log n)$. Thus the sum of the complexities of the t summands 2^m is also $O(\log\log n)$. Moreover, f when expressed in base-two notation has $\log k + O(1)$ fewer bit positions on the left than n does. Hence the complexity of f is $H(n) + O(1)$. In summary, we have $O(k)$ quantities n_i with the following properties:

$$n = \sum n_i, \quad n_i \le n/k, \quad \sum H(n_i) \le H(n) + O(\log\log n).$$

Thus $H_d(R) \le H(R) + O(\log H(R))$.

Proof that: $H_d(R) \ge H(R) + O(\log H(R))$

This follows immediately from the fact that $H_{|R|}(R) = H(R) + O(1)$ and $H_d(R)$ increases monotonically as d decreases.

Theorem 3: ("Twins")
For convenience assume n is even. Suppose that the region R consists of two repetitions of an $O(\log n)$-random $n/2$ -bit string u. Consider $d = n/k$, where n is large, and k is fixed and greater than unity. Then

$$H(R) = n/2 + O(\log n), \quad \text{and} \quad H_d(R) = 2\,H(R) + O(\log H(R)).$$

Proof that: $H_d(R) \le 2\,H(R) + O(\log H(R))$

The reasoning is the same as in the case of the "gas" (Theorem 1). Partition R into k successive strings of size floor($|R|/k$), with one (possibly null) string of size less than k left over at the end.

Proof that: $H_d(R) \ge 2\,H(R) + O(\log H(R))$

By the definition of $H_d(R)$, there is a partition $\Theta\,R_i$ of R into nonoverlapping regions which has the property that

$$H_d(R) = \alpha + \sum H(R_i), \quad \alpha = H(R \mid \Theta\,R_i) + H(k), \quad |R_i| \le d.$$

Classify the non-null R_i into three mutually exclusive sets A, B, and C: A is the set of all non-null R_i which come from the left half of R ("the first twin"), B is the (empty or singleton) set of all non-null R_i which come from both halves of R ("straddles the twins"), and C is the set of all non-null R_i which come from the right half of R ("the second twin"). Let $\dot{A}$, $\dot{B}$, and $\dot{C}$ be the sets of indices i of the regions R_i in A, B, and C, respectively. And let $\ddot{A}$, $\ddot{B}$, and $\ddot{C}$ be the three portions of R which contained the pieces in A, B, and C, respectively. Using the idea of Lemma 1, one sees that

$$H(\ddot{A}) \leq O(1) + H(\#(A)) + \sum_{i \in \dot{A}} H(R_i),$$

$$H(\ddot{B}) \leq O(1) + H(\#(B)) + \sum_{i \in \dot{B}} H(R_i),$$

$$H(\ddot{C}) \leq O(1) + H(\#(C)) + \sum_{i \in \dot{C}} H(R_i).$$

Here $\#$ denotes the cardinality of a set. Now $\ddot{A}$, $\ddot{B}$, and $\ddot{C}$ are each a substring of an $O(\log n)$-random $n/2$-bit string. This assertion holds for $\ddot{B}$ for the following two reasons: the $n/2$-bit string is considered to be a loop, and $|\ddot{B}| \leq d = n/k \leq n/2$ since k is assumed to be greater than 1. Hence, applying Lemma 2, one obtains the following inequalities:

$$|\ddot{A}| + O(\log n) \leq H(\ddot{A}), \quad |\ddot{B}| + O(\log n) \leq H(\ddot{B}), \quad |\ddot{C}| + O(\log n) \leq H(\ddot{C}).$$

Adding both of the above sets of three inequalities and using the facts that

$$|\ddot{A}| + |\ddot{B}| + |\ddot{C}| = |R| = n, \quad \#(A) \leq n/2, \quad \#(B) \leq 1, \quad \#(C) \leq n/2,$$

and that $H(m) = O(\log m)$, one sees that

$$n + O(\log n) \leq H(\ddot{A}) + H(\ddot{B}) + H(\ddot{C})$$
$$\leq O(1) + H(\#(A)) + H(\#(B)) + H(\#(C)) + \sum \{H(R_i) : i \in \dot{A} \cup \dot{B} \cup \dot{C}\}$$
$$\leq O(\log n) + \sum H(R_i).$$

Hence

$$H_d(R) \geq \sum H(R_i) \geq n + O(\log n) = 2 H(R) + O(\log H(R)).$$

Theorem 4: ("Bilateral Symmetry")
For convenience assume n is even. Suppose that the region R consists of an $O(\log n)$-random $n/2$-bit string u concatenated with its reversal. Consider $d = n/k$, where n is large, and k is fixed and greater than zero. Then

$$H(R) = n/2 + O(\log n), \quad \text{and} \quad H_d(R) = (2 - k^{-1}) H(R) + O(\log H(R)).$$

Proof: The proof is along the lines of that of Theorem 3, with one new idea. In the previous proof we considered $\ddot{B}$ which is the region R_i in the partition of R that straddles R's midpoint. Before $\ddot{B}$ was $O(\log |R|)$-random, but now it can be compressed into a program about half its size, i.e. about $|\ddot{B}|/2$ bits long. Hence the maximum departure from randomness for $\ddot{B}$ is for it to only be $O(\log |R|) + (|R|/2k)$-random, and this is attained by making $\ddot{B}$ as large as possible and having its midpoint coincide with that of R.

Theorem 5: ("Hierarchy")
For convenience assume n is a power of two. Suppose that the region R is constructed in the following fashion. Consider an $O(1)$-random $\log n$-bit string s. Start with the one-bit string 1, and successively concatenate the string with itself or with its bit by bit complement, so that its size doubles at each stage. At the ith stage, the string or its complement is chosen depending on whether the ith

TOWARD A MATHEMATICAL DEFINITION OF "LIFE"

bit of s is a 0 or a 1, respectively. Consider the resulting n-bit string R and $d = n/k$, where n is large, and k is fixed and greater than zero. Then

$$H(R) = \log n + O(\log\log n), \quad \text{and} \quad H_d(R) = k H(R) + O(\log H(R)).$$

Proof that: $H_d(R) \leq k H(R) + O(\log H(R))$

The reasoning is similar to the case of the upper bounds on $H_d(R)$ in Theorems 1 and 3. Partition R into k successive strings of size floor($|R|/k$), with one (possibly null) string of size less than k left over at the end.

Proof that: $H_d(R) \geq k H(R) + O(\log H(R))$

Proceeding as in the proof of Theorem 3, one considers a partition ΘR_i of R that realizes $H_d(R)$. Using Lemma 3, one can easily see that the following lower bound holds for any substring R_i of R:

$$H(R_i) \geq \max \{1, \log |R_i| - c \log\log |R_i|\}.$$

The max $\{1, \dots\}$ is because H is always greater than or equal to unity; otherwise U would have only a single output. Hence the following expression is a lower bound on $H_d(R)$:

$$\sum \Phi(|R_i|), \tag{7}$$

where

$$\Phi(x) = \max \{1, \log x - c \log\log x\}, \quad \sum |R_i| = |R| = n, \quad |R_i| \leq d.$$

It follows that one obtains a lower bound on (7) and thus on $H_d(R)$ by solving the following minimization problem: Minimize

$$\sum \Phi(n_i) \tag{8}$$

subject to the following constraints:

$$\sum n_i = n, \quad n_i \leq n/k, \quad n \text{ large}, \quad k \text{ fixed}.$$

Now to do the minimization. Note that as x goes to infinity, $\Phi(x)/x$ goes to the limit zero. Furthermore, the limit is never attained, i.e. $\Phi(x)/x$ is never equal to zero. Moreover, for x and y sufficiently large and x less than y, $\Phi(x)/x$ is greater than $\Phi(y)/y$. It follows that a sum of the form (8) with the n_i constrained as indicated is minimized by making the n_i as large as possible. Clearly this is achieved by taking all but one of the n_i equal to floor(n/k), with the last n_i equal to remainder(n/k). For this choice of n_i the value of (8) is

$$k[\log n + O(\log\log n)] + \Phi(\text{remainder}(n/k))$$
$$= k \log n + O(\log\log n) = k H(R) + O(\log H(R)).$$

Theorem 6: For convenience assume n is a perfect square. Suppose that the region R is an n-bit string consisting of $\sqrt{n}$ repetitions of an $O(\log n)$-random $\sqrt{n}$ bit string u. Consider $d = n/k$, where n is large, and k is fixed and greater than zero. Then

$$H(R) = \sqrt{n} + O(\log n), \quad \text{and} \quad H_d(R) = k H(R) + O(\log H(R)).$$

Proof that: $H_d(R) \leq k H(R) + O(\log H(R))$

The reasoning is identical to the case of the upper bound on $H_d(R)$ in Theorem 5.

Proof that: $H_d(R) \geq k H(R) + O(\log H(R))$

Proceeding as in the proof of Theorem 5, one considers a partition ΘR_i of R that realizes $H_d(R)$. Using Lemma 2, one can easily see that the following lower bound holds for any substring R_i of R:

$$H(R_i) \geq \max \{1, -c \log n + \min \{\sqrt{n}, |R_i|\}\}.$$

Hence the following expression is a lower bound on $H_d(R)$:

$$\sum \Phi_n(|R_i|), \tag{9}$$

where

$$\Phi_n(x) = \max \{1, -c \log n + \min \{\sqrt{n}, x\}\}, \quad \sum |R_i| = |R| = n, \quad |R_i| \leq d.$$

It follows that one obtains a lower bound on (9) and thus on $H_d(R)$ by solving the following minimization problem: Minimize

$$\sum \Phi_n(n_i) \tag{10}$$

subject to the following constraints:

$$\sum n_i = n, \quad n_i \leq n/k, \quad n \text{ large}, \quad k \text{ fixed}.$$

Now to do the minimization. Consider $\Phi_n(x)/x$ as x goes from 1 to n. It is easy to see that this ratio is much smaller, on the order of $1/\sqrt{n}$, for x near to n than it is for x anywhere else in the interval from 1 to n. Also, for x and y both greater than $\sqrt{n}$ and x less than y, $\Phi_n(x)/x$ is greater than $\Phi_n(y)/y$. It follows that a sum of the form (10) with the n_i constrained as indicated is minimized by making the n_i as large as possible. Clearly this is achieved by taking all but one of the n_i equal to $\text{floor}(n/k)$, with the last n_i equal to $\text{remainder}(n/k)$. For this choice of n_i the value of (10) is

$$k[\sqrt{n} + O(\log n)] + \Phi_n(\text{remainder}(n/k))$$
$$= k\sqrt{n} + O(\log n) = k H(R) + O(\log H(R)).$$

6. Determining Boundaries of Geometrical Patterns

What happens to the structures of Theorems 3 to 6 if they are imbedded in a gas or crystal, i.e. in a random or constant 0 background? And what about scenes with several independent structures imbedded in them – do their degrees of organization sum together? Is our definition sufficiently robust to work properly in these circumstances?

This raises the issue of determining the boundaries of structures. It is easy to pick out the hierarchy of Theorem 5 from an unstructured background. Any two "spheres" of diameter δ will have a high mutual information given δ^* if and only if they are both in the hierarchy instead of in the background. Here we are using the notion of the mutual information of X and Y given Z, which is denoted $H(X:Y|Z)$, and is defined to be $H(X|Z) + H(Y|Z) - H(\langle X,Y \rangle|Z)$. The special case of this concept that we are interested in, however, can be expressed more simply: for if X and Y are both strings of length n, then it can be shown that $H(X:Y|n^*) = H(X|Y) - H(n)$. This is done by using the de-

TOWARD A MATHEMATICAL DEFINITION OF "LIFE"

composition (4) and the fact that since X and Y are both of length n, $H(\langle n,X\rangle) = H(X) + O(1)$, $H(\langle n,Y\rangle) = H(Y) + O(1)$, and $H(\langle n, \langle X,Y\rangle\rangle) = H(\langle X,Y\rangle) + O(1)$, and thus

$$H(X \mid n^*) = H(X) - H(n) + O(1),$$
$$H(Y \mid n^*) = H(Y) - H(n) + O(1),$$
$$H(\langle X,Y\rangle \mid n^*) = H(\langle X,Y\rangle) - H(n) + O(1).$$

How can one dissect a structure from a comparatively unorganized background in the other cases, the structures of Theorems 3, 4, and 6? The following definition is an attempt to provide a tool for doing this: An ε, δ-pattern R is a maximal region ("maximal" means not extensible, not contained in a bigger region R' which is also an ε, δ-pattern) with the property that for any δ-diameter sphere R_1 in R there is a disjoint δ-diameter sphere R_2 in R such that

$$H(R_1:R_2 \mid \delta^*) \geq \varepsilon.$$

The following questions immediately arise: What is the probability of having an ε, δ-pattern in an n-bit string, i.e. what proportion of the n-bit strings contain an ε, δ-pattern? This is similar to asking what is the probability that an n-bit string s satisfies

$$H_{n/k}(s) - H(s) > x.$$

A small upper bound on the latter probability can be derived from Theorem 1.

7. Two and Higher Dimension Geometrical Patterns

We make a few brief remarks.

In the general case, to say that a geometrical object O is "random" means $H(O \mid \text{shape}(O)^*) \approx \text{volume}(O)$, or $H(O) \approx \text{volume}(O) + H(\text{shape}(O))$. Here shape($O$) denotes the object O with all the 1's that it contains in its unit cubes changed to 0's. Here are some examples: A random n by n square has complexity

$$n^2 + H(n) + O(1).$$

A random n by m rectangle doesn't have complexity $nm + H(n) + H(m) + O(1)$, for if $m = n$ this states that a random n by n square has complexity

$$n^2 + 2H(n) + O(1),$$

which is false. Instead a random n by m rectangle has complexity $nm + H(\langle n,m\rangle) + O(1) = nm + H(n) + H(m \mid n^*) + O(1)$, which gives the right answer for $m = n$, since $H(n \mid n^*) = O(1)$. One can show that most n by m rectangles have complexity $nm + H(\langle n,m\rangle) + O(1)$, and less than two raised to the $nm - k + O(1)$ have complexity less than $nm + H(\langle n,m\rangle) - k$.

Here is a two-dimensional version of Lemma 2: Any large chunk of a random square which has a shape that is easy to describe, must itself be random.

8. Common Information

We should mention some new concepts that are closely related to the notion of mutual information. They are called measures of common information. Here are three different expressions defining the common information content of two strings X and Y. In them the parameter ε denotes a small tolerance, and as before $H(X:Y \mid Z)$ denotes $H(X \mid Z) + H(Y \mid Z) - H(\langle X,Y\rangle \mid Z)$.

$$\max \ \{H(Z) : H(Z \mid X^*) < \varepsilon \ \& \ H(Z \mid Y^*) < \varepsilon\}$$
$$\min \ \{H(\langle X,Y \rangle{:}Z) : H(X{:}Y \mid Z^*) < \varepsilon\}$$
$$\min \ \{H(Z) : H(X{:}Y \mid Z^*) < \varepsilon\}$$

Thus the first expression for the common information of two strings defines it to be the maximum information content of a string that can be extracted easily from both, the second defines it to be the minimum of the mutual information of the given strings and any string in the light of which the given strings look nearly independent, and the third defines it to be the minimum information content of a string in the light of which the given strings appear nearly independent. Essentially these definitions of common information are given in [17-19]. [17] considers an algorithmic formulation of its common information measure, while [18] and [19] deal exclusively with the classical ensemble setting.

Appendix 1: Errors in [5]

...The definition of the d-diameter complexity given in [5] has a basic flaw which invalidates the entries for $R = R_2$, R_3, and R_4 and $d = n/k$ the table in [5]: It is insensitive to changes in the diameter d ...

There is also another error in the table in [5], even if we forget the flaw in the definition of the d-diameter complexity. The entry for the crystal is wrong, and should read $\log n$ rather than $k \log n$ (see Theorem 2 in Section 5 of this paper).

Appendix 2: An Information-Theoretic Proof That There Are Infinitely Many Primes

It is of methodological interest to use widely differing techniques in elementary proofs of Euclid's theorem that there are infinitely many primes. For example, see Chapter II of Hardy and Wright [20], and also [21-23]. Recently Billingsley [24] has given an information-theoretic proof of Euclid's theorem. The purpose of this appendix is to point out that there is an information-theoretic proof of Euclid's theorem that utilizes ideas from algorithmic information theory instead of the classical measure-theoretic setting employed by Billingsley. We consider the algorithmic entropy $H(n)$, which applies to individual natural numbers n instead of to ensembles.

The proof is by *reductio ad absurdum*. Suppose on the contrary that there are only finitely many primes $p_1, \ldots, p_k$. Then one way to specify algorithmically an arbitrary natural number

$$n = \prod p_i^{e_i}$$

is by giving the k-tuple $\langle e_1, \ldots, e_k \rangle$ of exponents in any of its prime factorizations (we pretend not to know that the prime factorization is unique). Thus we have

$$H(n) \leq H(\langle e_1, \ldots, e_k \rangle) + O(1).$$

By the subadditivity of algorithmic entropy we have

$$H(n) \leq \sum H(e_i) + O(1).$$

Let us examine this inequality. Most n are algorithmically random and so the left-hand side is usually $\log n + O(\log\log n)$. As for the right-hand side, since

$$n \geq p_i^{e_i} \geq 2^{e_i},$$

each e_i is $\leq \log n$. Thus $H(e_i) \leq \log\log n + O(\log\log\log n)$. So for random n we have

$$\log n + O(\log\log n) \leq k[\log\log n + O(\log\log\log n)],$$

where k is the assumed finite number of primes. This last inequality is false for large n, as it assuredly is not the case that $\log n = O(\log\log n)$. Thus our initial assumption that there are only k primes is refuted, and there must in fact be infinitely many primes.

This proof is merely a formalization of the observation that if there were only finitely many primes, the prime factorization of a number would usually be a much more compact representation for it than its base-two numeral, which is absurd. This proof appears, formulated as a counting argument, in Section 2.6 of the 1938 edition of Hardy and Wright [20]; we believe that it is also quite natural to present it in an information-theoretic setting.

Bibliography

1. L. E. Orgel, *The Origins of Life: Molecules and Natural Selection,* Wiley, New York, 1973, pp. 187-197.

2. P. H. A. Sneath, *Planets and Life,* Funk and Wagnalls, New York, 1970, pp. 54-71.

3. G. J. Chaitin, "Information-Theoretic Computational Complexity," *IEEE Trans. Info. Theor.* IT-20 (1974), pp. 10-15.

4. G. J. Chaitin, "Randomness and Mathematical Proof," *Sci. Amer.* 232, No. 5 (May 1975), pp. 47-52.

5. G. J. Chaitin, "To a Mathematical Definition of 'Life'," *ACM SICACT News* 4 (Jan. 1970), pp. 12-18.

6. J. von Neumann, "The General and Logical Theory of Automata," *John von Neumann – Collected Works, Volume V,* A. H. Taub (ed.), Macmillan, New York, 1963, pp. 288-328.

7. J. von Neumann, *Theory of Self-Reproducing Automata,* Univ. Illinois Press, Urbana, 1966, pp. 74-87; edited and completed by A. W. Burks.

8. R. J. Solomonoff, "A Formal Theory of Inductive Inference," *Info. & Contr.* 7 (1964), pp. 1-22, 224-254.

9. G. J. Chaitin and J. T. Schwartz, "A Note on Monte Carlo Primality Tests and Algorithmic Information Theory," *Comm. Pure & Appl. Math.,* to appear.

10. C. E. Shannon and W. Weaver, *The Mathematical Theory of Communication,* Univ. Illinois Press, Urbana, 1949.

11. H. A. Simon, *The Sciences of the Artificial,* MIT Press, Cambridge, MA, 1969, pp. 90-97, 114-117.

12. J. von Neumann, *The Computer and the Brain,* Silliman Lectures Series, Yale Univ. Press, New Haven, CT, 1958.

13. C. Sagan, *The Dragons of Eden – Speculations on the Evolution of Human Intelligence,* Random House, New York, 1977, pp. 19-47.

14. G. J. Chaitin, "A Theory of Program Size Formally Identical to Information Theory," *J. ACM* 22 (1975), pp. 329-340.

15. G. J. Chaitin, "Algorithmic Information Theory," *IBM J. Res. Develop.* 21 (1977), pp. 350-359, 496.

16. R. M. Solovay, "On Random R.E. Sets," *Non-Classical Logics, Model Theory, and Computability,* A. I. Arruda, N. C. A. da Costa, and R. Chuaqui (eds.), North-Holland, Amsterdam, 1977, pp. 283-307.

17. P. Gács and J. Körner, "Common Information Is Far Less Than Mutual Information," *Prob. Contr. & Info. Theor.* 2, No. 2 (1973), pp. 149-162.

18. A. D. Wyner, "The Common Information of Two Dependent Random Variables," *IEEE Trans. Info. Theor.* IT-21 (1975), pp. 163-179.

19. H. S. Witsenhausen, "Values and Bounds for the Common Information of Two Discrete Random Variables," *SIAM J. Appl. Math.* 31 (1976), pp. 313-333.

20. G. H. Hardy and E. M. Wright, *An Introduction to the Theory of Numbers,* Clarendon Press, Oxford, 1962.

21. G. H. Hardy, *A Mathematician's Apology,* Cambridge University Press, 1967.

22. G. H. Hardy, *Ramanujan – Twelve Lectures on Subjects Suggested by His Life and Work,* Chelsea, New York, 1959.

23. H. Rademacher and O. Toeplitz, *The Enjoyment of Mathematics,* Princeton University Press, 1957.

24. P. Billingsley, "The Probability Theory of Additive Arithmetic Functions," *Ann. of Prob.* 2 (1974), pp. 749-791.

25. A. W. Burks (ed.), *Essays on Cellular Automata,* Univ. Illinois Press, Urbana, 1970.

26. M. Eigen, "The Origin of Biological Information," *The Physicist's Conception of Nature,* J. Mehra (ed.), D. Reidel Publishing Co., Dordrecht-Holland, 1973, pp. 594-632.

27. R. Landauer, "Fundamental Limitations in the Computational Process," *Ber. Bunsenges. Physik. Chem.* 80 (1976), pp. 1048-1059.

28. H. P. Yockey, "A Calculation of the Probability of Spontaneous Biogenesis by Information Theory," *J. Theor. Biol.* 67 (1977), pp. 377-398.

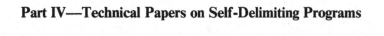

Part IV—Technical Papers on Self-Delimiting Programs

A THEORY OF PROGRAM SIZE FORMALLY IDENTICAL TO IN-FORMATION THEORY

Journal of the ACM 22 (1975), pp. 329-340.

GREGORY J. CHAITIN[27]
IBM Thomas J. Watson Research Center, Yorktown Heights, New York

Abstract

A new definition of program-size complexity is made. $H(A,B/C,D)$ is defined to be the size in bits of the shortest self-delimiting program for calculating strings A and B if one is given a minimal-size self-delimiting program for calculating strings C and D. This differs from previous definitions: (1) programs are required to be self-delimiting, i.e. no program is a prefix of another, and (2) instead of being given C and D directly, one is given a program for calculating them that is minimal in size. Unlike previous definitions, this one has precisely the formal properties of the entropy concept of information theory. For example, $H(A,B) = H(A) + H(B/A) + O(1)$. Also, if a program of length k is assigned measure 2^{-k}, then $H(A) = -\log_2$ (the probability that the standard universal computer will calculate A) + $O(1)$.

Key Words and Phrases

computational complexity, entropy, information theory, instantaneous code, Kraft inequality, minimal program, probability theory, program size, random string, recursive function theory, Turing machine

CR Categories

5.25, 5.26, 5.27, 5.5, 5.6

1. Introduction

There is a persuasive analogy between the entropy concept of information theory and the size of programs. This was realized by the first workers in the field of program-size complexity, Solomonoff [1], Kolmogorov [2], and Chaitin [3,4], and it accounts for the large measure of success of subsequent work in this area. However, it is often the case that results are cumbersome and have unpleasant error terms. These ideas cannot be a tool for general use until they are clothed in a powerful formalism like that of information theory.

---

[27]   Copyright © 1975, Association for Computing Machinery, Inc. General permission to republish, but not for profit, all or part of this material is granted provided that ACM's copyright notice is given and that reference is made to the publication, to its date of issue, and to the fact that reprinting privileges were granted by permission of the Association for Computing Machinery.

This paper was written while the author was a visitor at the IBM Thomas J. Watson Research Center, Yorktown Heights, New York, and was presented at the IEEE International Symposium on Information Theory, Notre Dame, Indiana, October 1974.

Author's present address: Rivadavia 3580, Dpto. 10A, Buenos Aires, Argentina.

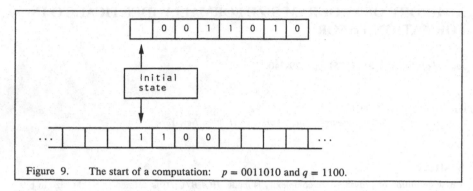

Figure 9. The start of a computation: $p = 0011010$ and $q = 1100$.

This opinion is apparently not shared by all workers in this field (see Kolmogorov [5]), but it has led others to formulate alternative definitions of program-size complexity, for example, Loveland's uniform complexity [6] and Schnorr's process complexity [7]. In this paper we present a new concept of program-size complexity. What train of thought led us to it?

Following [8, Sec. VI, p.7], think of a computer as decoding equipment at the receiving end of a noiseless binary communications channel. Think of its programs as code words, and of the result of the computation as the decoded message. Then it is natural to require that the programs/code words form what is called an "instantaneous code," so that successive messages sent across the channel (e.g. subroutines) can be separated. Instantaneous codes are well understood by information theorists [9-12]; they are governed by the Kraft inequality, which therefore plays a fundamental role in this paper.

One is thus led to define the relative complexity $H(A,B/C,D)$ of A and B with respect to C and D to be the size of the shortest self-delimiting program for producing A and B from C and D. However, this is still not quite right. Guided by the analogy with information theory, one would like

$$H(A,B) = H(A) + H(B/A) + \Delta$$

to hold with an error term Δ bounded in absolute value. But, as is shown in the Appendix, $|\Delta|$ is unbounded. So we stipulate instead that $H(A,B/C,D)$ is the size of the smallest self-delimiting program that produces A and B when it is given *a minimal-size self-delimiting program* for C and D. Then it can be shown that $|\Delta|$ is bounded.

In Sections 2-4 we define this new concept formally, establish the basic identities, and briefly consider the resulting concept of randomness or maximal entropy.

We recommend reading Willis [13]. In retrospect it is clear that he was aware of some of the basic ideas of this paper, though he developed them in a different direction. Chaitin's study [3,4] of the state complexity of Turing machines may be of interest, because in his formalism programs can also be concatenated. To compare the properties of our entropy function H with those it has in information theory, see [9-12]; to contrast its properties with those of previous definitions of program-size complexity, see [14]. Cover [15] and Gewirtz [16] use our new definition. See [17-32] for other applications of information/entropy concepts.

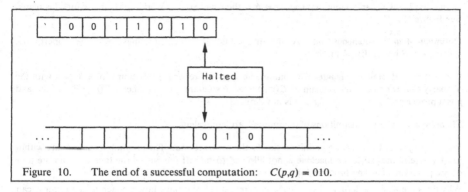

Figure 10.   The end of a successful computation: $C(p,q) = 010$.

2. Definitions

$X = \{\Lambda,0,1,00,01,10,11,000, \dots \}$ is the set of finite binary strings, and X^∞ is the set of infinite binary strings. Henceforth we shall merely say "string" instead of "binary string," and a string will be understood to be finite unless the contrary is explicitly stated. X is ordered as indicated, and $|s|$ is the length of the string s. The variables p, q, s, and t denote strings. The variables α and ω denote infinite strings. α_n is the prefix of α of length n. $N = \{0, 1, 2, \dots \}$ is the set of natural numbers. The variables c, i, j, k, m, and n denote natural numbers. R is the set of positive rationals. The variable r denotes an element of R. We write "r.e." instead of "recursively enumerable," "lg" instead of "$\log_2$, " and sometimes "$2\uparrow(x)$" instead of "2^x." $\#(S)$ is the cardinality of the set S.

Concrete Definition of a Computer: A computer C is a Turing machine with two tapes, a program tape and a work tape. The program tape is finite in length. Its leftmost square is called the dummy square and always contains a blank. Each of its remaining squares contains either a 0 or a 1. It is a read-only tape, and has one read head on it which can move only to the right. The work tape is two-way infinite and each of its squares contains either a 0, a 1, or a blank. It has one read-write head on it.

At the start of a computation the machine is in its initial state, the program p occupies the whole program tape except for the dummy square, and the read head is scanning the dummy square. The work tape is blank except for a single string q whose leftmost symbol is being scanned by the read-write head. Note that q can be equal to Λ. In that case the read-write head initially scans a blank square. p can also be equal to Λ. In that case the program tape consists solely of the dummy square. See Figure 9 on page 104.

During each cycle of operation the machine may halt, move the read head of the program tape one square to the right, move the read-write head of the work tape one square to the left or to the right, erase the square of the work tape being scanned, or write a 0 or a 1 on the square of the work tape being scanned. The the machine changes state. The action performed and the next state are both functions of the present state and the contents of the two squares being scanned, and are indicated in two finite tables with nine columns and as many rows as there are states.

If the Turing machine eventually halts with the read head of the program tape scanning its rightmost square, then the computation is a success. If not, the computation is a failure. $C(p,q)$ denotes the result of the computation. If the computation is a failure, then $C(p,q)$ is undefined. If it is a success, then $C(p,q)$ is the string extending to the right from the square of the work tape that is being scanned to the first blank square. Note that $C(p,q) = \Lambda$ if the square of the work tape being scanned is blank. See Figure 10.

Definition of an Instantaneous Code: An instantaneous code is a set of strings S with the property that no string in S is a prefix of another.

Abstract Definition of a Computer: A computer is a partial recursive function $C:X \times X \rightarrow X$ with the property that for each q the domain of $C(., q)$ is an instantaneous code; i.e. if $C(p,q)$ is defined and p is a proper prefix of p', then $C(p', q)$ is not defined.

Theorem 2.1: The two definitions of a computer are equivalent.

Proof: Why does the concrete definition satisfy the abstract one? The program must indicate within itself where it ends since the machine is not allowed to run off the end of the tape or to ignore part of the program. Thus no program for a successful computation is the prefix of another.

Why does the abstract definition satisfy the concrete one? We show how a concrete computer C can simulate an abstract computer C'. The idea is that C should read another square of its program tape only when it is sure that this is necessary.

Suppose C found the string q on its work tape. C then generates the r.e. set $S = \{p \mid C'(p,q) \text{ is defined}\}$ on its work tape.

As it generates S, C continually checks whether or not that part p of the program that it has already read is a prefix of some known element s of S. Note that initially $p = \Lambda$.

Whenever C finds that p is a prefix of an $s \in S$, it does the following. If p is a proper prefix of s, C reads another square of the program tape. And if $p = s$, C calculates $C'(p,q)$ and halts, indicating this to be the result of the computation. Q.E.D.

Definition of an Optimal Universal Computer: U is an optimal universal computer iff for each computer C there is a constant sim(C) with the following property: if $C(p,q)$ is defined, then there is a p' such that $U(p', q) = C(p,q)$ and $|p'| \leq |p| + \text{sim}(C)$.

Theorem 2.2: There is an optimal universal computer U.

Proof: U reads its program tape until it gets to the first 1. If U has read i 0's, it then simulates C_i, the i th computer (i.e. the computer with the ith pair of tables in a recursive enumeration of all possible pairs of defining tables), using the remainder of the program tape as the program for C_i. Thus if $C_i(p,q)$ is defined, then $U(0^i 1 p,q) = C_i(p,q)$. Hence U satisfies the definition of an optimal universal computer with sim(C_i) $= i + 1$. Q.E.D.

We somehow pick out a particular optimal universal computer U as the standard one for use throughout the rest of this paper.

Definition of Canonical Programs, Complexities, and Probabilities:

(a) The canonical program.

- $s^* = \min p \ (U(p, \Lambda) = s)$.

I.e. s^* is the first element in the ordered set X of all strings that is a program for U to calculate s.

A THEORY OF PROGRAM SIZE FORMALLY IDENTICAL TO INFORMATION THEORY

(b) Complexities.

- $H_C(s) = \min |p| \ (C(p, \Lambda) = s)$ (may be ∞),
- $H(s) = H_U(s)$,
- $H_C(s/t) = \min |p| \ (C(p,t^*) = s)$ (may be ∞),
- $H(s/t) = H_U(s/t)$.

(c) Probabilities.

- $P_C(s) = \Sigma \, 2^{-|p|} \ (C(p, \Lambda) = s)$,
- $P(s) = P_U(s)$,
- $P_C(s/t) = \Sigma \, 2^{-|p|} \ (C(p,t^*) = s)$,
- $P(s/t) = P_U(s/t)$.

Remark on Nomenclature: There are two different sets of terminology for these concepts, one derived from computational complexity and the other from information theory. $H(s)$ may be referred to as the information-theoretic or program-size complexity, and $H(s/t)$ may be referred to as the relative information-theoretic or program-size complexity. Or $H(s)$ and $H(s/t)$ may be termed the algorithmic entropy and the conditional algorithmic entropy, respectively. Similarly, this field might be referred to as "information-theoretic complexity" or as "algorithmic information theory."

Remark on the Definition of Probabilities: There is a very intuitive way of looking at the definition of P_C. Change the definition of the computer C so that the program tape is infinite to the right, and remove the (now impossible) requirement for a computation to be successful that the rightmost square of the program tape is being scanned when C halts. Imagine each square of the program tape except for the dummy square to be filled with a 0 or a 1 by a separate toss of a fair coin. Then the probability that the result s is obtained when the work tape is initially blank is $P_C(s)$, and the probability that the result s is obtained when the work tape initially has t^* on it is $P_C(s/t)$.

Theorem 2.3:
(a) $H(s) \leq H_C(s) + \text{sim}(C)$,
(b) $H(s/t) \leq H_C(s/t) + \text{sim}(C)$,
(c) $s^* \neq \Lambda$,
(d) $s = U(s^*, \Lambda)$,
(e) $H(s) = |s^*|$,
(f) $H(s) \neq \infty$,
(g) $H(s/t) \neq \infty$,
(h) $0 \leq P_C(s) \leq 1$,
(i) $0 \leq P_C(s/t) \leq 1$,
(j) $1 \geq \Sigma_s P_C(s)$,
(k) $1 \geq \Sigma_s P_C(s/t)$,
(l) $P_C(s) \geq 2\dagger(-H_C(s))$,
(m) $P_C(s/t) \geq 2\dagger(-H_C(s/t))$,
(n) $0 < P(s) < 1$,
(o) $0 < P(s/t) < 1$,
(p) $\#(\{s \mid H_C(s) < n\}) < 2^n$,
(q) $\#(\{s \mid H_C(s,t) < n\}) < 2^n$,
(r) $\#(\{s \mid P_C(s) > r\}) < 1/r$,
(s) $\#(\{s \mid P_C(s/t) > r\}) < 1/r$.

Proof: These are immediate consequences of the definitions. Q.E.D.

Definition of Tuples of Strings: Somehow pick out a particular recursive bijection $b{:}X \times X \to X$ for use throughout the rest of this paper. The 1-tuple $\langle s_1 \rangle$ is defined to be the string s_1. For $n \geq 2$ the n-tuple $\langle s_1, \ldots, s_n \rangle$ is defined to be the string $b(\langle s_1, \ldots, s_{n-1} \rangle, s_n)$.

Extensions of the Previous Concepts to Tuples of Strings ($n \geq 1,\ m \geq 1$).

- $H_C(s_1, \ldots, s_n) = H_C(\langle s_1, \ldots, s_n \rangle)$,
- $H_C(s_1, \ldots, s_n/t_1, \ldots, t_m) = H_C(\langle s_1, \ldots, s_n \rangle / \langle t_1, \ldots, t_m \rangle)$,
- $H(s_1, \ldots, s_n) = H_U(s_1, \ldots, s_n)$,
- $H(s_1, \ldots, s_n/t_1, \ldots, t_m) = H_U(s_1, \ldots, s_n/t_1, \ldots, t_m)$,
- $P_C(s_1, \ldots, s_n) = P_C(\langle s_1, \ldots, s_n \rangle)$,
- $P_C(s_1, \ldots, s_n/t_1, \ldots, t_m) = P_C(\langle s_1, \ldots, s_n \rangle / \langle t_1, \ldots, t_m \rangle)$,
- $P(s_1, \ldots, s_n) = P_U(s_1, \ldots, s_n)$,
- $P(s_1, \ldots, s_n/t_1, \ldots, t_m) = P_U(s_1, \ldots, s_n/t_1, \ldots, t_m)$.

Definition of the Information in One Tuple of Strings About Another ($n \geq 1,\ m \geq 1$).

- $I_C(s_1, \ldots, s_n{:}t_1, \ldots, t_m) = H_C(t_1, \ldots, t_m) - H_C(t_1, \ldots, t_m/s_1, \ldots, s_n)$,
- $I(s_1, \ldots, s_n{:}t_1, \ldots, t_m) = I_U(s_1, \ldots, s_n{:}t_1, \ldots, t_m)$.

Extensions of the Previous Concepts to Natural Numbers: We have defined H, P, and I for tuples of strings. This is now extended to tuples each of whose elements may either be a string or a natural number. We do this by identifying the natural number n with the nth string ($n = 0,1,2, \ldots$). Thus, for example, "$H(n)$" signifies "H(the nth element of X), " and "$U(p, \Lambda) = n$" stands for "$U(p, \Lambda)$ = the nth element of X."

3. Basic Identities

This section has two objectives. The first is to show that H and I satisfy the fundamental inequalities and identities of information theory to within error terms of the order of unity. For example, the information in s about t is nearly symmetrical. The second objective is to show that P is approximately a conditional probability measure: $P(t/s)$ and $P(s,t)/P(s)$ are within a constant multiplicative factor of each other.

The following notation is convenient for expressing these approximate relationships. $O(1)$ denotes a function whose absolute value is less than or equal to c for all values of its arguments. And $f \approx g$ means that the functions f and g satisfy the inequalities $cf \geq g$ and $f \leq cg$ for all values of their arguments. In both cases $c \in N$ is an unspecified constant.

Theorem 3.1:

(a) $H(s,t) = H(t,s) + O(1)$,

(b) $H(s/s) = O(1)$,

(c) $H(H(s)/s) = O(1)$,

(d) $H(s) \leq H(s,t) + O(1)$,

(e) $H(s/t) \leq H(s) + O(1)$,

(f) $H(s,t) \leq H(s) + H(t/s) + O(1)$,

(g) $H(s,t) \leq H(s) + H(t) + O(1)$,

(h) $I(s{:}t) \geq O(1)$,

A THEORY OF PROGRAM SIZE FORMALLY IDENTICAL TO INFORMATION THEORY

(i) $I(s{:}t) \le H(s) + H(t) - H(s,t) + O(1)$,

(j) $I(s{:}s) = H(s) + O(1)$,

(k) $I(\Lambda{:}s) = O(1)$,

(l) $I(s{:}\Lambda) = O(1)$.

Proof: These are easy consequences of the definitions. The proof of Theorem 3.1(f) is especially interesting, and is given in full below. Also, note that Theorem 3.1(g) follows immediately from Theorem 3.1(f,e), and Theorem 3.1(i) follows immediately from Theorem 3.1(f) and the definition of I.

Now for the proof of Theorem 3.1(f). We claim that there is a computer C with the following property. If $U(p,s^*) = t$ and $|p| = H(t/s)$ (i.e. if p is a minimal-size program for calculating t from s^*), then $C(s^*p, \Lambda) = \langle s, t \rangle$. By using Theorem 2.3(e,a) we see that $H_C(s,t) \le |s^*p| = |s^*| + |p| = H(s) + H(t/s)$, and $H(s,t) \le H_C(s,t) + \text{sim}(C) \le H(s) + H(t/s) + O(1)$.

It remains to verify the claim that there is such a computer. C does the following when it is given the program s^*p on its program tape and the string Λ on its work tape. First it simulates the computation that U performs when given the same program and work tapes. In this manner C reads the program s^* and calculates s. Then it simulates the computation that U performs when given s^* on its work tape and the remaining portion of C's program tape. In this manner C reads the program p and calculates t from s^*. The entire program tape has now been read, and both s and t have been calculated. C finally forms the pair $\langle s,t \rangle$ and halts, indicating this to be the result of the computation. Q.E.D.

Remark: The rest of this section is devoted to showing that the " $\le$ " in Theorem 3.1(f) and 3.1(i) can be replaced by "=." The arguments used to do this are more probabilistic than information-theoretic in nature.

Theorem 3.2: (Extension of the Kraft inequality condition for the existence of an instantaneous code).

Hypothesis. Consider an effectively given list of finitely or infinitely many "requirements" $\langle s_k, n_k \rangle$ ($k = 0,1,2, \dots$) for the construction of a computer. The requirements are said to be "consistent" if $1 \ge \Sigma_k 2\dagger(-n_k)$, and we assume that they are consistent. Each requirement $\langle s_k, n_k \rangle$ requests that a program of length n_k be "assigned" to the result s_k. A computer C is said to "satisfy" the requirements if there are precisely as many programs p of length n such that $C(p, \Lambda) = s$ as there are pairs $\langle s,n \rangle$ in the list of requirements. Such a C must have the property that $P_C(s) = \Sigma 2\dagger(-n_k) (s_k = s)$ and $H_C(s) = \min n_k (s_k = s)$.

Conclusion. There are computers that satisfy these requirements. Moreover, if we are given the requirements one by one, then we can simulate a computer that satisfies them. Hereafter we refer to the particular computer that the proof of this theorem shows how to simulate as the one that is "determined" by the requirements.

Proof:

(a)   First we give what we claim is the (abstract) definition of a particular computer C that satisfies the requirements. In the second part of the proof we justify this claim.

As we are given the requirements, we assign programs to results. Initially all programs for C are available. When we are given the requirement $\langle s_k, n_k \rangle$ we assign *the first available program* of length n_k to the result s_k (first in the ordering which X was defined to have in Section 2). As each program is assigned, it and all its prefixes and extensions become unavailable for future assignments. Note that a result can have many programs assigned to it (of the same or different lengths) if there are many requirements involving it.

How can we simulate C? As we are given the requirements, we make the above assignments, and we simulate C by using the technique that was given in the proof of Theorem 2.1 for a concrete computer to simulate an abstract one.

(b) Now to justify the claim. We must show that the above rule for making assignments never fails, i.e. we must show that it is never the case that all programs of the requested length are unavailable. The proof we sketch is due to N. J. Pippenger.

A geometrical interpretation is necessary. Consider the unit interval $[0,1)$. The kth program of length n ($0 \leq k < 2^n$) corresponds to the interval $[k2^{-n}, (k+1)2^{-n})$. Assigning a program corresponds to assigning all the points in its interval. The condition that the set of assigned programs must be an instantaneous code corresponds to the rule that an interval is available for assignment iff no point in it has already been assigned. The rule we gave above for making assignments is to assign that interval $[k2^{-n}, (k+1)2^{-n})$ of the requested length 2^{-n} that is available that has the smallest possible k. Using this rule for making assignments gives rise to the following fact.

Fact. The set of those points in $[0,1)$ that are unassigned can always be expressed as the union of a finite number of intervals $[k_i 2\dagger(-n_i), (k_i + 1)2\dagger(-n_i))$ with the following properties: $n_i > n_{i+1}$, and

$$(k_i + 1)2\dagger(-n_i) \leq k_{i+1}2\dagger(-n_{i+1}).$$

I.e. these intervals are disjoint, their lengths are *distinct* powers of 2, and they appear in $[0,1)$ in order of increasing length.

We leave to the reader the verification that this fact is always the case and that it implies that an assignment is impossible only if the interval requested is longer than the total length of the unassigned part of $[0,1)$, i.e. only if the requirements are inconsistent. Q.E.D.

Theorem 3.3: (Recursive "estimates" for H_C and P_C).

Consider a computer C.
(a) The set of all true propositions of the form "$H_C(s) \leq n$" is r.e. Given t^* one can recursively enumerate the set of all true propositions of the form "$H_C(s/t) \leq n$."
(b) The set of all true propositions of the form "$P_C(s) > r$" is r.e. Given t^* one can recursively enumerate the set of all true propositions of the form "$P_C(s/t) > r$."

Proof: This is an easy consequence of the fact that the domain of C is an r.e. set. Q.E.D.

Remark: The set of all true propositions of the form "$H(s/t) \leq n$" is not r.e.; for if it were r.e., it would easily follow from Theorems 3.1(c) and 2.3(q) that Theorem 5.1(f) is false, which is a contradiction.

Theorem 3.4:

For each computer C there is a constant c such that
(a) $H(s) \leq - \lg P_C(s) + c$,
(b) $H(s/t) \leq - \lg P_C(s/t) + c$.

Proof: It follows from Theorem 3.3(b) that the set T of all true propositions of the form "$P_C(s) > 2^{-n}$" is r.e., and that given t^* one can recursively enumerate the set T_t of all true propositions

A THEORY OF PROGRAM SIZE FORMALLY IDENTICAL TO INFORMATION THEORY

of the form "$P_C(s/t) > 2^{-n}$." This will enable us to use Theorem 3.2 to show that there is a computer C' with these properties:

$$H_{C'}(s) = \lceil -\lg P_C(s) \rceil + 1, \quad P_{C'}(s) = 2\dagger(-\lceil -\lg P_C(s) \rceil), \tag{1}$$

$$H_{C'}(s/t) = \lceil -\lg P_C(s/t) \rceil + 1, \quad P_{C'}(s/t) = 2\dagger(-\lceil -\lg P_C(s/t) \rceil), \tag{2}$$

Here $\lceil x \rceil$ denotes the least integer greater than x. By applying Theorem 2.3(a,b) to (1) and (2), we see that Theorem 3.4 holds with $c = \text{sim}(C') + 2$.

How does the computer C' work? First of all, it checks whether it has been given Λ or t^* on its work tape. These two cases can be distinguished, for by Theorem 2.3(c) it is impossible for t^* to be equal to Λ.

(a) If C' has been given Λ on its work tape, it enumerates T and simulates the computer determined by all requirements of the form

$$\langle s, \ n + 1 \rangle \ (\text{``}P_C(s) > 2^{-n}\text{''} \ \epsilon \ T). \tag{3}$$

Thus $\langle s,n \rangle$ is taken as a requirement iff $n \geq \lceil -\lg P_C(s) \rceil + 1$. Hence the number of programs p of length n such that $C'(p, \Lambda) = s$ is 1 if $n \geq \lceil -\lg P_C(s) \rceil + 1$ and is 0 otherwise, which immediately yields (1).

However, we must check that the requirements (3) are consistent. $\Sigma \ 2^{-|p|}$ (over all programs p we wish to assign to the result s) $= 2\dagger(-\lceil -\lg P_C(s) \rceil) < P_C(s)$. Hence $\Sigma \ 2^{-|p|}$ (over all p we wish to assign) $< \Sigma_s P_C(s) \leq 1$ by Theorem 2.3(j). Thus the hypothesis of Theorem 3.2 is satisfied, the requirements (3) indeed determine a computer, and the proof of (1) and Theorem 3.4(a) is complete.

(b) If C' has been given t^* on its work tape, it enumerates T_t and simulates the computer determined by all requirements of the form

$$\langle s, \ n + 1 \rangle \ (\text{``}P_C(s/t) > 2^{-n}\text{''} \ \epsilon \ T_t). \tag{4}$$

Thus $\langle s,n \rangle$ is taken as a requirement iff $n \geq \lceil -\lg P_C(s/t) \rceil + 1$. Hence the number of programs p of length n such that $C'(p,t^*) = s$ is 1 if $n \geq \lceil -\lg P_C(s/t) \rceil + 1$ and is 0 otherwise, which immediately yields (2).

However, we must check that the requirements (4) are consistent. $\Sigma \ 2^{-|p|}$ (over all programs p we wish to assign to the result s) $= 2\dagger(-\lceil -\lg P_C(s/t) \rceil) < P_C(s/t)$. Hence $\Sigma \ 2^{-|p|}$ (over all p we wish to assign) $< \Sigma_t P_C(s/t) \leq 1$ by Theorem 2.3(k). Thus the hypothesis of Theorem 3.2 is satisfied, the requirements (4) indeed determine a computer, and the proof of (2) and Theorem 3.4(b) is complete. Q.E.D.

Theorem 3.5:

(a) For each computer C there is a constant c such that $P(s) \geq 2^{-c}P_C(s)$, $P(s/t) \geq 2^{-c}P_C(s/t)$.
(b) $H(s) = -\lg P(s) + O(1)$, $H(s/t) = -\lg P(s/t) + O(1)$.

Proof: Theorem 3.5(a) follows immediately from Theorem 3.4 using the fact that $P(s) \geq 2\dagger(-H(s))$ and $P(s/t) \geq 2\dagger(-H(s/t))$ (Theorem 2.3(l,m)). Theorem 3.5(b) is obtained by taking $C = U$ in Theorem 3.4 and also using these two inequalities. Q.E.D.

Remark: Theorem 3.4(a) extends Theorem 2.3(a,b) to probabilities. Note that Theorem 3.5(a) is not an immediate consequence of our weak definition of an optimal universal computer.

Theorem 3.5(b) enables one to reformulate results about H as results concerning P, and vice versa; it is the first member of a trio of formulas that will be completed with Theorem 3.9(e,f). These formulas are closely analogous to expressions in information theory for the information content of *individual* events or symbols [10, Secs. 2.3, 2.6, pp. 27-28, 34-37].

Theorem 3.6:

(a) $\#(\{p \mid U(p, \Lambda) = s \ \& \ |p| \leq H(s) + n\}) \leq 2^{\dagger}(n + O(1))$.

(b) $\#(\{p \mid U(p, t^*) = s \ \& \ |p| \leq H(s/t) + n\}) \leq 2^{\dagger}(n + O(1))$.

Proof: This follows immediately from Theorem 3.5(b). Q.E.D.

Theorem 3.7: $P(s) \approx \Sigma_t P(s,t)$.

Proof: On the one hand, there is a computer C such that $C(p, \Lambda) = s$ if $U(p, \Lambda) = \langle s,t \rangle$. Thus $P_C(s) \geq \Sigma_t P(s,t)$. Using Theorem 3.5(a), we see that $P(s) \geq 2^{-c} \Sigma_t P(s,t)$.

On the other hand, there is a computer C such that $C(p, \Lambda) = \langle s,s \rangle$ if $U(p, \Lambda) = s$. Thus $\Sigma_t P_C(s,t) \geq P_C(s,s) \geq P(s)$. Using Theorem 3.5(a), we see that $\Sigma_t P(s,t) \geq 2^{-c}P(s)$. Q.E.D.

Theorem 3.8: There is a computer C and a constant c such that $H_C(t/s) = H(s,t) - H(s) + c$.

Proof: The set of all programs p such that $U(p, \Lambda)$ is defined is r.e. Let p_k be the kth program in a particular recursive enumeration of this set, and define s_k and t_k by $\langle s_k, t_k \rangle = U(p_k, \Lambda)$. By Theorems 3.7 and 3.5(b) there is a c such that $2^{\dagger}(H(s) - c) \Sigma_t P(s,t) \leq 1$ for all s. Given s^* on its work tape, C simulates the computer C_s determined by the requirements $\langle t_k, |p_k| - |s^*| + c \rangle$ for $k = 0,1,2, \ldots$ such that $s_k = U(s^*, \Lambda)$. Recall Theorem 2.3(d,e). Thus for each p such that $U(p, \Lambda) = \langle s,t \rangle$ there is a corresponding p' such that $C(p', s^*) = C_s(p', \Lambda) = t$ and $|p'| = |p| - H(s) + c$. Hence

$$H_C(t/s) = H(s,t) - H(s) + c.$$

However, we must check that the requirements for C_s are consistent. $\Sigma \, 2^{\dagger}(-|p'|)$ (over all programs p' we wish to assign to any result t) $= \Sigma \, 2^{\dagger}(-|p| + H(s) - c)$ (over all p such that $U(p, \Lambda) = \langle s,t \rangle$) $= 2^{\dagger}(H(s) - c) \Sigma_t P(s,t) \leq 1$ because of the way c was chosen. Thus the hypothesis of Theorem 3.2 is satisfied, and these requirements indeed determine C_s. Q.E.D.

Theorem 3.9:

(a) $H(s,t) = H(s) + H(t/s) + O(1)$,

(b) $I(s:t) = H(s) + H(t) - H(s,t) + O(1)$,

(c) $I(s:t) = I(t:s) + O(1)$,

(d) $P(t/s) \approx P(s,t)/P(s)$,

(e) $H(t/s) = \lg P(s)/P(s,t) + O(1)$,

(f) $I(s:t) = \lg P(s,t)/P(s)P(t) + O(1)$.

Proof:

* Theorem 3.9(a) follows immediately from Theorems 3.8, 2.3(b), and 3.1(f).
* Theorem 3.9(b) follows immediately from Theorem 3.9(a) and the definition of $I(s:t)$.
* Theorem 3.9(c) follows immediately from Theorems 3.9(b) and 3.1(a).

- Theorem 3.9(d,e) follows immediately from Theorems 3.9(a) and 3.5(b).
- Theorem 3.9(f) follows immediately from Theorems 3.9(b) and 3.5(b). Q.E.D.

Remark: We thus have at our disposal essentially the entire formalism of information theory. Results such as these can now be obtained effortlessly:

- $H(s_1) \leq H(s_1/s_2) + H(s_2/s_3) + H(s_3/s_4) + H(s_4) + O(1)$,
- $H(s_1, s_2, s_3, s_4) = H(s_1/s_2, s_3, s_4) + H(s_2/s_3, s_4) + H(s_3/s_4) + H(s_4) + O(1)$.

However, there is an interesting class of identities satisfied by our H function that has no parallel in information theory. The simplest of these is $H(H(s)/s) = O(1)$ (Theorem 3.1(c)), which with Theorem 3.9(a) immediately yields $H(s,H(s)) = H(s) + O(1)$. This is just one pair of a large family of identities, as we now proceed to show.

Keeping Theorem 3.9(a) in mind, consider modifying the computer C used in the proof of Theorem 3.1(f) so that it also measures the lengths $H(s)$ and $H(t/s)$ of its subroutines s^* and p, and halts indicating $\langle s,t,H(s),H(t/s)\rangle$ to be the result of the computation instead of $\langle s,t\rangle$. It follows that $H(s,t) = H(s,t,H(s),H(t/s)) + O(1)$ and $H(H(s),H(t/s)/s,t) = O(1)$. In fact, it is easy to see that $H(H(s),H(t),H(t/s),H(s/t),H(s,t)/s,t) = O(1)$, which implies $H(I(s:t)/s,t) = O(1)$. And of course these identities generalize to tuples of three or more strings.

4. A Random Infinite String

The undecidability of the halting problem is a fundamental theorem of recursive function theory. In algorithmic information theory the corresponding theorem is as follows: The base-two representation of the probability that U halts is a random (i.e. maximally complex) infinite string. In this section we formulate this statement precisely and prove it.

Theorem 4.1: (Bounds on the complexity of natural numbers).
(a) $\sum_n 2^{-H(n)} \leq 1$.
- Consider a recursive function $f : N \to N$.
(b) If $\sum_n 2^{-f(n)}$ diverges, then $H(n) > f(n)$ infinitely often.
(c) If $\sum_n 2^{-f(n)}$ converges, then $H(n) \leq f(n) + O(1)$.

Proof:

(a) By Theorem 2.3(l,j), $\sum_n 2^{-H(n)} \leq \sum_n P(n) \leq 1$.

(b) If $\sum_n 2^{-f(n)}$ diverges, and $H(n) \leq f(n)$ held for all but finitely many values of n, then $\sum_n 2^{-H(n)}$ would also diverge. But this would contradict Theorem 4.1(a), and thus $H(n) > f(n)$ infinitely often.

(c) If $\sum_n 2^{-f(n)}$ converges, there is an n_0 such that $\sum_{n \geq n_0} 2^{-f(n)} \leq 1$. By Theorem 3.2 there is a computer C determined by the requirements $\langle n, f(n)\rangle$ $(n \geq n_0)$. Thus $H(n) \leq f(n) + \text{sim}(C)$ for all $n \geq n_0$. Q.E.D.

Theorem 4.2: (Maximal complexity finite and infinite strings).
(a) $\max H(s)\,(\,|s| = n\,) = n + H(n) + O(1)$.
(b) $\#(\{s \mid |s| = n \ \& \ H(s) \leq n + H(n) - k\}) \leq 2^{\uparrow}(n - k + O(1))$.
(c) Imagine that the infinite string α is generated by tossing a fair coin once for each if its bits. Then, with probability one, $H(\alpha_n) > n$ for all but finitely many n.

Proof:

- Consider a string s of length n. By Theorem 3.9(a), $H(s) = H(n,s) + O(1)$ $= H(n) + H(s/n) + O(1)$. We now obtain Theorem 4.2(a,b) from this estimate for $H(s)$.

  There is a computer C such that $C(p, |p|*) = p$ for all p. Thus $H(s/n) \leq n + \text{sim}(C)$, and $H(s) \leq n + H(n) + O(1)$. On the other hand, by Theorem 2.3(q), fewer than 2^{n-k} of the s satisfy $H(s/n) < n - k$. Hence fewer than 2^{n-k} of the s satisfy $H(s) < n - k + H(n) + O(1)$. Thus we have obtained Theorem 4.2(a,b).

- Now for the proof of Theorem 4.2(c). By Theorem 4.2(b), at most a fraction of $2\dagger(-H(n) + c)$ of the strings s of length n satisfy $H(s) \leq n$. Thus the probability that α satisfies $H(\alpha_n) \leq n$ is $\leq 2\dagger(-H(n) + c)$. By Theorem 4.1(a), $\Sigma_n 2\dagger(-H(n) + c)$ converges. Invoking the Borel-Cantelli lemma, we obtain Theorem 4.2(c). Q.E.D.

Definition of Randomness: A string s is random iff $H(s)$ is approximately equal to $|s| + H(|s|)$. An infinite string α is random iff $\exists c \, \forall n \, H(\alpha_n) > n - c$.

Remark: In the case of infinite strings there is a sharp distinction between randomness and nonrandomness. In the case of finite strings it is a matter of degree. To the question "How random is s?" one must reply indicating how close $H(s)$ is to $|s| + H(|s|)$.

C. P. Schnorr (private communication) has shown that this complexity-based definition of a random infinite string and P. Martin-Löf's statistical definition of this concept [7, pp. 379-380] are equivalent.

Definition of Base-Two Representations: The base-two representation of a real number $x \in (0,1]$ is that unique string $b_1 b_2 b_3 \ldots$ *with infinitely many 1's* such that $x = \Sigma_k b_k 2^{-k}$.

Definition of the Probability ω that U Halts: $\omega = \Sigma_s P(s) = \Sigma \, 2^{-|p|}$ ($U(p, \Lambda)$ is defined).

By Theorem 2.3(j,n), $\omega \in (0,1]$. Therefore the real number ω has a base-two representation. Henceforth ω denotes both the real number and its base-two representation. Similarly, ω_n denotes a string of length n and a rational number $m/2^n$ with the property that $\omega > \omega_n$ and $\omega - \omega_n \leq 2^{-n}$.

Theorem 4.3: (Construction of a random infinite string).

(a) There is a recursive function $w:N \to R$ such that $w(n) \leq w(n + 1)$ and $\omega = \lim_{n\to\infty} w(n)$.
(b) ω is random.
(c) There is a recursive predicate $D:N \times N \times N \to \{true, false\}$ such that the k-th bit of ω is a 1 iff $\exists i \, \forall j \, D(i,j,k)$ ($k = 0,1,2, \ldots$).

Proof:

(a) $\{p \mid U(p, \Lambda) \text{ is defined}\}$ is r.e. Let p_k ($k = 0,1,2, \ldots$) denote the kth p in a particular recursive enumeration of this set. Let $w(n) = \Sigma_{k \leq n} 2\dagger(-|p_k|)$. $w(n)$ tends monotonically to ω from below, which proves Theorem 4.3(a).

(b) In view of the fact that $\omega > \omega_n \geq \omega - 2^{-n}$ (see the definition of ω), if one is given ω_n one can find an m such that $\omega \geq w(m) > \omega_n \geq \omega - 2^{-n}$. Thus $\omega - w(m) < 2^{-n}$, and $\{p_k \mid k \leq m\}$ contains all programs p of length less than or equal to n such that $U(p, \Lambda)$ is defined. Hence $\{U(p_k, \Lambda) \mid k \leq m \, \& \, |p_k| \leq n\} = \{s \mid H(s) \leq n\}$. It follows there is a computer C with the property that if $U(p, \Lambda) = \omega_n$, then $C(p, \Lambda)$ equals the first string s such that $H(s) > n$. Thus $n < H(s) \leq H(\omega_n) + \text{sim}(C)$, which proves Theorem 4.3(b).

(c)   To prove Theorem 4.3(c), define D as follows: $D(i,j,k)$ iff $j \geq i$ implies the kth bit of the base-two representation of $w(j)$ is a 1. Q.E.D.

Appendix. The Traditional Concept of Relative Complexity

In this Appendix programs are required to be self-delimiting, but the relative complexity $H(s/t)$ of s with respect to t will now mean that one is directly given t, instead of being given a minimal-size program for t.

The standard optimal universal computer U remains the same as before. H and P are redefined as follows:

- $H_C(s/t) = \min |p| \ (C(p,t) = s)$ (may be ∞),
- $H_C(s) = H_C(s/\Lambda)$,
- $H(s/t) = H_U(s/t)$,
- $H(s) = H_U(s)$,
- $P_C(s/t) = \Sigma\ 2^{-|p|} \ (C(p,t) = s)$,
- $P_C(s) = P_C(s/\Lambda)$,
- $P(s/t) = P_U(s/t)$,
- $P(s) = P_U(s)$.

These concepts are extended to tuples of strings and natural numbers as before. Finally, $\Delta(s,t)$ is defined as follows:

- $H(s,t) = H(s) + H(t/s) + \Delta(s,t)$.

Theorem 5.1:

(a) $H(s,H(s)) = H(s) + O(1)$,
(b) $H(s,t) = H(s) + H(t/s,H(s)) + O(1)$,
(c) $-H(H(s)/s) - O(1) \leq \Delta(s,t) \leq O(1)$,
(d) $\Delta(s,s) = O(1)$,
(e) $\Delta(s,H(s)) = -H(H(s)/s) + O(1)$,
(f) $H(H(s)/s) \neq O(1)$.

Proof:

(a)   On the one hand, $H(s,H(s)) \leq H(s) + c$ because a minimal-size program for s also tells one its length $H(s)$, i.e. because there is a computer C such that $C(p, \Lambda) = \langle U(p, \Lambda), |p| \rangle$ if $U(p, \Lambda)$ is defined. On the other hand, obviously $H(s) \leq H(s,H(s)) + c$.

(b)   On the one hand, $H(s,t) \leq H(s) + H(t/s,H(s)) + c$ follows from Theorem 5.1(a) and the obvious inequality $H(s,t) \leq H(s,H(s)) + H(t/s,H(s)) + c$.   On the other hand, $H(s,t) \geq H(s) + H(t/s,H(s)) - c$ follows from the inequality $H(t/s,H(s)) \leq H(s,t) - H(s) + c$ analogous to Theorem 3.8 and obtained by adapting the methods of Section 3 to the present setting.

(c)   This follows from Theorem 5.1(b) and the obvious inequality $H(t/s,H(s)) - c \leq H(t/s) \leq H(H(s)/s) + H(t/s,H(s)) + c$.

(d)   If $t = s$, $H(s,t) - H(s) - H(t/s) = H(s,s) - H(s) - H(s/s) = H(s) - H(s) + O(1) = O(1)$, for obviously $H(s,s) = H(s) + O(1)$ and $H(s/s) = O(1)$.

(e) If $t = H(s)$, $H(s,t) - H(s) - H(t/s) = H(s,H(s)) - H(s) - H(H(s)/s) = -H(H(s)/s) + O(1)$ by Theorem 5.1(a).

(f) The proof is by *reductio ad absurdum*. Suppose on the contrary that $H(H(s)/s) < c$ for all s. First we adapt an idea of A. R. Meyer and D. W. Loveland [6, pp. 525-526] to show that there is a partial recursive function $f:X \to N$ with the property that if $f(s)$ is defined it is equal to $H(s)$ and this occurs for infinitely many values of s. Then we obtain the desired contradiction by showing that such a function f cannot exist.

Consider the set K_s of all natural numbers k such that $H(k/s) < c$ and $H(s) \leq k$. Note that $\min K_s = H(s)$, $\#(K_s) < 2^c$, and given s one can recursively enumerate K_s. Also, given s and $\#(K_s)$ one can recursively enumerate K_s until one finds all its elements, and, in particular, its smallest element, which is $H(s)$. Let $m = \lim \sup \#(K_s)$, and let n be such that $|s| \geq n$ implies $\#(K_s) \leq m$.

Knowing m and n one calculates $f(s)$ as follows. First one checks if $|s| < n$. If so, $f(s)$ is undefined. If not, one recursively enumerates K_s until m of its elements are found. Because of the way n was chosen, K_s cannot have more than m elements. If it has less than m, one never finishes searching for m of them, and so $f(s)$ is undefined. However, if $\#(K_s) = m$, which occurs for infinitely many values of s, then one eventually realizes all of them have been found, including $f(s) = \min K_s = H(s)$. Thus $f(s)$ is defined and equal to $H(s)$ for infinitely many values of s.

It remains to show that such an f is impossible. As the length of s increases, $H(s)$ tends to infinity, and so f is unbounded. Thus given n and $H(n)$ one can calculate a string s_n such that $H(n) + n < f(s_n) = H(s_n)$, and so $H(s_n/n,H(n))$ is bounded. Using Theorem 5.1(b) we obtain $H(n) + n < H(s_n) \leq H(n, s_n) + c' \leq H(n) + H(s_n/n,H(n)) + c'' \leq H(n) + c'''$, which is impossible for $n \geq c''''$. Thus f cannot exist, and our initial assumption that $H(H(s)/s) < c$ for all s must be false. Q.E.D.

Remark: Theorem 5.1 makes it clear that the fact that $H(H(s)/s)$ is unbounded implies that $H(t/s)$ is less convenient to use than $H(t/s,H(s))$. In fact, R. Solovay (private communication) has announced that max $H(H(s)/s)$ taken over all strings s of length n is asymptotic to $\lg n$. The definition of the relative complexity of s with respect to t given in Section 2 is equivalent to $H(s/t,H(t))$.

Acknowledgments

The author is grateful to the following for conversations that helped to crystallize these ideas: C. H. Bennett, T. M. Cover, R. P. Daley, M. Davis, P. Elias, T. L. Fine, W. L. Gewirtz, D. W. Loveland, A. R. Meyer, M. Minsky, N. J. Pippenger, R. J. Solomonoff, and S. Winograd. The author also wishes to thank the referees for their comments.

References

(Note. Reference [33] is not cited in the text.)

1. SOLOMONOFF, R. J. A formal theory of inductive inference. *Inform. and Contr. 7* (1964), 1-22, 224-254.

2. KOLMOGOROV, A. N. Three approaches to the quantitative definition of information. *Problems of Inform. Transmission 1*, 1 (Jan.-March 1965), 1-7.

3. CHAITIN, G. J. On the length of programs for computing finite binary sequences. *J. ACM 13*, 4 (Oct. 1966), 547-569.

4. CHAITIN, G. J. On the length of programs for computing finite binary sequences: Statistical considerations. *J. ACM 16,* 1 (Jan. 1969), 145-159.

5. KOLMOGOROV, A. N. On the logical foundations of information theory and probability theory. *Problems of Inform. Transmission 5,* 3 (July-Sept. 1969), 1-4.

6. LOVELAND, D. W. A variant of the Kolmogorov concept of complexity. *Inform. and Contr. 15* (1969), 510-526.

7. SCHNORR, C. P. Process complexity and effective random tests. *J. Comput. and Syst. Scis. 7* (1973), 376-388.

8. CHAITIN, G. J. On the difficulty of computations. *IEEE Trans. IT-16* (1970), 5-9.

9. FEINSTEIN, A. *Foundations of Information Theory.* McGraw-Hill, New York, 1958.

10. FANO, R. M. *Transmission of Information.* Wiley, New York, 1961.

11. ABRAMSON, N. *Information Theory and Coding.* McGraw-Hill, New York, 1963.

12. ASH, R. *Information Theory.* Wiley-Interscience, New York, 1965.

13. WILLIS, D. G. Computational complexity and probability constructions. *J. ACM 17,* 2 (April 1970), 241-259.

14. ZVONKIN, A. K. AND LEVIN, L. A. The complexity of finite objects and the development of the concepts of information and randomness by means of the theory of algorithms. *Russ. Math. Survs. 25,* 6 (Nov.-Dec. 1970), 83-124.

15. COVER, T. M. Universal gambling schemes and the complexity measures of Kolmogorov and Chaitin. Rep. No. 12, Statistics Dep., Stanford U., Stanford, Calif., 1974. Submitted to *Ann. Statist.*

16. GEWIRTZ, W. L. Investigations in the theory of descriptive complexity. Ph.D. Thesis, New York University, 1974 (to be published as a Courant Institute rep.).

17. WEISS, B. The isomorphism problem in ergodic theory. *Bull. Amer. Math. Soc. 78* (1972), 668-684.

18. RÉNYI, A. *Foundations of Probability.* Holden-Day, San Francisco, 1970.

19. FINE, T. L. *Theories of Probability: An Examination of Foundations.* Academic Press, New York, 1973.

20. COVER, T. M. On determining the irrationality of the mean of a random variable. *Ann. Statist. 1* (1973), 862-871.

21. CHAITIN, G. J. Information-theoretic computational complexity. *IEEE Trans. IT-20* (1974), 10-15.

22. LEVIN, M. Mathematical Logic for Computer Scientists. Rep. TR-131, M.I.T. Project MAC, 1974, pp. 145-147,153.

23. CHAITIN, G. J. Information-theoretic limitations of formal systems. *J. ACM 21,* 3 (July 1974), 403-424.

24. MINSKY, M. L. *Computation: Finite and Infinite Machines.* Prentice-Hall, Englewood Cliffs, N.J., 1967, pp. 54, 55, 66.

25. MINSKY, M. AND PAPERT, S. *Perceptrons: An Introduction to Computational Geometry.* M.I.T. Press, Cambridge, Mass. 1969, pp. 150-153.

26. SCHWARTZ, J. T. On Programming: An Interim Report on the SETL Project. Installment I: Generalities. Lecture Notes, Courant Institute, New York University, New York, 1973, pp. 1-20.

27. BENNETT, C. H. Logical reversibility of computation. *IBM J. Res. Develop. 17* (1973), 525-532.

28. DALEY, R. P. The extent and density of sequences within the minimal-program complexity hierarchies. *J. Comput. and Syst. Scis.* (to appear).

29. CHAITIN, G. J. Information-theoretic characterizations of recursive infinite strings. Submitted to *Theoretical Comput. Sci.*

30. ELIAS, P. Minimum times and memories needed to compute the values of a function. *J. Comput. and Syst. Scis.* (to appear).

31. ELIAS, P. Universal codeword sets and representations of the integers. *IEEE Trans. IT* (to appear).

32. HELLMAN, M. E. The information theoretic approach to cryptography. Center for Systems Research, Stanford U., Stanford, Calif., 1974.

33. CHAITIN, G. J. Randomness and mathematical proof. *Sci. Amer. 232,* 5 (May 1975), in press.

RECEIVED APRIL 1974; REVISED DECEMBER 1974

INCOMPLETENESS THEOREMS FOR RANDOM REALS

Advances in Applied Mathematics 8 (1987), pp. 119-146.

G. J. Chaitin

IBM Thomas J. Watson Research Center, P. O. Box 218,
Yorktown Heights, New York 10598

Abstract

We obtain some dramatic results using statistical mechanics–thermodynamics kinds of arguments concerning randomness, chaos, unpredictability, and uncertainty in mathematics. We construct an equation involving only whole numbers and addition, multiplication, and exponentiation, with the property that if one varies a parameter and asks whether the number of solutions is finite or infinite, the answer to this question is indistinguishable from the result of independent tosses of a fair coin. This yields a number of powerful Gödel incompleteness-type results concerning the limitations of the axiomatic method, in which entropy–information measures are used. © 1987 Academic Press, Inc.

1. Introduction

It is now half a century since Turing published his remarkable paper *On Computable Numbers, with an Application to the Entscheidungsproblem* (Turing [15]). In that paper Turing constructs a universal Turing machine that can simulate any other Turing machine. He also uses Cantor's method to diagonalize over the countable set of computable real numbers and construct an uncomputable real, from which he deduces the unsolvability of the halting problem and as a corollary a form of Gödel's incompleteness theorem. This paper has penetrated into our thinking to such a point that it is now regarded as obvious, a fate which is suffered by only the most basic conceptual contributions. Speaking as a mathematician, I cannot help noting with pride that the idea of a general purpose electronic digital computer was invented in order to cast light on a fundamental question regarding the foundations of mathematics, years before such objects were actually constructed. Of course, this is an enormous simplification of the complex genesis of the computer, to which many contributed, but there is as much truth in this remark as there is in many other historical "facts."

In another paper [5], I used ideas from algorithmic information theory to construct a diophantine equation whose solutions are in a sense random. In the present paper I shall try to give a relatively self-contained exposition of this result via another route, starting from Turing's original construction of an uncomputable real number.

Following Turing, consider an enumeration $r_1, r_2, r_3, \ldots$ of all computable real numbers between zero and one. We may suppose that r_k is the real number, if any, computed by the kth computer program. Let $.d_{k1}d_{k2}d_{k3} \ldots$ be the successive digits in the decimal expansion of r_k. Following Cantor, consider the diagonal of the array of r_k,

$$r_1 = .d_{11}d_{12}d_{13} \cdots$$
$$r_2 = .d_{21}d_{22}d_{23} \cdots$$
$$r_3 = .d_{31}d_{32}d_{33} \cdots .$$

This gives us a new real number with decimal expansion $.d_{11}d_{22}d_{33} \ldots$. Now change each of these digits, avoiding the digits zero and nine. The result is an uncomputable real number, because its first digit is different from the first digit of the first computable real, its second digit is different from the second digit of the second computable real, etc. It is necessary to avoid zero and nine, because real numbers with different digit sequences can be equal to each other if one of them ends with an infinite sequence of zeros and the other ends with an infinite sequence of nines, for example, .3999999... = .4000000... .

Having constructed an uncomputable real number by diagonalizing over the computable reals, Turing points out that it follows that the halting problem is unsolvable. In particular, there can be no way of deciding if the kth computer program ever outputs a kth digit. Because if there were, one could actually calculate the successive digits of the uncomputable real number defined above, which is impossible. Turing also notes that a version of Gödel's incompleteness theorem is an immediate corollary, because if there cannot be an algorithm for deciding if the kth computer program ever outputs a kth digit, there also cannot be a formal axiomatic system which would always enable one to prove which of these possibilities is the case, for in principle one could run through all possible proofs to decide. Using the powerful techniques which were developed in order to solve Hilbert's tenth problem (see Davis *et al.* [7] and Jones and Matijasevic [11]), it is possible to encode the unsolvability of the halting problem as a statement about an exponential diophantine equation. An exponential diophantine equation is one of the form

$$P(x_1, \ldots, x_m) = P'(x_1, \ldots, x_m),$$

where the variables $x_1, \ldots, x_m$ range over natural numbers and P and P' are functions built up from these variables and natural number constants by the operations of addition, multiplication, and exponentiation. The result of this encoding is an exponential diophantine equation $P = P'$ in $m + 1$ variables $n, x_1, \ldots, x_m$ with the property that

$$P(n, x_1, \ldots, x_m) = P'(n, x_1, \ldots, x_m)$$

has a solution in natural numbers $x_1, \ldots, x_m$ if and only if the nth computer program ever outputs an nth digit. It follows that there can be no algorithm for deciding as a function of n whether or not $P = P'$ has a solution, and thus there cannot be any complete proof system for settling such questions either.

Up to now we have followed Turing's original approach, but now we will set off into new territory. Our point of departure is a remark of Courant and Robbins [6] that another way of obtaining a real number that is not on the list $r_1, r_2, r_3, \ldots$ is by tossing a coin. Here is their measure-theoretic argument that the real numbers are uncountable. Recall that $r_1, r_2, r_3, \ldots$ are the computable reals between zero and one. Cover r_1 with an interval of length $\varepsilon/2$, cover r_2 with an interval of length $\varepsilon/4$, cover r_3 with an interval of length $\varepsilon/8$, and in general cover r_k with an interval of length $\varepsilon/2^k$. Thus all computable reals in the unit interval are covered by this infinite set of intervals, and the total length of the covering intervals is

$$\sum_{k=1}^{\infty} \frac{\varepsilon}{2^k} = \varepsilon.$$

Hence if we take ε sufficiently small, the total length of the covering is arbitrarily small. In summary, the reals between zero and one constitute an interval of length one, and the subset that are computable can be covered by intervals whose total length is arbitrarily small. In other words, the computable reals are a set of measure zero, and if we choose a real in the unit interval at random, the probability that it is computable is zero. Thus one way to get an uncomputable real with probability one is to flip a fair coin, using independent tosses to obtain each bit of the binary expansion of its base-two representation.

If this train of thought is pursued, it leads one to the notion of a random real number, which can never be a computable real. Following Martin-Löf [12], we give a definition of a random real using constructive measure theory. We say that a set of real numbers X is a constructive measure zero set if there is an algorithm A which given n generates a (possibly infinite) set of intervals whose total length is less than or equal to 2^{-n} and which covers the set X. More precisely, the covering is in the form of a set C of finite binary strings s such that

$$\sum_{s \in C} 2^{-|s|} \leq 2^{-n}$$

(here $|s|$ denotes the length of the string s), and each real in the covered set X has a member of C as the initial part of its base-two expansion. In other words, we consider sets of real numbers with the property that there is an algorithm A for producing arbitrarily small coverings of the set. Such sets of reals are constructively of measure zero. Since there are only countably many algorithms A for constructively covering measure zero sets, it follows that almost all real numbers are not contained in any set of constructive measure zero. Such reals are called (Martin-Löf) random reals. In fact, if the successive bits of a real number are chosen by coin flipping, with probability one it will not be contained in any set of constructive measure zero, and hence will be a random real number.

Note that no computable real number r is random. Here is how we get a constructive covering of arbitrarily small measure. The covering algorithm, given n, yields the n-bit initial sequence of the binary digits of r. This covers r and has total length or measure equal to 2^{-n}. Thus there is an algorithm for obtaining arbitrarily small coverings of the set consisting of the computable real r, and r is not a random real number. We leave to the reader the adaptation of the argument in Feller [9] proving the strong law of large numbers to show that reals in which all digits do not have equal limiting frequency have constructive measure zero. It follows that random reals are normal in Borel's sense, that is, in any base all digits have equal limiting frequency.

Let us consider the real number p whose nth bit in base-two notation is a zero or a one depending on whether or not the exponential diophantine equation

$$P(n, x_1, \ldots, x_m) = P'(n, x_1, \ldots, x_m)$$

has a solution in natural numbers $x_1, \ldots, x_m$. We will show that p is not a random real. In fact, we will give an algorithm for producing coverings of measure $(n + 1)2^{-n}$, which can obviously be changed to one for producing coverings of measure not greater than 2^{-n}. Consider the first N values of the parameter n. If one knows for how many of these values of n, $P = P'$ has a solution, then one can find for which values of $n < N$ there are solutions. This is because the set of solutions of $P = P'$ is recursively enumerable, that is, one can try more and more solutions and eventually find each value of the parameter n for which there is a solution. The only problem is to decide when to give up further searches because all values of $n < N$ for which there are solutions have been found. But if one is told how many such n there are, then one knows when to stop searching for solutions. So one can assume each of the $N + 1$ possibilities ranging from p has all of its initial N bits off to p has all of them on, and each one of these assumptions determines the actual values of the first N bits of p. Thus we have determined $N + 1$ different possibilities for the first N bits of p, that is, the real number p is covered by a set of intervals of total length $(N + 1)2^{-N}$, and hence is a set of constructive measure zero, and p cannot be a random real number.

Thus asking whether an exponential diophantine equation has a solution as a function of a parameter cannot give us a random real number. However asking whether or not the number of solutions is infinite can give us a random real. In particular, there is a exponential diophantine equation

$Q = Q'$ such that the real number q is random whose nth bit is a zero or a one depending on whether or not there are infinitely many natural numbers $x_1, \ldots, x_m$ such that

$$Q(n, x_1, \ldots, x_m) = Q'(n, x_1, \ldots, x_m).$$

The equation $P = P'$ that we considered before encoded the halting problem, that is, the nth bit of the real number p was zero or one depending on whether the nth computer program ever outputs an nth digit. To construct an equation $Q = Q'$ such that q is random is somewhat more difficult; we shall limit ourselves to giving an outline of the proof:[28]

1. First show that if one had an oracle for solving the halting problem, then one could compute the successive bits of the base-two representation of a particular random real number q.

2. Then show that if a real number q can be computed using an oracle for the halting problem, it can be obtained without using an oracle as the limit of a computable sequence of dyadic rational numbers (rationals of the form $K/2^L$).

3. Finally show that any real number q that is the limit of a computable sequence of dyadic rational numbers can be encoded into an exponential diophantine equation $Q = Q'$ in such a manner that

$$Q(n, x_1, \ldots, x_m) = Q'(n, x_1, \ldots, x_m)$$

has infinitely many solutions $x_1, \ldots, x_m$ if and only if the nth bit of the real number q is a one. This is done using the fact "that every r.e. set has a *singlefold* exponential diophantine representation" (Jones and Matijasevic [11]).

$Q = Q'$ is quite a remarkable equation, as it shows that there is a kind of uncertainty principle even in pure mathematics, in fact, even in the theory of whole numbers. Whether or not $Q = Q'$ has infinitely many solutions jumps around in a completely unpredictable manner as the parameter n varies. It may be said that the truth or falsity of the assertion that there are infinitely many solutions is indistinguishable from the result of independent tosses of a fair coin. In other words, these are independent mathematical facts with probability one-half! This is where our search for a probabilistic proof of Turing's theorem that there are uncomputable real numbers has led us, to a dramatic version of Gödel's incompleteness theorem.

In Section 2 we define the real number Ω, and we develop as much of algorithmic information theory as we shall need in the rest of the paper. In Section 3 we compare a number of definitions of randomness, we show that Ω is random, and we show that Ω can be encoded into an exponential diophantine equation. In Section 4 we develop incompleteness theorems for Ω and for its exponential diophantine equation.

2. Algorithmic Information Theory [3]

First a piece of notation. By $\log x$ we mean the integer part of the base-two logarithm of x. That is, if $2^n \le x < 2^{n+1}$, then $\log x = n$. Thus $2^{\log x} \le x$, even if $x < 1$.

Our point of departure is the observation that the series

$$\sum \frac{1}{n}, \qquad \sum \frac{1}{n \log n}, \qquad \sum \frac{1}{n \log n \, \log\log n} \cdots$$

---

[28]   The full proof is given later in this paper (Theorems R6 and R7), but is slightly different; it uses a particular random real number, Ω, that arises naturally in algorithmic information theory.

all diverge. On the other hand,

$$\sum \frac{1}{n^2}, \quad \sum \frac{1}{n(\log n)^2}, \quad \sum \frac{1}{n \log n (\log\log n)^2} \cdots$$

all converge. To show this we use the Cauchy condensation test (Hardy [10]): if $\phi(n)$ is a nonincreasing function of n, then the series $\Sigma\phi(n)$ is convergent or divergent according as $\Sigma 2^n\phi(2^n)$ is convergent or divergent.

Here is a proof of the Cauchy condensation test

$$\sum\phi(k) \geq \sum\left[\phi(2^n+1) + \cdots + \phi(2^{n+1})\right] \geq \sum 2^n\phi(2^{n+1}) = \frac{1}{2}\sum 2^{n+1}\phi(2^{n+1}).$$

$$\sum\phi(k) \leq \sum\left[\phi(2^n) + \cdots + \phi(2^{n+1}-1)\right] \leq \sum 2^n\phi(2^n).$$

Thus

$$\sum \frac{1}{n} \text{ behaves the same as } \sum 2^n \frac{1}{2^n} = \sum 1, \quad \text{which diverges.}$$

$$\sum \frac{1}{n \log n} \text{ behaves the same as } \sum 2^n \frac{1}{2^n n} = \sum \frac{1}{n}, \quad \text{which diverges.}$$

$$\sum \frac{1}{n \log n \log\log n} \text{ behaves the same as } \sum 2^n \frac{1}{2^n n \log n} = \sum \frac{1}{n \log n},$$

which diverges, etc.

On the other hand,

$$\sum \frac{1}{n^2} \text{ behaves the same as } \sum 2^n \frac{1}{2^{2n}} = \sum \frac{1}{2^n}, \quad \text{which converges.}$$

$$\sum \frac{1}{n(\log n)^2} \text{ behaves the same as } \sum 2^n \frac{1}{2^n n^2} = \sum \frac{1}{n^2}, \quad \text{which converges.}$$

$$\sum \frac{1}{n \log n (\log\log n)^2} \text{ behaves the same as } \sum 2^n \frac{1}{2^n n (\log n)^2} = \sum \frac{1}{n(\log n)^2},$$

which converges, etc.

For the purposes of this paper, it is best to think of the algorithmic information content H, which we shall now define, as the borderline between $\Sigma 2^{-f(n)}$ converging and diverging!

Definition. Define an *information content measure* $H(n)$ to be a function of the natural number n having the property that

(1) $$\Omega \equiv \sum 2^{-H(n)} \leq 1,$$

and that $H(n)$ is computable as a limit from above, so that the set

(2) $$\{ \text{ "}H(n) \leq k\text{"} \}$$

of all upper bounds is r.e. We also allow $H(n) = +\infty$, which contributes zero to the sum (1) since $2^{-\infty} = 0$. It contributes no elements to the set of upper bounds (2).

Note. If H is an information content measure, then it follows immediately from $\Sigma 2^{-H(n)} = \Omega \leq 1$ that

$$\#\{n \mid H(n) \leq n\} \leq 2^n.$$

That is, there are at most 2^n natural numbers with information content less than or equal to n.

Theorem I. There is a minimal information content measure H, i.e., an information content measure with the property that for any other information content measure H', there exists a constant c depending only on H and H' but not on n such that

$$H(n) \leq H'(n) + c.$$

That is, H is smaller, within $O(1)$, than any other information content measure.

Proof. Define H as

(3) $$H(n) = \min_{k \geq 1}\Big[H_k(n) + k \Big],$$

where H_k denotes the information content measure resulting from taking the kth ($k \geq 1$) computer algorithm and patching it, if necessary, so that it gives limits from above and does not violate the $\Omega \leq 1$ condition (1). Then (3) gives H as a computable limit from above, and

$$\Omega = \sum_n 2^{-H(n)} \leq \sum_{k \geq 1}\left(2^{-k} \sum_n 2^{-H_k(n)} \right) \leq \sum_{k \geq 1} 2^{-k} = 1.$$

Q.E.D.

Definition. Henceforth we use this minimal information content measure H, and we refer to $H(n)$ as the *information content* of n. We also consider each natural number n to correspond to a bit string s and vice versa, so that H is defined for strings as well as numbers.[29] In addition, let $\langle n, m \rangle$ denote a fixed computable one-to-one correspondence between natural numbers and ordered pairs of natural numbers. We define the *joint information content* of n and m to be $H(\langle n, m \rangle)$. Thus H is defined for ordered pairs of natural numbers as well as individual natural numbers. We define the *relative information content* $H(m \mid n)$ of m relative to n by the equation

$$H(\langle n, m \rangle) \equiv H(n) + H(m \mid n).$$

That is,

$$H(m \mid n) \equiv H(\langle n, m \rangle) - H(n).$$

And we define the *mutual information content* $I(n : m)$ of n and m by the equation

$$I(n : m) \equiv H(m) - H(m \mid n) \equiv H(n) + H(m) - H(\langle n, m \rangle).$$

---

[29] It is important to distinguish between the length of a string and its information content! However, a possible source of confusion is the fact that the "natural unit" for both length and information content is the "bit." Thus one often speaks of an n-bit string, and also of a string whose information content is $\leq n$ bits.

     INCOMPLETENESS THEOREMS FOR RANDOM REALS

Note. $\Omega = \Sigma 2^{-H(n)}$ is just on the borderline between convergence and divergence:

- $\Sigma 2^{-H(n)}$ converges.
- If $f(n)$ is computable and unbounded, then $\Sigma 2^{-H(n)+f(n)}$ diverges.
- If $f(n)$ is computable and $\Sigma 2^{-f(n)}$ converges, then $H(n) \leq f(n) + O(1)$.
- If $f(n)$ is computable and $\Sigma 2^{-f(n)}$ diverges, then $H(n) \geq f(n)$ infinitely often.

Let us look at a real-valued function $\rho(n)$ that is computable as a limit of rationals from below. And suppose that $\Sigma \rho(n) \leq 1$. Then $H(n) \leq - \log \rho(n) + O(1)$. So $2^{-H(n)}$ can be thought of as a *maximal* function $\rho(n)$ that is computable in the limit from *below* and has $\Sigma \rho(n) \leq 1$, instead of thinking of $H(n)$ as a *minimal* function $f(n)$ that is computable in the limit from *above* and has $\Sigma 2^{-f(n)} \leq 1$.

Lemma I. For all n,

$$
\begin{aligned}
H(n) & \leq & 2 \log n + c, \\
& \leq & \log n + 2 \log\log n + c', \\
& \leq & \log n + \log\log n + 2 \log\log\log n + c'' \dots .
\end{aligned}
$$

For infinitely many values of n,

$$
\begin{aligned}
H(n) & \geq & \log n, \\
& \geq & \log n + \log\log n, \\
& \geq & \log n + \log\log n + \log\log\log n \dots .
\end{aligned}
$$

Lemma I2. $H(s) \leq |s| + H(|s|) + O(1)$. $|s| = $ the length in bits of the string s.

Proof.

$$
1 \geq \Omega = \sum_n 2^{-H(n)} = \sum_n \left(2^{-H(n)} \sum_{|s|=n} 2^{-n} \right)
$$

$$
= \sum_n \sum_{|s|=n} 2^{-[n+H(n)]} = \sum_s 2^{-[|s|+H(|s|)]}.
$$

The lemma follows by the minimality of H. Q.E.D.

Lemma I3. There are $< 2^{n-k+c}$ n-bit strings s such that $H(s) < n + H(n) - k$. Thus there are $< 2^{n-H(n)-k+c}$ n-bit strings s such that $H(s) < n - k$.

Proof.

$$
\sum_n \sum_{|s|=n} 2^{-H(s)} = \sum_s 2^{-H(s)} = \Omega \leq 1.
$$

Hence by the minimality of H

$$
2^{-H(n)+c} \geq \sum_{|s|=n} 2^{-H(s)},
$$

which yields the lemma. Q.E.D.

Lemma I4. If $\psi(n)$ is a computable partial function, then

$$H(\psi(n)) \leq H(n) + c_\psi.$$

Proof.

$$1 \geq \Omega = \sum_n 2^{-H(n)} \geq \sum_y \sum_{\psi(x) = y} 2^{-H(x)}.$$

Note that

(4)
$$2^{-a} \geq \sum_i 2^{-b_i} \Rightarrow a \leq \min_i b_i .$$

The lemma follows by the minimality of H. Q.E.D.

Lemma I5. $H(\langle n, m \rangle) = H(\langle m, n \rangle) + O(1).$

Proof.

$$\sum_{\langle n, m \rangle} 2^{-H(\langle n, m \rangle)} = \sum_{\langle m, n \rangle} 2^{-H(\langle n, m \rangle)} = \Omega \leq 1.$$

The lemma follows by using the minimality of H in both directions. Q.E.D.

Lemma I6. $H(\langle n, m \rangle) \leq H(n) + H(m) + O(1).$

Proof.

$$\sum_{\langle n, m \rangle} 2^{-[H(n)+H(m)]} = \Omega^2 \leq 1^2 \leq 1.$$

The lemma follows by the minimality of H. Q.E.D.

Lemma I7. $H(n) \leq H(\langle n, m \rangle) + O(1).$

Proof.

$$\sum_n \sum_{\langle n, m \rangle} 2^{-H(\langle n, m \rangle)} = \sum_{\langle n, m \rangle} 2^{-H(\langle n, m \rangle)} = \Omega \leq 1.$$

The lemma follows from (4) and the minimality of H. Q.E.D.

Lemma I8. $H(\langle n, H(n) \rangle) = H(n) + O(1).$

Proof. By Lemma I7,

$$H(n) \leq H(\langle n, H(n) \rangle) + O(1).$$

On the other hand, consider

$$\sum_{\substack{\langle n, i \rangle \\ H(n) \leq i}} 2^{-i-1} = \sum_{\langle n, H(n)+j \rangle} 2^{-H(n)-j-1} = \sum_n \sum_{k \geq 1} 2^{-H(n)-k} = \sum_n 2^{-H(n)} = \Omega \leq 1.$$

By the minimality of H,

$$H(\langle n, H(n) + j \rangle) \leq H(n) + j + O(1).$$

Take $j = 0$. Q.E.D.

Lemma I9. $H(\langle n, n \rangle) = H(n) + O(1)$.

Proof. By Lemma I7,

$$H(n) \leq H(\langle n, n \rangle) + O(1).$$

On the other hand, consider $\psi(n) = \langle n, n \rangle$. By Lemma I4,

$$H(\psi(n)) \leq H(n) + c_\psi.$$

That is,

$$H(\langle n, n \rangle) \leq H(n) + O(1).$$

Q.E.D.

Lemma I10. $H(\langle n, 0 \rangle) = H(n) + O(1)$.

Proof. By Lemma I7,

$$H(n) \leq H(\langle n, 0 \rangle) + O(1).$$

On the other hand, consider $\psi(n) = \langle n, 0 \rangle$. By Lemma I4,

$$H(\psi(n)) \leq H(n) + c_\psi.$$

That is,

$$H(\langle n, 0 \rangle) \leq H(n) + O(1).$$

Q.E.D.

Lemma I11. $H(m \mid n) \equiv H(\langle n, m \rangle) - H(n) \geq -c$.
(Proof: use Lemma I7.)

Lemma I12. $I(n : m) \equiv H(n) + H(m) - H(\langle n, m \rangle) \geq -c$.
(Proof: use Lemma I6.)

Lemma I13. $I(n : m) = I(m : n) + O(1)$.
(Proof: use Lemma I5.)

Lemma I14. $I(n : n) = H(n) + O(1)$.
(Proof: use Lemma I9.)

Lemma I15. $I(n : 0) = O(1)$.
(Proof: use Lemma I10.)

Note. The further development of this algorithmic version of information theory[30] requires the notion of the size in bits of a self-delimiting computer program (Chaitin [3]), which, however, we can do without in this paper.

3. Random Reals

Definition (Martin-Löf [12]). Speaking geometrically, a real r is Martin-Löf random if it is never the case that it is contained in each set of an r.e. infinite sequence A_i of sets of intervals with the property that the measure[31] of the ith set is always less than or equal to 2^{-i},

---

[30]   Compare the original ensemble version of information theory given in Shannon and Weaver [13].
[31]   I.e., the sum of the lengths of the intervals, being careful to avoid counting overlapping intervals twice.

(5) $$\mu(A_i) \leq 2^{-i}.$$

Here is the definition of a Martin-Löf random real r in a more compact notation:

$$\forall i \left[\mu(A_i) \leq 2^{-i} \right] \Rightarrow \neg \forall i \left[r \in A_i \right].$$

An equivalent definition, if we restrict ourselves to reals in the unit interval $0 \leq r \leq 1$, may be formulated in terms of bit strings rather than geometrical notions, as follows. Define a *covering* to be an r.e. set of ordered pairs consisting of a natural number i and a bit string s,

$$\text{Covering} = \{\langle i, s \rangle\},$$

with the property that if $\langle i, s \rangle \in$ Covering and $\langle i, s' \rangle \in$ Covering, then it is not the case that s is an extension of s' or that s' is an extension of s.[32] We simultaneously consider A_i to be a set of (finite) bit strings

$$\{s \mid \langle i, s \rangle \in \text{Covering}\}$$

and to be a set of real numbers, namely those which in base-two notation have a bit string in A_i as an initial segment.[33] Then condition (5) becomes

(6) $$\mu(A_i) = \sum_{\langle i, s \rangle \in \text{Covering}} 2^{-|s|} \leq 2^{-i},$$

where $|s| =$ the length in bits of the string s.

Note. This is equivalent to stipulating the existence of an arbitrary "regulator of convergence" $f \to \infty$ that is computable and nondecreasing such that $\mu(A_i) \leq 2^{-f(i)}$. A_0 is only required to have measure ≤ 1 and is sort of useless, since we are working within the unit interval $0 \leq r \leq 1$.[34]

Any real number, considered as a singleton set, is a set of measure zero, but not constructively so! Similarly, the notion of a von Mises' collective,[35] which is an infinite bit string such that any place selection rule based on the preceding bits picks out a substring with the same limiting frequency of 0's and 1's as the whole string has, is contradictory. But Alonzo Church's idea, to allow only computable place selection rules, saves the concept.

Definition (Solovay [14]). A real r is Solovay random if for any r.e. infinite sequence A_i of sets of intervals with the property that the sum of the measures of the A_i converges

$$\sum \mu(A_i) < \infty,$$

r is contained in at most finitely many of the A_i. In other words,

---

32   This is to avoid overlapping intervals and enable us to use the formula (6). It is easy to convert a covering which does not have this property into one that covers exactly the same set and does have this property. How this is done depends on the order in which overlaps are discovered: intervals which are subsets of ones which have already been included in the enumeration of A_i are eliminated, and intervals which are supersets of ones which have already been included in the enumeration must be split into disjoint subintervals, and the common portion must be thrown away.

33   I.e., the geometrical statement that a point is covered by (the union of) a set of intervals, corresponds in bit string language to the statement that an initial segment of an infinite bit string is contained in a set of finite bit strings. The fact that some reals correspond to two infinite bit strings, e.g., .100000... = .011111..., causes no problems. We are working with closed intervals, which include their endpoints.

34   It makes $\Sigma\mu(A_i) \leq 2$ instead of what it should be, namely, ≤ 1. So A_0 really ought to be abolished!

35   See Feller [9].

$$\sum \mu(A_i) < \infty \;\Rightarrow\; \exists N \; \forall (i > N) \left[r \notin A_i \right].$$

A real r is weakly Solovay random ("Solovay random with a regulator of convergence") if for any r.e. infinite sequence A_i of sets of intervals with the property that the sum of the measures of the A_i converges constructively, then r is contained in at most finitely many of the A_i. In other words, a real r is weakly Solovay random if the existence of a computable function $f(n)$ such that for each n,

$$\sum_{i \geq f(n)} \mu(A_i) \;\leq\; 2^{-n},$$

implies that r is contained in at most finitely many of the A_i. That is to say,

$$\forall n \left[\sum_{i \geq f(n)} \mu(A_i) \;\leq\; 2^{-n} \right] \;\Rightarrow\; \exists N \; \forall (i > N) \left[r \notin A_i \right].$$

Definition (Chaitin [3]). A real r is Chaitin random if (the information content of the initial segment r_n of length n of the base-two expansion of r) does not drop arbitrarily far below n: $\liminf H(r_n) - n > - \infty$.[36] In other words,

$$\exists c \; \forall n \left[H(r_n) \geq n - c \right].$$

A real r is strongly Chaitin random if (the information content of the initial segment r_n of length n of the base-two expansion of r) eventually becomes and remains arbitrarily greater than n: $\liminf H(r_n) - n = \infty$. In other words,

$$\forall k \; \exists N_k \; \forall (n \geq N_k) \left[H(r_n) \geq n + k \right].$$

Note. All these definitions hold with probability one (see Theorem R4).

Theorem R1. Martin-Löf random $\iff$ Chaitin random.

Proof. $\neg$ Martin-Löf $\Rightarrow \neg$ Chaitin. Suppose that a real number r has the property that

$$\forall i \left[\mu(A_i) \leq 2^{-i} \; \& \; r \in A_i \right].$$

The series

$$\sum 2^n / 2^{n^2} \;=\; \sum 2^{-n^2 + n} \;=\; 2^{-0} + 2^{-0} + 2^{-2} + 2^{-6} + 2^{-12} + 2^{-20} + \cdots$$

obviously converges, and define N so that

$$\sum_{n \geq N} 2^{-n^2 + n} \;\leq\; 1.$$

(In fact, we can take $N = 2$.) Let the variable s range over bit strings, and consider

---

[36] Thus

$$\begin{aligned} n - c \;\leq\; H(r_n) \;&\leq\; n + H(n) + c' \\ &\leq\; n + \log n + 2 \log\log n + c'' \end{aligned}$$

by Lemmas 12 and I.

$$\sum_{n \geq N} \sum_{s \in A_{n^2}} 2^{-\lceil |s| - n \rceil} = \sum_{n \geq N} 2^n \mu(A_{n^2}) \leq \sum_{n \geq N} 2^{-n^2 + n} \leq 1.$$

It follows from the minimality of H that

$$s \in A_{n^2} \quad \text{and} \quad n \geq N \quad \Rightarrow \quad H(s) \leq |s| - n + c.$$

Thus, since $r \in A_{n^2}$ for all $n \geq N$, there will be infinitely many initial segments r_k of length k of the base-two expansion of r with the property that $r_k \in A_{n^2}$ and $n \geq N$, and for each of these r_k we have

$$H(r_k) \leq |r_k| - n + c.$$

Thus the information content of an initial segment of the base-two expansion of r can drop arbitrarily far below its length.

Proof. $\neg$Chaitin $\Rightarrow$ $\neg$Martin-Löf. Suppose that $H(r_n) - n$ can go arbitrarily negative. There are $< 2^{n-k+c}$ n-bit strings s such that $H(s) < n + H(n) - k$ (Lemma I3). Thus there are $< 2^{n-H(n)-k}$ n-bit strings s such that $H(s) < n - k - c$. That is, the probability that an n-bit string s has $H(s) < n - k - c$ is $< 2^{-H(n)-k}$. Summing this over all n, we get

$$\sum_n 2^{-H(n)-k} = 2^{-k} \sum_n 2^{-H(n)} = 2^{-k} \Omega \leq 2^{-k},$$

since $\Omega \leq 1$. Thus if a real r has the property that $H(r_n)$ dips below $n - k - c$ for even one value of n, then r is covered by an r.e. set A_k of intervals & $\mu(A_k) \leq 2^{-k}$. Thus if $H(r_n) - n$ goes arbitrarily negative, for each k we can compute an A_k with $\mu(A_k) \leq 2^{-k}$ & $r \in A_k$, and r is not Martin-Löf random. Q.E.D.

Theorem R2. Solovay random $\Longleftrightarrow$ strong Chaitin random.

Proof. $\neg$Solovay $\Rightarrow$ $\neg$(strong Chaitin). Suppose that a real number r has the property that it is in infinitely many A_i and

$$\sum \mu(A_i) < \infty.$$

Then there must be an N such that

$$\sum_{i \geq N} \mu(A_i) \leq 1.$$

Hence

$$\sum_{i \geq N} \sum_{s \in A_i} 2^{-|s|} = \sum_{i \geq N} \mu(A_i) \leq 1.$$

It follows from the minimality of H that

$$s \in A_i \quad \text{and} \quad i \geq N \quad \Rightarrow \quad H(s) \leq |s| + c,$$

i.e., if a bit string s is in A_i and $i \geq N$, then its information content is less than or equal to its size in bits $+ c$. Thus $H(r_n) \leq |r_n| + c = n + c$ for infinitely many initial segments r_n of length n of the base-two expansion of r, and it is not the case that $H(r_n) - n \to \infty$.

Proof. ¬(strong Chaitin) ⇒ ¬Solovay. ¬(strong Chaitin) says that there is a k such that for infinitely many values of n we have $H(r_n) - n < k$. The probability that an n-bit string s has $H(s) < n + k$ is $< 2^{-H(n)+k+c}$ (Lemma I3). Let A_n be the r.e. set of all n-bit strings s such that $H(s) < n + k$.

$$\sum \mu(A_n) \leq \sum_n 2^{-H(n)+k+c} = 2^{k+c} \sum 2^{-H(n)} = 2^{k+c}\, \Omega \leq 2^{k+c},$$

since $\Omega \leq 1$. Hence $\sum \mu(A_n) < \infty$ and r is in infinitely many of the A_n, and thus r is not Solovay random. Q.E.D.

Theorem R3. Martin-Löf random ⟺ weak Solovay random.

Proof. ¬Martin-Löf ⇒ ¬(weak Solovay). We are given that $\forall i\, \big[r \in A_i\big]$ and $\forall i\, \big[\mu(A_i) \leq 2^{-i}\big]$. Hence $\sum \mu(A_i)$ converges and the inequality

$$\sum_{i > N} \mu(A_i) \leq 2^{-N}$$

gives us a regulator of convergence.

Proof. ¬(weak Solovay) ⇒ ¬Martin-Löf. Suppose

$$\sum_{i \geq f(n)} \mu(A_i) \leq 2^{-n}$$

and the real number r is in infinitely many of the A_i. Let

$$B_n = \bigcup_{i \geq f(n)} A_i .$$

Then $\mu(B_n) \leq 2^{-n}$ and $r \in B_n$, so r is not Martin-Löf random. Q.E.D.

Note. In summary, the five definitions of randomness reduce to at most two:

- Martin-Löf random ⟺ Chaitin random ⟺ weak Solovay random.[37]

- Solovay random ⟺ strong Chaitin random.[38]

- Solovay random ⇒ Martin-Löf random.[39]

- Martin-Löf random ⇒ Solovay random???

Theorem R4. With probability one, a real number r is Martin-Löf random and Solovay random.

Proof 1. Since Solovay random ⇒ Martin-Löf random (is the converse true?), it is sufficient to show that r is Solovay random with probability one. Suppose

$$\sum \mu(A_i) < \infty,$$

where the A_i are an r.e. infinite sequence of sets of intervals. Then (this is the Borel–Cantelli lemma (Feller [9])),

---

[37] Theorems R1 and R3.
[38] Theorem R2.
[39] Because strong Chaitin ⇒ Chaitin.

$$\lim_{N \to \infty} \Pr\{\bigcup_{i \ge N} A_i\} \le \lim_{N \to \infty} \sum_{i \ge N} \mu(A_i) = 0,$$

and the probability is zero that a real r is in infinitely many of the A_i. But there are only countably many choices for the r.e. sequence of A_i, since there are only countably many algorithms. Since the union of a countable number of sets of measure zero is also of measure zero, it follows that with probability one r is Solovay random.

Proof 2. We use the Borel–Cantelli lemma again. This time we show that the strong Chaitin criterion for randomness, which is equivalent to the Solovay criterion, is true with probability one. Since for each k,

$$\sum_n \Pr\{H(r_n) < n + k\} \le 2^{k+c}$$

and thus converges,[40] it follows that for each k with probability one $H(r_n) < n + k$ only finitely often. Thus, with probability one,

$$\lim_{n \to \infty} H(r_n) - n = \infty.$$

Q.E.D.

Theorem R5. r Martin-Löf random $\Rightarrow H(r_n) - n$ is unbounded. (Does r Martin-Löf random $\Rightarrow \lim H(r_n) - n = \infty$?)

Proof. We shall prove the theorem by assuming that $H(r_n) - n < c$ for all n and deducing that r cannot be Martin-Löf random. Let c' be the constant of Lemma I3, so that the number of k-bit strings s with $H(s) < k + H(k) - i$ is $< 2^{k-i+c'}$.

Consider r_k for $k = 1$ to $2^{n+c+c'}$. We claim that the probability of the event A_n that r simultaneously satisfies the $2^{n+c+c'}$ inequalities

$$H(r_k) < k + c \quad (k = 1, \dots, 2^{n+c+c'})$$

is $\le 2^{-n}$. (See the next paragraph for the proof of this claim.) Thus we have an r.e. infinite sequence A_n of sets of intervals with measure $\mu(A_n) \le 2^{-n}$ which all contain r. Hence r is not Martin-Löf random.

Proof of Claim. Since $\sum 2^{-H(k)} = \Omega \le 1$, there is a k between 1 and $2^{n+c+c'}$ such that $H(k) \ge n + c + c'$. For this value of k,

$$\Pr\{H(r_k) < k + c\} \le 2^{-H(k)+c+c'} \le 2^{-n},$$

since the number of k-bit strings s with $H(s) < k + H(k) - i$ is $< 2^{k-i+c'}$ (Lemma I3). Q.E.D.

Theorem R6. Ω is a Martin-Löf–Chaitin–weak Solovay random real number. More generally, if N is an infinite r.e. set of natural numbers, then

$$\theta = \sum_{n \in N} 2^{-H(n)}$$

---

[40] See the second half of the proof of Theorem R2.

is a Martin-Löf–Chaitin–weak Solovay random real.[41]

Proof. Since $H(n)$ can be computed as a limit from above, $2^{-H(n)}$ can be computed as a limit from below. It follows that given θ_k, the first k bits of the base-two expansion *without infinitely many consecutive trailing zeros*[42] of the real number θ, one can calculate the finite set of all $n \in N$ such that $H(n) \leq k$, and then, since N is infinite, one can calculate an $n \in N$ with $H(n) > k$. That is, there is a computable partial function ψ such that

$$\psi(\theta_k) = \text{a natural number } n \text{ with } H(n) > k.$$

But by Lemma I4,

$$H(\psi(\theta_k)) \leq H(\theta_k) + c_\psi.$$

Hence

$$k < H(\psi(\theta_k)) \leq H(\theta_k) + c_\psi$$

and

$$H(\theta_k) > k - c_\psi.$$

Thus θ is Chaitin random, and by Theorems R1 and R3 it is also Martin-Löf random and weakly Solovay random. Q.E.D.

Theorem R7. There is an exponential diophantine equation

$$L(n, x_1, \dots, x_m) = R(n, x_1, \dots, x_m)$$

which has only finitely many solutions $x_1, \dots, x_m$ if the nth bit of Ω is a 0, and which has infinitely many solutions $x_1, \dots, x_m$ if the nth bit of Ω is a 1.

Proof. Since $H(n)$ can be computed as a limit from above, $2^{-H(n)}$ can be computed as a limit from below. It follows that

$$\Omega = \sum 2^{-H(n)}$$

is the limit from below of a computable sequence $\omega_1 \leq \omega_2 \leq \omega_3 \leq \cdots$ of rational numbers

$$\Omega = \lim_{k \to \infty} \omega_k.$$

This sequence converges extremely slowly! The exponential diophantine equation $L = R$ is constructed from the sequence ω_k by using the theorem that "every r.e. relation has a *singlefold* exponential diophantine representation" (Jones and Matijasevic [11]). Since the assertion that

"the nth bit of ω_k is a 1"

---

[41] Incidentally, this implies that θ is not a computable real number, from which it follows that $0 < \theta < 1$, that θ is irrational, and even that θ is transcendental.

[42] If there is a choice between ending the base-two expansion of θ with infinitely many consecutive zeros or with infinitely many consecutive ones (i.e., if θ is a dyadic rational), then we must choose the infinity of consecutive ones. This is to ensure that considered as real numbers

$$\theta_k < \theta < \theta_k + 2^{-k}.$$

Of course, it will follow from this theorem that θ must be an irrational number, so this situation cannot actually occur, but we don't know that yet!

is an r.e. relation between n and k (in fact, it is a recursive relation), the theorem of Jones and Matijasevic yields an equation

$$L(n, k, x_2, \dots, x_m) \;=\; R(n, k, x_2, \dots, x_m)$$

involving only additions, multiplications, and exponentiations of natural number constants and variables, and this equation has exactly one solution $x_2, \dots, x_m$ in natural numbers if the nth bit of the base-two expansion of ω_k is a 1, and it has no solution $x_2, \dots, x_m$ in natural numbers if the nth bit of the base-two expansion of ω_k is a 0. The number of different m-tuples $x_1, \dots, x_m$ of natural numbers which are solutions of the equation

$$L(n, x_1, \dots, x_m) \;=\; R(n, x_1, \dots, x_m)$$

is therefore infinite if the nth bit of the base-two expansion of Ω is a 1, and it is finite if the nth bit of the base-two expansion of Ω is a 0. Q.E.D.

4. Incompleteness Theorems

Having developed the necessary information-theoretic formalism in Section 2, and having studied the notion of a random real in Section 3, we can now begin to derive incompleteness theorems.

The setup is as follows. The axioms of a formal theory are considered to be encoded as a single finite bit string, the rules of inference are considered to be an algorithm for enumerating the theorems given the axioms, and in general we shall fix the rules of inference and vary the axioms. More formally, the rules of inference F may be considered to be an r.e. set of propositions of the form

$$\text{``Axioms } |-_F \text{ Theorem.''}$$

The r.e. set of theorems deduced from the axiom A is determined by selecting from the set F the theorems in those propositions which have the axiom A as an antecedent. In general we will consider the rules of inference F to be fixed and study what happens as we vary the axioms A. By an n-bit theory we shall mean the set of theorems deduced from an n-bit axiom.

4.1. Incompleteness Theorems for Lower Bounds on Information Content

Let us start by rederiving within our current formalism an old and very basic result, which states that even though most strings are random, one can never prove that a specific string has this property.

If one produces a bit string s by tossing a coin n times, 99.9% of the time it will be the case that $H(s) \approx n + H(n)$ (Lemmas I2 and I3). In fact, if one lets n go to infinity, with probability one $H(s) > n$ for all but finitely many n (Theorem R4). However,

Theorem LB (Chaitin [1,2,4]). Consider a formal theory all of whose theorems are assumed to be true. Within such a formal theory a specific string cannot be proven to have information content more than $O(1)$ greater than the information content of the axioms of the theory. That is, if "$H(s) \geq n$" is a theorem only if it is true, then it is a theorem only if $n \leq H(\text{axioms}) + O(1)$. Conversely, there are formal theories whose axioms have information content $n + O(1)$ in which it is possible to establish all true propositions of the form "$H(s) \geq n$" and of the form "$H(s) = k$" with $k < n$.

Proof. Consider the enumeration of the theorems of the formal axiomatic theory in order of the size of their proofs. For each natural number k, let s^* be the string in the theorem of the form "$H(s) \geq n$" with $n > H(\text{axioms}) + k$ which appears first in the enumeration. On the one hand, if all theorems are true, then

$$H(\text{axioms}) + k \quad < \quad H(s^*).$$

On the other hand, the above prescription for calculating s^* shows that

$$s^* \quad = \quad \psi(\langle\langle\text{axioms}, \ H(\text{axioms})\rangle, \ k\rangle) \qquad (\psi \text{ partial recursive}),$$

and thus

$$H(s^*) \quad \leq \quad H(\langle\langle\text{axioms}, H(\text{axioms})\rangle, k\rangle) + c_\psi \quad \leq \quad H(\text{axioms}) + H(k) + O(1).$$

Here we have used the subadditivity of information $H(\langle s, t\rangle) \leq H(s) + H(t) + O(1)$ (Lemma I6) and the fact that $H(\langle s, H(s)\rangle) \leq H(s) + O(1)$ (Lemma I8). It follows that

$$H(\text{axioms}) + k \quad < \quad H(s^*) \quad \leq \quad H(\text{axioms}) + H(k) + O(1),$$

and thus

$$k \quad < \quad H(k) + O(1).$$

However, this inequality is false for all $k \geq k_0$, where k_0 depends only on the rules of inference. A contradiction is avoided only if s^* does not exist for $k = k_0$, i.e., it is impossible to prove in the formal theory that a specific string has H greater than $H(\text{axioms}) + k_0$.

Proof of Converse. The set T of all true propositions of the form "$H(s) \leq k$" is r.e. Choose a fixed enumeration of T without repetitions, and for each natural number n, let s^* be the string in the last proposition of the form "$H(s) \leq k$" with $k < n$ in the enumeration. Let

$$\Delta \quad = \quad n - H(s^*) \quad > \quad 0.$$

Then from s^*, $H(s^*)$, & Δ we can calculate $n = H(s^*) + \Delta$, then all strings s with $H(s) < n$, and then a string s_n with $H(s_n) \geq n$. Thus

$$n \quad \leq \quad H(s_n) \quad = \quad H(\psi(\langle\langle s^*, \ H(s^*)\rangle, \ \Delta\rangle)) \qquad (\psi \text{ partial recursive}),$$

and so

(7) $$n \quad \leq \quad H(\langle\langle s^*, \ H(s^*)\rangle, \ \Delta\rangle) + c_\psi \quad \begin{aligned} &\leq \quad H(s^*) + H(\Delta) + O(1) \\ &\leq \quad n + H(\Delta) + O(1) \end{aligned}$$

by Lemmas I6 and I8. The first line of (7) implies that

$$\Delta \quad \equiv \quad n - H(s^*) \quad \leq \quad H(\Delta) + O(1),$$

which implies that Δ and $H(\Delta)$ are both bounded. Then the second line of (7) implies that

$$H(\langle\langle s^*, \ H(s^*)\rangle, \ \Delta\rangle) \quad = \quad n + O(1).$$

The triple $\langle\langle s^*, \ H(s^*)\rangle, \ \Delta\rangle$ is the desired axiom: it has information content $n + O(1)$, and by enumerating T until all true propositions of the form "$H(s) \leq k$" with $k < n$ have been discovered, one can immediately deduce all true propositions of the form "$H(s) \geq n$" and of the form "$H(s) = k$" with $k < n$. Q.E.D.

4.2. Incompleteness Theorems for Random Reals: First Approach

In this section we begin our study of incompleteness theorems for random reals. We show that any particular formal theory can enable one to determine at most a finite number of bits of Ω. In the following sections (4.3 and 4.4) we express the upper bound on the number of bits of Ω which can be determined, in terms of the axioms of the theory; for now, we just show that an upper bound exists. We shall not use any ideas from algorithmic information theory until Section 4.4; for now (Sections 4.2 and 4.3) we only make use of the fact that Ω is Martin-Löf random.

If one tries to guess the bits of a random sequence, the average number of correct guesses before failing is exactly 1 guess! Reason: if we use the fact that the expected value of a sum is equal to the sum of the expected values, the answer is the sum of the chance of getting the first guess right, plus the chance of getting the first and the second guesses right, plus the chance of getting the first, second and third guesses right, etc.,

$$\frac{1}{2} + \frac{1}{4} + \frac{1}{8} + \frac{1}{16} + \cdots = 1.$$

Or if we directly calculate the expected value as the sum of (the # right till first failure) × (the probability),

$$0 \times \frac{1}{2} + 1 \times \frac{1}{4} + 2 \times \frac{1}{8} + 3 \times \frac{1}{16} + 4 \times \frac{1}{32} + \cdots$$

$$= 1 \times \sum_{k>1} 2^{-k} + 1 \times \sum_{k>2} 2^{-k} + 1 \times \sum_{k>3} 2^{-k} + \cdots$$

$$= \frac{1}{2} + \frac{1}{4} + \frac{1}{8} + \cdots = 1.$$

On the other hand (see the next section), if we are allowed to try 2^n times a series of n guesses, one of them will always get it right, if we try all 2^n different possible series of n guesses.

Theorem X. Any given formal theory T can yield only finitely many (scattered) bits of (the base-two expansion of) Ω.

When we say that a theory yields a bit of Ω, we mean that it enables us to determine its position and its $0/1$ value.

Proof. Consider a theory T, an r.e. set of true assertions of the form

"The nth bit of Ω is 0."
"The nth bit of Ω is 1."

Here n denotes specific natural numbers.

If T provides k different (scattered) bits of Ω, then that gives us a covering A_k of measure 2^{-k} which includes Ω: Enumerate T until k bits of Ω are determined, then the covering is all bit strings up to the last determined bit with all determined bits okay. If n is the last determined bit, this covering will consist of 2^{n-k} n-bit strings, and will have measure $2^{n-k}/2^n = 2^{-k}$.

It follows that if T yields infinitely many different bits of Ω, then for any k we can produce by running through all possible proofs in T a covering A_k of measure 2^{-k} which includes Ω. But this contradicts the fact that Ω is Martin-Löf random. Hence T yields only finitely many bits of Ω. Q.E.D.

Corollary X. Since by Theorem R7 Ω can be encoded into an exponential diophantine equation

$$(8) \qquad\qquad L(n, x_1, \ldots, x_m) = R(n, x_1, \ldots, x_m),$$

it follows that any given formal theory can permit one to determine whether (8) has finitely or infinitely many solutions $x_1, \ldots, x_m$, for only finitely many specific values of the parameter n.

4.3. Incompleteness Theorems for Random Reals: | Axioms |

Theorem A. If $\Sigma 2^{-f(n)} \leq 1$ and f is computable, then there is a constant c_f with the property that no n-bit theory ever yields more than $n + f(n) + c_f$ bits of Ω.

Proof. Let A_k be the event that there is at least one n such that there is an n-bit theory that yields $n + f(n) + k$ or more bits of Ω.

$$\Pr\{A_k\} \leq \sum_n \left[\binom{2^n}{\substack{n\text{-bit} \\ \text{theories}}} \binom{2^{-\lceil n+f(n)+k\rceil}}{\substack{\text{probability that yields} \\ n + f(n) + k \text{ bits of } \Omega}} \right] = 2^{-k} \sum_n 2^{-f(n)} \leq 2^{-k}$$

since $\Sigma 2^{-f(n)} \leq 1$. Hence $\Pr\{A_k\} \leq 2^{-k}$, and $\Sigma \Pr\{A_k\}$ also converges. Thus only finitely many of the A_k occur (Borel–Cantelli lemma (Feller [9])). That is,

$$\lim_{N \to \infty} \Pr\{ \bigcup_{k > N} A_k \} \leq \sum_{k > N} \Pr\{A_k\} \leq 2^{-N} \to 0.$$

More Detailed Proof. Assume the opposite of what we want to prove, namely that for every k there is at least one n-bit theory that yields $n + f(n) + k$ bits of Ω. From this we shall deduce that Ω cannot be Martin-Löf random, which is impossible.

To get a covering A_k of Ω with measure $\leq 2^{-k}$, consider a specific n and all n-bit theories. Start generating theorems in each n-bit theory until it yields $n + f(n) + k$ bits of Ω (it does not matter if some of these bits are wrong). The measure of the set of possibilities for Ω covered by the n-bit theories is thus $\leq 2^n 2^{-n-f(n)-k} = 2^{-f(n)-k}$. The measure $\mu(A_k)$ of the union of the set of possibilities for Ω covered by n-bit theories with any n is thus

$$\leq \sum_n 2^{-f(n)-k} = 2^{-k} \sum_n 2^{-f(n)} \leq 2^{-k} \text{ (since } \sum 2^{-f(n)} \leq 1\text{).}$$

Thus Ω is covered by A_k and $\mu(A_k) \leq 2^{-k}$ for every k if there is always an n-bit theory that yields $n + f(n) + k$ bits of Ω, which is impossible. Q.E.D.

Corollary A. If $\Sigma 2^{-f(n)}$ converges and f is computable, then there is a constant c_f with the property that no n-bit theory ever yields more than $n + f(n) + c_f$ bits of Ω.

Proof. Choose c so that $\Sigma 2^{-f(n)} \leq 2^c$. Then $\Sigma 2^{-\lceil f(n)+c\rceil} \leq 1$, and we can apply Theorem A to $f'(n) = f(n) + c$. Q.E.D.

Corollary A2. Let $\Sigma 2^{-f(n)}$ converge and f be computable as before. If $g(n)$ is computable, then there is a constant $c_{f, g}$ with the property that no $g(n)$-bit theory ever yields more than $g(n) + f(n) + c_{f, g}$ bits of Ω. For example, consider N of the form 2^{2^n}. For such N, no N-bit theory ever yields more than $N + f(\log\log N) + c_{f, g}$ bits of Ω.

Note. Thus for n of special form, i.e., which have concise descriptions, we get better upper bounds on the number of bits of Ω which are yielded by n-bit theories. This is a foretaste of the way algorithmic information theory will be used in Theorem C and Corollary C2 (Sect. 4.4).

Lemma for Second Borel–Cantelli Lemma! For any finite set $\{x_k\}$ of non-negative real numbers,

$$\prod (1 - x_k) \leq \frac{1}{\sum x_k}.$$

Proof. If x is a real number, then

$$1 - x \leq \frac{1}{1 + x}.$$

Thus

$$\prod(1 - x_k) \leq \frac{1}{\prod(1 + x_k)} \leq \frac{1}{\sum x_k},$$

since if all the x_k are non-negative

$$\prod(1 + x_k) \geq \sum x_k.$$

Q.E.D.

Second Borel–Cantelli Lemma (Feller [9]). Suppose that the events A_n have the property that it is possible to determine whether or not the event A_n occurs by examining the first $f(n)$ bits of Ω, where f is a computable function. If the events A_n are mutually independent and $\sum \Pr\{A_n\}$ diverges, then Ω has the property that infinitely many of the A_n must occur.

Proof. Suppose on the contrary that Ω has the property that only finitely many of the events A_n occur. Then there is an N such that the event A_n does not occur if $n \geq N$. The probability that none of the events $A_N, A_{N+1}, \ldots, A_{N+k}$ occur is, since the A_n are mutually independent, precisely

$$\prod_{i=0}^{k}(1 - \Pr\{A_{N+i}\}) \leq \frac{1}{\left[\displaystyle\sum_{i=0}^{k}\Pr\{A_{N+i}\}\right]},$$

which goes to zero as k goes to infinity. This would give us arbitrarily small covers for Ω, which contradicts the fact that Ω is Martin-Löf random. Q.E.D.

Theorem B. If $\sum 2^{n-f(n)}$ diverges and f is computable, then infinitely often there is a run of $f(n)$ zeros between bits 2^n & 2^{n+1} of Ω ($2^n \leq$ bit $< 2^{n+1}$). Hence there are rules of inference which have the property that there are infinitely many N-bit theories that yield (the first) $N + f(\log N)$ bits of Ω.

Proof. We wish to prove that infinitely often Ω must have a run of $k = f(n)$ consecutive zeros between its 2^nth & its 2^{n+1}th bit position. There are 2^n bits in the range in question. Divide this into nonoverlapping blocks of $2k$ bits each, giving a total of $2^n/2k$ blocks. The chance of having a run of k consecutive zeros in each block of $2k$ bits is

$$(9) \qquad \geq \frac{k\, 2^{k-2}}{2^{2k}}.$$

Reason:

- There are $2k - k + 1 \geq k$ different possible choices for where to put the run of k zeros in the block of $2k$ bits.

- Then there must be a 1 at each end of the run of 0's, but the remaining $2k - k - 2 = k - 2$ bits can be anything.

- This may be an underestimate if the run of 0's is at the beginning or end of the $2k$ bits, and there is no room for endmarker 1's.

- There is no room for another $10^k 1$ to fit in the block of $2k$ bits, so we are not overestimating the probability by counting anything twice.

Summing (9) over all $2^n/2k$ blocks and over all n, we get

$$\geq \sum_n \left[\frac{k \, 2^{k-2}}{2^{2k}} \cdot \frac{2^n}{2k} \right] = \frac{1}{8} \sum_n 2^{n-k} = \frac{1}{8} \sum 2^{n-f(n)} = \infty.$$

Invoking the second Borel–Cantelli lemma (if the events A_i are independent and $\sum \Pr\{A_i\}$ diverges, then infinitely many of the A_i must occur), we are finished. Q.E.D.

Corollary B. If $\sum 2^{-f(n)}$ diverges and f is computable and nondecreasing, then infinitely often there is a run of $f(2^{n+1})$ zeros between bits 2^n & 2^{n+1} of Ω ($2^n \leq$ bit $< 2^{n+1}$). Hence there are infinitely many N-bit theories that yield (the first) $N + f(N)$ bits of Ω.

Proof. If $\sum 2^{-f(n)}$ diverges and f is computable and nondecreasing, then by the Cauchy condensation test

$$\sum 2^n \, 2^{-f(2^n)}$$

also diverges, and therefore so does

$$\sum 2^n \, 2^{-f(2^{n+1})}.$$

Hence, by Theorem B, infinitely often there is a run of $f(2^{n+1})$ zeros between bits 2^n and 2^{n+1}. Q.E.D.

Corollary B2. If $\sum 2^{-f(n)}$ diverges and f is computable, then infinitely often there is a run of $n + f(n)$ zeros between bits 2^n & 2^{n+1} of Ω ($2^n \leq$ bit $< 2^{n+1}$). Hence there are infinitely many N-bit theories that yield (the first) $N + \log N + f(\log N)$ bits of Ω.

Proof. Take $f(n) = n + f'(n)$ in Theorem B. Q.E.D.

Theorem AB. (a) There is a c with the property that no n-bit theory ever yields more than $n + \log n + 2 \log\log n + c$ (scattered) bits of Ω.

(b) There are infinitely many n-bit theories that yield (the first) $n + \log n + \log\log n$ bits of Ω.

Proof. Using the Cauchy condensation test, we have seen (beginning of Sect. 2) that

(a) $\displaystyle \sum \frac{1}{n \, (\log n)^2}$ converges and (b) $\displaystyle \sum \frac{1}{n \log n}$ diverges.

The theorem follows immediately from Corollaries A and B. Q.E.D.

4.4. Incompleteness Theorems for Random Reals: H(Axioms)

Theorem C is a remarkable extension of Theorem R6:

- We have seen that the information content of [knowing the first n bits of Ω] is $\geq n - c$.

- Now we show that the information content of [knowing any n bits of Ω (their positions and $0/1$ values)] is $\geq n - c$.

Lemma C.

$$\sum_n \#\{s \mid H(s) < n\} \, 2^{-n} \;\; = \;\; \Omega \;\; \le \;\; 1.$$

Proof.

$$1 \;\; \ge \;\; \Omega \;\; = \;\; \sum_s 2^{-H(s)} \;\; = \;\; \sum_n \#\{s \mid H(s) = n\} \, 2^{-n} \;\; = \;\; \sum_n \#\{s \mid H(s) = n\} \, 2^{-n} \sum_{k \ge 1} 2^{-k} \;\; =$$

$$\sum_n \sum_{k \ge 1} \#\{s \mid H(s) = n\} \, 2^{-n-k} \;\; = \;\; \sum_n \#\{s \mid H(s) < n\} \, 2^{-n}.$$

Q.E.D.

Theorem C. If a theory has $H(\text{axiom}) < n$, then it can yield at most $n + c$ (scattered) bits of Ω.

Proof. Consider a particular k and n. If there is an axiom with $H(\text{axiom}) < n$ which yields $n + k$ scattered bits of Ω, then even without knowing which axiom it is, we can cover Ω with an r.e. set of intervals of measure

$$\le \;\; \begin{pmatrix} \#\{s \mid H(s) < n\} \\ \text{\# of axioms} \\ \text{with } H < n \end{pmatrix} \begin{pmatrix} 2^{-n-k} \\ \text{measure of set of} \\ \text{possibilities for } \Omega \end{pmatrix} \;\; = \;\; \#\{s \mid H(s) < n\} \, 2^{-n-k}.$$

But by the preceding lemma, we see that

$$\sum_n \#\{s \mid H(s) < n\} \, 2^{-n-k} \;\; = \;\; 2^{-k} \sum_n \#\{s \mid H(s) < n\} \, 2^{-n} \;\; \le \;\; 2^{-k}.$$

Thus if even one theory with $H < n$ yields $n + k$ bits of Ω, for any n, we get a cover for Ω of measure $\le 2^{-k}$. This can only be true for finitely many values of k, or Ω would not be Martin-Löf random. Q.E.D.

Corollary C. No n-bit theory ever yields more than $n + H(n) + c$ bits of Ω.
(Proof: Theorem C and by Lemma 12, $H(\text{axiom}) \le |\text{axiom}| + H(|\text{axiom}|) + c$.)

Lemma C2. If $g(n)$ is computable and unbounded, then $H(n) < g(n)$ for infinitely many values of n.

Proof. Define the inverse of g as

$$g^{-1}(n) \;\; = \;\; \min_{g(k) \ge n} k.$$

Then using Lemmas I and I4 we see that for all sufficiently large values of n,

$$H(g^{-1}(n)) \;\; \le \;\; H(n) + O(1) \;\; \le \;\; O(\log n) \;\; < \;\; n \;\; \le \;\; g(g^{-1}(n)).$$

That is, $H(k) < g(k)$ for all $k = g^{-1}(n)$ and n sufficiently large. Q.E.D.

Corollary C2. Let $g(n)$ be computable and unbounded. For infinitely many n, no n-bit theory yields more than $n + g(n) + c$ bits of Ω.

(Proof: Corollary C and Lemma C2.)

Note. In appraising Corollaries C and C2, the trivial formal systems in which there is always an n-bit axiom that yields the first n bits of Ω should be kept in mind. Also, compare Corollaries C and A, and Corollaries C2 and A2.

In summary,

Theorem D. There is an exponential diophantine equation

$$(10) \qquad L(n, x_1, \ldots, x_m) = R(n, x_1, \ldots, x_m)$$

which has only finitely many solutions $x_1, \ldots, x_m$ if the nth bit of Ω is a 0, and which has infinitely many solutions $x_1, \ldots, x_m$ if the nth bit of Ω is a 1. Let us say that a formal theory "settles k cases" if it enables one to prove that the number of solutions of (10) is finite or that it is infinite for k specific values (possibly scattered) of the parameter n. Let $f(n)$ and $g(n)$ be computable functions.

- $\sum 2^{-f(n)} < \infty \;\Rightarrow\;$ all n-bit theories settle $\leq n + f(n) + O(1)$ cases.

- $\sum 2^{-f(n)} = \infty$ and $f(n) \leq f(n+1) \;\Rightarrow\;$ for infinitely many n, there is an n-bit theory that settles $\geq n + f(n)$ cases.

- $H(\text{theory}) < n \;\Rightarrow\;$ it settles $\leq n + O(1)$ cases.

- n-bit theory $\;\Rightarrow\;$ it settles $\leq n + H(n) + O(1)$ cases.

- g unbounded $\;\Rightarrow\;$ for infinitely many n, all n-bit theories settle $\leq n + g(n) + O(1)$ cases.

Proof. The theorem combines Theorem R7, Corollaries A and B, Theorem C, and Corollaries C and C2. Q.E.D.

5. Conclusion

In conclusion, we have seen that proving whether particular exponential diophantine equations have finitely or infinitely many solutions, is absolutely intractable. Such questions escape the power of mathematical reasoning. This is a region in which mathematical truth has no discernible structure or pattern and appears to be completely random. These questions are completely beyond the power of human reasoning. Mathematics cannot deal with them.

Quantum physics has shown that there is randomness in nature. I believe that we have demonstrated in this paper that randomness is already present in pure mathematics. This does not mean that the universe and mathematics are lawless, it means that laws of a different kind apply: statistical laws.

References

1. G. J. CHAITIN, Information-theoretic computational complexity, *IEEE Trans. Inform. Theory* **20** (1974), 10–15.

2. G. J. CHAITIN, Randomness and mathematical proof, *Sci. Amer.* **232,** No. 5 (1975), 47–52.

3. G. J. CHAITIN, A theory of program size formally identical to information theory, *J. Assoc. Comput. Mach.* **22** (1975), 329–340.

4. G. J. CHAITIN, Gödel's theorem and information, *Internat. J. Theoret. Phys.* **22** (1982), 941–954.

5. G. J. CHAITIN, Randomness and Gödel's theorem, "Mondes en Développement," Vol. 14, No. 53, in press.

6. R. COURANT and H. ROBBINS, "What is Mathematics?," Oxford Univ. Press, London, 1941.

7. M. DAVIS, H. PUTNAM, and J. ROBINSON, The decision problem for exponential diophantine equations, *Ann. Math.* **74** (1961), 425–436.

8. M. DAVIS, "The Undecidable—Basic Papers on Undecidable Propositions, Unsolvable Problems and Computable Functions," Raven, New York, 1965.

9. W. FELLER, "An Introduction to Probability Theory and Its Applications, I," Wiley, New York, 1970.

10. G. H. HARDY, "A Course of Pure Mathematics," 10th ed., Cambridge Univ. Press, London, 1952.

11. J. P. JONES and Y. V. MATIJASEVIC, Register machine proof of the theorem on exponential diophantine representation of enumerable sets, *J. Symbolic Logic* **49** (1984), 818-829.

12. P. MARTIN-LÖF, The definition of random sequences, *Inform. Control* **9** (1966), 602-619.

13. C. E. SHANNON and W. WEAVER, "The Mathematical Theory of Communication," Univ. of Illinois Press, Urbana, 1949.

14. R. M. SOLOVAY, Private communication, 1975.

15. A. M. TURING, On computable numbers, with an application to the Entscheidungsproblem, *Proc. London Math. Soc.* **42** (1937), 230-265; also in [8].

ALGORITHMIC ENTROPY OF SETS

Computers & Mathematics with Applications 2 (1976), pp. 233-245.

GREGORY J. CHAITIN

IBM Thomas J. Watson Research Center, Yorktown Heights, NY 10598, U.S.A.

Communicated by J. T. Schwartz

(Received July 1976)

Abstract

In a previous paper a theory of program size formally identical to information theory was developed. The entropy of an individual finite object was defined to be the size in bits of the smallest program for calculating it. It was shown that this is $-\log_2$ of the probability that the object is obtained by means of a program whose successive bits are chosen by flipping an unbiased coin. Here a theory of the entropy of recursively enumerable sets of objects is proposed which includes the previous theory as the special case of sets having a single element. The primary concept in the generalized theory is the probability that a computing machine enumerates a given set when its program is manufactured by coin flipping. The entropy of a set is defined to be $-\log_2$ of this probability.

1. Introduction

In a classical paper on computability by probabilistic machines [1], de Leeuw *et al.* showed that if a machine with a random element can enumerate a specific set of natural numbers with positive probability, then there is a deterministic machine that also enumerates this set. We propose to throw further light on this matter by bringing into play the concepts of algorithmic information theory [2,3].

As in [3], we require a computing machine to read the successive bits of its program from a semi-infinite tape that has been filled with 0's and 1's by flipping an unbiased coin, and to decide by itself where to stop reading the program, for there is no endmarker. In [3] this convention has the important consequence that a program can be built up from subroutines by concatenating them.

In this paper we turn from finite computations to unending computations. The computer is used to enumerate a set of objects instead of a single one. An important difference between this paper and [3] is that here it is possible for the machine to read the entire program tape, so that in a sense infinite programs are permitted. However, following [1] it is better to think of these as cases in which a nondeterministic machine uses coin-flipping infinitely often.

Here, as in [3], we pick a universal computer that makes the probability of obtaining any given machine output as high as possible.

We are thus led to define three concepts: $P(A)$, the probability that the standard machine enumerates the set A, which may be called the algorithmic probability of the set A; $H(A)$, the entropy of the set A, which is $-\log_2$ of $P(A)$; and the amount of information that must be specified to enumerate A, denoted $I(A)$, which is the size in bits of the smallest program for A. In other words, $I(A)$ is the least number n such that for some program tape contents the standard machine enumerates the set A and in the process of doing so reads precisely n bits of the program tape.

One may also wish to use the standard machine to simultaneously enumerate two sets A and B, and this leads to the joint concepts $P(A, B)$, $H(A, B)$, and $I(A, B)$. In [3] programs could be concatenated, and this fact carries over here to programs that enumerate singleton sets (i.e. sets with a single element). What about arbitrary sets? Programs that enumerate arbitrary sets can be merged by interweaving their bits in the order that they are read when running at the same time, that is, in parallel. This implies that the joint probability $P(A, B)$ is not less than the product of the individual probabilities $P(A)$ and $P(B)$, from which it is easy to show that H has all the formal properties of the entropy concept of classical information theory [4]. This also implies that $I(A, B)$ is not greater than the sum of $I(A)$ and $I(B)$.

The purpose of this paper is to propose this new approach and to determine what is the number of sets A that have probability $P(A)$ greater than 2^{-n}, in other words, that have entropy $H(A)$ less than n. It must be emphasized that we do not present a complete theory. For example, the relationship between $H(A)$ and $I(A)$ requires further study. In [3] we proved that the difference between $H(A)$ and $I(A)$ is bounded for singleton sets A, but we shall show that even for finite A this is no longer the case.

2. Definitions and Their Elementary Properties

The formal definition of computing machine that we use is the Turing machine. However, we have made a few changes in the standard definition [5, pp. 13-16].

Our Turing machines have three tapes: a program tape, a work tape and an output tape. The program tape is only infinite to the right. It can be read by the machine and it can be shifted to the left. Each square of the program tape contains a 0 or a 1. The program tape is initially positioned at its leftmost square. The work tape is infinite in both directions, can be read, written and erased, and can be shifted in either direction. Each of its squares may contain a blank, a 0, or a 1. Initially all squares are blank. The output tape is infinite in both directions and it can be written on and shifted to the left. Each square may contain a blank or a $. Initially all squares are blank.

A Turing machine with n states, the first of which is its initial state, is defined in a table with $6n$ entries which is consulted each machine cycle. Each entry corresponds to one of the 6 possible contents of the 2 squares being read, and to one of the n states. All entries must be present, and each specifies an action to be performed and the next state. There are 8 possible actions: program tape left, output tape left, work tape left/right, write blank/0/1 on work tape and write $ on output tape.

Each way of filling this $6 \times n$ table produces a different n-state Turing machine M. We imagine M to be equipped with a clock that starts with time 1 and advances one unit each machine cycle. We call a unit of time a quantum. Starting at its initial state M carries out an unending computation, in the course of which it may read all or part of the program tape. The output from this computation is a set of natural numbers A. n is in A iff a $ is written by M on the output tape that is separated by exactly n blank squares from the previous $ on the tape. The time at which M outputs n is defined to be the clock reading when two $'s separated by n blanks appear on the output tape for the first time.

Let p be a finite binary sequence (henceforth *string*) or an infinite binary sequence (henceforth *sequence*). $M(p)$ denotes the set of natural numbers output *(enumerated)* by M with p as the contents of the program tape if p is a sequence, and with p written at the beginning of the program tape if p is a string. $M(p)$ is always defined if p is a sequence, but if p is a string and M reads beyond the end of p, then $M(p)$ is undefined. However, instead of saying that $M(p)$ is undefined, we shall say that $M(p)$ halts. Thus for any string p, $M(p)$ is either defined or halts. If $M(p)$ halts, the clock reading when M reads past the end of p is said to be the time at which $M(p)$ halts.

Definition:

- $P_M(A)$ is the probability that $M(p) = A$ if each bit of the sequence p is obtained by a separate toss of an unbiased coin. In other words, $P_M(A)$ is the probability that a program tape produced by coin flipping makes M enumerate A.
- $H_M(A) = -\log_2 P_M(A)$ ($= \infty$ if $P_M(A) = 0$).
- $I_M(A)$ is the number of bits in the smallest string p such that $M(p) = A$ ($=\infty$ if no such p exists).

We now pick a particular universal Turing machine U having the ability to simulate any other machine as the standard one for use throughout this paper. U has the property that for each M there is a string $\%_M$ such that for all sequences p, $U(\%_M p) = M(p)$ and U reads exactly as much of p as M does. To be more precise $\%_M = 0^g 1$, where g is the Gödel number for M. That is to say, g is the position of M in a standard list of all possible Turing machine defining tables.

Definition

- $P(A) = P_U(A)$ is the algorithmic probability of the set A.
- $H(A) = H_U(A)$ is the algorithmic entropy of the set A.
- $I(A) = I_U(A)$ is the algorithmic information of the set A.

The qualification "algorithmic" is usually omitted below.

We say that a string or sequence p is a program for A if $U(p) = A$. If $U(p) = A$ and p is a string of $I(A)$ bits, then p is said to be a minimal-size program for A. The recursively enumerable (r.e.) sets are defined to be those sets of natural numbers A for which $I(A) < \infty$. (This is equivalent to the standard definition [5, p. 58].) As there are nondenumerably many sets of natural numbers and only denumerably many r.e. sets, most A have $I(A) = \infty$.

The following theorem, whose proof is immediate, shows why U is a good machine to use. First some notation must be explained. $f(x) \lesssim g(x)$ means that $\exists c\, \forall x\, f(x) \le cg(x)$. $f(x) \gtrsim g(x)$ means that $g(x) \lesssim f(x)$. And $f(x) \approx g(x)$ means that $f(x) \lesssim g(x)$ and $f(x) \gtrsim g(x)$. $O(f(x))$ denotes an $F(x)$ with the property that there are constants c_1 and c_2 such that for all x, $|F(x)| \le |c_1 f(x)| + |c_2|$, where $f(x)$ is to be replaced by 0 if it is undefined for a particular value of x.

Theorem 1: $P(A) \gtrsim P_M(A)$, $H(A) \le H_M(A) + O(1)$, and $I(A) \le I_M(A) + O(1)$.

Definition:

- A join $B = \{2n : n \in A\} \cup \{2n + 1 : n \in B\}$. [5, pp. 81, 168]. Enumerating A join B is equivalent to simultaneously enumerating A and B.
- $P(A, B) = P(A \text{ join } B)$ (joint probability)
- $H(A, B) = H(A \text{ join } B)$ (joint entropy)
- $I(A, B) = I(A \text{ join } B)$ (joint information)
- $P(A/B) = P(A, B)/P(B)$ (conditional probability)
- $H(A/B) = -\log_2 P(A/B) = H(A, B) - H(B)$ (conditional entropy)
- $P(A : B) = P(A)P(B)/P(A, B)$ (mutual probability)
- $H(A : B) = -\log_2 P(A : B) = H(A) + H(B) - H(A, B)$ (mutual entropy).

Theorem 2:

(a) $P(A) \ge 2^{I(A)}$

(b) $H(A) \leq I(A)$

(c) For singleton A, $H(A) = I(A) + O(1)$.

(d) $H(A) < \infty$ implies $I(A) < \infty$.

Proof: (a) and (b) are immediate; (c) is Theorem 3.5(b) [3]; (d) follows from Theorem 2 [1].

Theorem 3:

(a) $P(A, B) \approx P(B, A)$

(b) $P(A, A) \approx P(A)$

(c) $P(A, \emptyset) \approx P(A)$

(d) $P(A/A) \approx 1$

(e) $P(A/\emptyset) \approx P(A)$

(f) $P(A, B) \gtrsim P(A)P(B)$

(g) $P(A/B) \gtrsim P(A)$

(h) $\Sigma_A P(A, B) \approx P(B)$

(i) $P(A, B) \lesssim P(B)$

(j) $\Sigma_A P(A/B) \approx 1$.

Proof: The proof is straightforward. For example, (f) was shown in Section 1. And (h) follows from the fact that there is a $\% = 0^g1$ such that $n \in U(\%p)$ iff $2n + 1 \in U(p)$. Thus $P(B) \geq 2^{-|\%|}\Sigma_A P(A, B)$, which taken together with (b) yields (h). Here, and henceforth, the absolute value $|s|$ of a string s signifies the number of bits in s.

The remainder of the proof is omitted.

Theorem 4:

(a) $H(A, B) = H(B, A) + O(1)$

(b) $H(A, A) = H(A) + O(1)$

(c) $H(A, \emptyset) = H(A) + O(1)$

(d) $H(A/A) = O(1)$

(e) $H(A/\emptyset) = H(A) + O(1)$

(f) $H(A, B) \leq H(A) + H(B) + O(1)$

(g) $H(A/B) \leq H(A) + O(1)$

(h) $H(A) \leq H(A, B) + O(1)$

(i) $H(A : \emptyset) = O(1)$

(j) $H(A : A) = H(A) + O(1)$

(k) $H(A : B) = H(B : A) + O(1)$

(l) $H(A : B) = H(A) - H(A/B) + O(1)$.

Theorem 5:

(a) $I(A, B) = I(B, A) + O(1)$

(b) $I(A, A) = I(A) + O(1)$

(c) $I(A, B) \leq I(A) + I(B) + O(1)$

(d) $I(A) \leq I(A, B) + O(1)$

(e) $I(A) = I(A, \{n : n < I(A)\}) + O(1)$.

The proofs of Theorems 4 and 5 are straightforward and are omitted.

ALGORITHMIC ENTROPY OF SETS

2'. The Oracle Machine U'

In order to study P, H, and I, which are defined in terms of U, we shall actually need to study a more powerful machine called U', which, unlike U, could never actually be built. U' is almost identical to U, but it cannot be built because it contains one additional feature, an oracle that gives U' *yes/no* answers to specific questions of the form "Does $U(p)$ halt?" U' can ask the oracle such questions whenever it likes. An oracle is needed because of a famous theorem on the undecidability of the halting problem [5, pp. 24-26], which states that there is no algorithm for answering these questions. U' is a special case of the general concept of relative recursiveness [5, pp. 128-134].

As a guide to intuition it should be stated that the properties of U' are precisely analogous to those of U; one simply imagines a universe exactly like ours except that sealed oracle boxes can be computer subsystems. We now indicate how to modify Section 2 so that it applies to U' instead of U.

One begins by allowing an oracle machine M to indicate in each entry of its table one of 9 possible actions (before there were 8). The new possibility is to ask the oracle if the string s currently being read on the work tape has the property that $U(s)$ halts. In response the oracle instantly writes a 1 on the work tape if the answer is *yes* and writes a 0 if the answer is *no*.

After defining an arbitrary oracle machine M, and P_M', H_M', and I_M', one then defines the standard oracle machine U' which can simulate any M. The next step is to define $P'(A)$, $H'(A)$, and $I'(A)$, which are the probability, entropy, and information of the set A relative to the halting problem. Furthermore, p is said to be an oracle program for A if $U'(p) = A$, and a minimal-size one if in addition $|p| = I'(A)$. Then A is defined to be r.e. in the halting problem if $I'(A) < \infty$. One sees as before that P' is maximal and H' and I' are minimal, and then defines the corresponding joint, conditional, and mutual concepts. Lastly one formulates the proves the corresponding Theorems 2', 3', 4', and 5'.

Theorem 6: $P'(A) \gtrsim P(A)$, $H'(A) \leq H(A) + O(1)$, and $I'(A) \leq I(A) + O(1)$.

Proof: There is a $\% = 0^g1$ such that for all sequences p, $U'(\%p) = U(p)$ and U' reads precisely as much of p as U does.

3. Summary and Discussion of Results

The remainder of this paper is devoted to counting the number of sets A of different kinds having information $I(A)$ less than n and having entropy $H(A)$ less than n. The kinds of A we shall consider are: singleton sets, consecutive sets, finite sets, cofinite sets, and arbitrary sets. A is consecutive if it is finite and $n + 1 \in A$ implies $n \in A$. A is cofinite if it contains all but finitely many natural numbers.

The following 4 pairs of estimates will be demonstrated in this paper. The first pair is due to Solovay [6]. $\#X$ denotes the cardinality of X. S_n denotes the Singleton set $\{n\}$. C_n denotes the Consecutive set $\{k : k < n\}$.

$$\begin{cases} \log_2 \#\{\text{singleton } A : I(A) < n\} = n - I(S_n) + O(1). \\ \log_2 \#\{\text{singleton } A : H(A) < n\} = n - I(S_n) + O(1). \end{cases}$$

$$\begin{cases} \log_2 \#\{\text{consecutive } A : I(A) < n\} = n - I(C_n) + O(\log I(C_n)). \\ \log_2 \#\{A : I(A) < n\} = n - I(C_n) + O(\log I(C_n)). \end{cases}$$

$$\begin{cases} \log_2 \#\{\text{consecutive } A : H(A) < n\} = n - I'(S_n) + O(1). \\ \log_2 \#\{\text{finite } A : H(A) < n\} = n - I'(S_n) + O(1). \end{cases}$$

$$\begin{cases} \log_2 \#\{\text{cofinite } A : H(A) < n\} = n - I'(C_n) + O(\log I'(C_n)). \\ \log_2 \#\{A : H(A) < n\} = n - I'(C_n) + O(\log I'(C_n)). \end{cases}$$

These estimates are expressed in terms of $I(S_n)$, $I(C_n)$, $I'(S_n)$ and $I'(C_n)$. These quantities are variations on a theme: specifying the natural number n in a more or less constructive manner. $I(S_n)$ is the number of bits of information needed to directly calculate n. $I(C_n)$ is the number of bits of information needed to obtain n in the limit from below. $I'(S_n)$ is the number of bits of information needed to directly calculate n using an oracle for the halting problem. And $I'(C_n)$ is the number of bits of information needed to obtain n in the limit from below using an oracle for the halting problem. The following theorem, whose straightforward proof is omitted, gives some facts about these quantities and the relationship between them.

Theorem 7: $I(S_n)$, $I(C_n)$, $I'(S_n)$ and $I'(C_n)$

(a) All four quantities vary smoothly. For example, $|I(S_n) - I(S_m)| \leq O(\log |n - m|)$, and the same inequality holds for the other three quantities.

(b) For most n all four quantities are $\log_2 n + O(\log\log n)$. Such n are said to be *random* because they are specified by table look-up without real computation.

(c) The four ways of specifying n are increasingly indirect:
$$I'(C_n) \leq I'(S_n) + O(1), \quad I'(S_n) \leq I(C_n) + O(1), \quad \text{and } I(C_n) \leq I(S_n) + O(1).$$

(d) Occasionally n is random with respect to one kind of specification, but has a great deal of pattern and its description can be considerably condensed if more indirect means of specification are allowed. For example, the least $n \geq 2^k$ such that $I(S_n) \geq k$ has the following properties: $n < 2^{k+1}$, $I(S_n) = k + O(\log k)$ and $I(C_n) \leq \log_2 k + O(\log\log k)$. This relationship between $I(S_n)$ and $I(C_n)$ also holds for $I(C_n)$ and $I'(S_n)$, and for $I'(S_n)$ and $I'(C_n)$.

We see from Theorem 7(b) that all 4 pairs of estimates for $\log_2 \#_n$ are usually $n - \log_2 n + O(\log\log n)$ and thus close to each other. But Theorem 7(c) shows that the 4 pairs are shown above in what is essentially ascending numerical order. In fact, by Theorem 7(d), for each k there is an n such that $k = \log_2 n + O(1)$ and one pair of estimates is that

$$\log_2 \#_n = n - \log_2 n + O(\log\log n)$$

while the next pair is that

$$\log_2 \#_n = n - (\text{a quantity} \leq \log_2 \log_2 n) + O(\log\log\log n).$$

Hence each pair of cardinalities can be an arbitrarily small fraction of the next pair.

Having examined the comparative magnitude of these cardinalities, we obtain two corollaries.

As was pointed out in Theorem 2(c), for singleton sets $I(A) = H(A) + O(1)$. Suppose consecutive sets also had this property. Then using the fifth estimate and Theorem 7(a) one would immediately conclude that #{consecutive $A : I(A) < n$} $\approx$ #{consecutive $A : H(A) < n$}. But we have seen that the first of these cardinalities can be an arbitrarily small fraction of the second one. This contradiction shows that consecutive sets do not have the property that $I(A) = H(A) + O(1)$. Nevertheless, in Section 5 it is shown that these sets do have the property that $I(A) = H(A) + O(\log H(A))$. Further research is needed to clarify the relationship between $I(A)$ and $H(A)$ for A that are neither singleton nor consecutive.

It is natural to ask what is the relationship between the probabilities of sets and the probabilities of their unions, intersections, and complements. $P(A \cup B) \gtrsim P(A, B) \gtrsim P(A)P(B)$, and the same inequality holds for $P(A \cap B)$. But is $P(\overline{A}) \gtrsim P(A)$? If this were the case, since the complement of a cofinite set is finite, using the sixth estimate and Theorem 7(a) it would immediately follow that #{finite $A : H(A) < n$} is $\gtrsim$ #{cofinite $A : H(A) < n$}. But we have seen that the first of these cardinalities can be an arbitrarily small fraction of the second. Hence it is not true that $P(\overline{A}) \gtrsim P(A)$. However in Section 7 it is shown that $P'(\overline{A}) \gtrsim P(A)$.

Corollary 1:

(a) For consecutive A it is not true that $I(A) = H(A) + O(1)$.
(b) For cofinite A it is not true that $P(\overline{A}) \gtrsim P(A)$.

4. The Estimates Involving Singleton Sets

The following theorem and its proof are due to Solovay [6], who formulated them in a string-entropy setting.

Definition: Consider a program p for a singleton set A. The bits of p which have not been read by U by the time the element of A is output are said to be *superfluous*.

Theorem 8:

(a) $\log_2$ #{singleton $A : I(A) < n$} $= n - I(S_n) + O(1)$.
(b) $\log_2$ #{singleton $A : H(A) < n$} $= n - I(S_n) + O(1)$.

Proof: (b) follows immediately from (a) by using Theorems 2(c) and 7(a). To prove (a) we break it up into two assertions: an upper bound on $\log_2$ #, and a lower bound.

Let us start by explaining how to *mend* a minimal-size program for a singleton set. The program is mended by replacing each of its superfluous bits by a 0 and adding an endmarker 1 bit.

There is a $\% = 0^g1$ such that if p is a mended minimal-size program for S_j, then $U(\%p) = \{ |p| - 1 \} = \{I(S_j)\}$. $\%$ accomplishes this by instructing U to execute p in a special way: when U would normally output the first number, it instead immediately advances the program tape to the endmarker 1 bit, outputs the amount of tape that has been read, and goes to sleep.

The crux of the matter is that with this $\%$

$$P(S_m) \geq \#\{j : I(S_j) = m\} \, 2^{-|\%|-m-1},$$

and so

$$\#\{j : I(S_j) = m\} \lesssim P(S_m) \, 2^m.$$

Substituting $n - k$ for m and summing over all k from 1 to n, we obtain

$$\#\{j : I(S_j) < n\} \lesssim P(S_n) \, 2^n \sum_{k=1}^{n} (P(S_{n-k})/P(S_n)) \, 2^{-k}.$$

It is easy to see that $P(S_n) \gtrsim P(S_k)P(S_{n-k})$ and so $P(S_{n-k})/P(S_n) \lesssim 1/P(S_k) \lesssim k^2$. Hence the above summation is $\lesssim$

$$\sum_{k=1}^{n} k^2 \, 2^{-k},$$

which converges for $n = \infty$. Thus

$$\#\{j : I(S_j) < n\} \lesssim P(S_n) \, 2^n.$$

Taking logarithms of both sides and using Theorem 2(c) we finally obtain

$$\log_2 \#\{j : I(S_j) < n\} \leq n - I(S_n) + O(1).$$

This upper bound is the first half of the proof of (a). To complete the proof we now obtain the corresponding lower bound.

There is a $\% = 0^s1$ with the following property. Concatenate $\%$, a minimal-size program p for S_n with all superfluous bits deleted, and an arbitrary string s that brings the total number of bits up to $n - 1$. $\%$ is chosen so that $U(\%p) = S_k$, where k has the property that s is a binary numeral for it.

$\%$ instructs U to proceed as follows with the rest of its program, which consists of the subroutine p followed by $n - 1 - |\%p|$ bits of data s. First U executes p to obtain n. Then U calculates the size of s, reads s, converts s to a natural number k, outputs k, and goes to sleep.

The reason for considering this $\%$ is that $\log_2$ of the number of possible choices for s is $|s|$ $= |\%ps| - |\%p| = n - 1 - |\%| - |p| \geq n - 1 - |\%| - I(S_n)$. And each choice of s yields a different singleton set $S_k = U(\%ps)$ such that $I(S_k) \leq |\%ps| = n - 1$. Hence

$$\log_2 \#\{k : I(S_k) < n\} \geq n - 1 - |\%| - I(S_n) = n - I(S_n) + O(1).$$

The proof of (a), and thus of (b), is now complete.

Theorem 8':

(a) $\log_2 \#\{\text{singleton } A : I'(A) < n\} = n - I'(S_n) + O(1)$.
(b) $\log_2 \#\{\text{singleton } A : H'(A) < n\} = n - I'(S_n) + O(1)$.

Proof: Imagine that the proofs of Theorem 8 and its auxiliary theorems refer to U' instead of U.

5. The Remaining Estimates Involving I(A)

Definition:

- $Q(n) = \Sigma \, P(A) \, (\#A < n)$ is the probability that a set has less than n elements.
- $Q(n)^t$ is the probability that with a program tape produced by coin flipping, U outputs less than n different numbers by time t. Note that $Q(n)^t$ can be calculated from n and t, and is a rational number of the form $k/2^t$ because U can read at most t bits of program by time t.

Lemma 1:

(a) $Q(0) = 0$, $Q(n) \leq Q(n + 1)$, $\lim_{n \to \infty} Q(n) < 1$.

(b) $Q(0)' = 0$, $Q(n)' \leq Q(n + 1)'$, $\lim_{n \to \infty} Q(n)' = 1$.
(c) For $n > 0$, $Q(n)^0 = 1$, $Q(n)' \geq Q(n)^{t+1}$, $\lim_{t \to \infty} Q(n)^t = Q(n)$.
(d) If A is finite, then $Q(\#A + 1) - Q(\#A) \geq P(A)$.

Theorem 9: If A is consecutive and $P(A) > 2^{-n}$, then $I(A) \leq n + I(C_n) + O(1)$.

Proof: There is a $\% = 0^g1$ with the following property. After reading $\%$, U expects to find on its program tape a string of length $I(C_n) + n$ which consists of a minimal-size program p for C_n appropriately merged with the binary expansion of a rational number $x = j/2^n$ $(0 \leq j < 2^n)$. In parallel U executes p to obtain C_n, reads x, and outputs a consecutive set. This is done in stages.

U begins stage t $(t = 1, 2, 3, \ldots)$ by simulating one more time quantum of the computation that yields C_n. During this simulation, whenever it is necessary to read another bit of the program U supplies this bit by reading the next square of the actual program tape. And whenever the simulated computation produces a new output (this will occur n times), U instead takes this as a signal to read the next bit of x from the program tape. Let x_t denote the value of x based on what U has read from its program tape by stage t. Note that $0 \leq x_t \leq x_{t+1}$ and $\lim_{t \to \infty} x_t = x < 1$.

In the remaining portion of stage t U does the following. It calculates $Q(k)'$ for $k = 0, 1, 2, \ldots$ until $Q(k)' = 1$. Then it determines m_t which is the greatest value of k for which $Q(k)' \leq x_t$. Note that since $Q(0)' = 0$ there is always such a k. Also, since $Q(k)'$ is monotone decreasing in t, and x_t is monotone increasing, it follows that m_t is also monotone increasing in t. Finally U outputs the m_t natural numbers less than m_t, and proceeds to stage $t + 1$.

This concludes the description of the instructions incorporated in $\%$. $\%$ is now used to prove the theorem by showing that if A is consecutive and $P(A) > 2^{-n}$, then $I(A) \leq n + I(C_n) + |\%|$.

As pointed out in the lemma, $Q(\#A + 1) - Q(\#A) \geq P(A) > 2^{-n}$. It follows that the open interval of real numbers between $Q(\#A)$ and $Q(\#A + 1)$ contains a rational number x of the form $j/2^n$ $(0 \leq j < 2^n)$. It is not difficult to see that one obtains a program for A that is $|\%| + I(C_n) + n$ bits long by concatenating $\%$ and the result of merging in an appropriate fashion a minimal-size program for C_n with the binary expansion of x. Hence $I(A) \leq n + I(C_n) + |\%| = n + I(C_n) + O(1)$.

Theorem 10: If A is consecutive $I(A) = H(A) + O(\log H(A))$ and $H(A) = I(A) + O(\log I(A))$.

Proof: Consider a consecutive set A. By Theorem 7(a), $I(C_n) = O(\log n)$. Restating Theorem 9, if $H(A) < n$ then $I(A) \leq n + I(C_n) + O(1) = n + O(\log n)$. Taking $n = H(A) + 1$, we see that $I(A) \leq H(A) + O(\log H(A))$. Moreover, $H(A) \leq I(A)$ (Theorem 2(b)). Hence $I(A) = H(A) + O(\log H(A))$, and thus $I(A) = H(A) + O(\log I(A))$.

Theorem 11: $\log_2 \#\{A : I(A) < n\} \leq n - I(C_n) + O(\log I(C_n))$.

Proof: There is a $\% = 0^g1$ with the following property. Let p be an arbitrary sequence, and suppose that U reads precisely m bits of the program p. Then $U(\%p) = C_m$. $\%$ accomplishes this by instructing U to execute p in a special way: normal output is replaced by a continually updated indication of how many bits of program have been read.

The crux of the matter is that with this $\%$

$$P(C_m) \geq \#\{A : I(A) = m\} \, 2^{-|\%|-m},$$

and so

$$\#\{A : I(A) = m\} \leq P(C_m) \, 2^m.$$

Replacing m by $n - k$ and summing over all k from 1 to n, we obtain

$$\#\{A : I(A) < n\} \leq P(C_n)\, 2^n \sum_{k=1}^{n} (P(C_{n-k})/P(C_n))\, 2^{-k}.$$

It is easy to see that $P(C_n) \geq P(C_k)P(C_{n-k})$ and so $P(C_{n-k})/P(C_n) \leq 1/P(C_k) \lesssim k^2$. Hence the above summation is $\leq$

$$\sum_{k=1}^{n} k^2\, 2^{-k},$$

which converges for $n = \infty$. Thus

$$\#\{A : I(A) < n\} \leq P(C_n)\, 2^n.$$

Taking logarithms of both sides and using $\log_2 P(C_n) = -I(C_n) + O(\log I(C_n))$ (Theorem 10), we finally obtain

$$\log_2 \#\{A : I(A) < n\} \leq n - I(C_n) + O(\log I(C_n)).$$

Theorem 12: $\log_2 \#\{\text{consecutive } A : I(A) < n\} \geq n - I(C_n) + O(\log I(C_n))$.

Proof: There is a $\% = 0^s1$ that is used in the following manner. Concatenate these strings: $\%$, a minimal-size program for $\{I(C_n)\}$ with all superfluous bits deleted, a minimal-size program for C_n, and an arbitrary string s of size sufficient to bring the total number of bits up to $n - 1$. Call the resulting $(n - 1)$-bit string p. Note that s is at least $n - 1 - |\%| - I(\{I(C_n)\}) - I(C_n)$ bits long. Hence $\log_2$ of the number of possible choices for s is, taking Theorem 7(a) into account, at least $n - I(C_n) + O(\log I(C_n))$.

$\%$ instructs U to proceed as follows with the rest of p, which consists of two subroutines and the data s. First U executes the first subroutine in order to calculate the size of the second subroutine and know where s begins. Then U executes the second subroutine, and uses each new number output by it as a signal to read another bit of the data s. Note that U will never know when it has finished reading s. As U reads the string s, it interprets s as the reversal of the binary numeral for a natural number m. And U contrives to enumerate the set C_m by outputting 2^k consecutive natural numbers each time the kth bit of s that is read is a 1.

To recapitulate, for each choice of s one obtains an $(n - 1)$-bit program p for a different consecutive set (in fact, the set C_m, where s is the reversal of a binary numeral for m). In as much as $\log_2$ of the number of possible choices for s was shown to be at least $n - I(C_n) + O(\log I(C_n))$, we conclude that $\log_2 \#\{\text{consecutive } A : I(A) < n\}$ is $\geq n - I(C_n) + O(\log I(C_n))$.

Theorem 13:

(a) $\log_2 \#\{\text{consecutive } A : I(A) < n\} = n - I(C_n) + O(\log I(C_n))$.
(b) $\log_2 \#\{A : I(A) < n\} = n - I(C_n) + O(\log I(C_n))$.

Proof: Since $\#\{\text{consecutive } A : I(A) < n\} \leq \#\{A : I(A) < n\}$, this follows immediately from Theorems 12 and 11.

Theorem 13':

(a) $\log_2 \#\{\text{consecutive } A : I'(A) < n\} = n - I'(C_n) + O(\log I'(C_n))$.
(b) $\log_2 \#\{A : I'(A) < n\} = n - I'(C_n) + O(\log I'(C_n))$.

Proof: Imagine that the proofs of Theorem 13 and its auxiliary theorems refer to U' instead of U.

6. The Remaining Lower Bounds

In this section we construct many consecutive sets and cofinite sets with probability greater than 2^{-n}. To do this, computations using an oracle for the halting problem are simulated using a fake oracle that answers that $U(p)$ halts iff it does so within time t. As t goes to infinity, any finite set of questions will eventually be answered correctly by the fake oracle. This *simulation in the limit* % is used to: (a) take any n-bit oracle program p for a singleton set and construct from it a consecutive set $U(\%px)$ with probability greater than or equal to $2^{-|x|-n}$, and (b) take any n-bit oracle program p for a consecutive set and construct from it a cofinite set $U(\%px)$ with probability greater than or equal to $2^{-|x|-n}$.

The critical feature of the simulation in the limit that accomplishes (a) and (b) can best be explained in terms of two notions: *harmless overshoot* and *erasure*. The crux of the matter is that although in the limit the fake oracle realizes its mistakes and changes its mind, U may already have read beyond p into x. This is called overshoot, and could make the probability of the constructed set fall far below $2^{-|x|-n}$. But the construction process contrives to make overshoot harmless by eventually forgetting bits in x and by erasing its mistakes. In case (a) erasure is accomplished by moving the end of the consecutive set. In case (b) erasure is accomplished by filling in holes that were left in the cofinite set. As a result bits in x do not affect which set is enumerated; they can only affect the time at which its elements are output.

Lemma 2: With our Turing machine model, if k is output at time $\leq t$, then $k < t < 2^t$.

Theorem 14: There is a $\% = 0^g 1$ with the following property. Suppose the string p is an oracle program for S_k. Let t_1 be the time at which k is output. Consider the finite set of questions that are asked to the oracle during these t_1 time quanta. Let t_2 be the maximum time taken to halt by any program that the oracle is asked about. ($t_2 = 0$ if none of them halt or if no questions are asked.) Finally, let $t = \max t_1,\ t_2$. Then for all sequences x, $\%px$ is a program for the set C_l, where $l = 2^t + k$. By the lemma k can be recovered from l.

Proof: % instructs U to act as follows on px. Initially U sets $i = 0$. Then U works in stages. At stage t ($t = 1, 2, 3, \dots$) U simulates t time quanta of the computation $U'(px)$, but truncates the simulation immediately if U' outputs a number. U fakes the halting-problem oracle used by U' by answering that a program halts iff it takes $\leq t$ time quanta to do so. Did an output k occur during the simulated computation? If not, nothing more is done at this stage. If so, U does the following. First it sets $i = i + 1$. Let L_i be the chronological list of *yes/no* answers given by the fake oracle during the simulation. U checks whether $i = 1$ or $L_{i-1} \neq L_i$. (Note that $L_{i-1} = L_i$ iff the same questions were asked in the same order and all the answers are the same.) If $i > 1$ and $L_{i-1} = L_i$, U does nothing at this stage. If $i = 1$ or $L_{i-1} \neq L_i$, U outputs all natural numbers less than $2^t + k$, and proceeds to stage $t + 1$.

It is not difficult to see that this % proves the theorem.

Theorem 15: $\log_2 \#\{\text{consecutive } A : H(A) < n\} \geq n - I'(S_n) + O(1)$.

Proof: By Theorem 14, $c = |\%|$ has the property that for each singleton set S_k such that $I'(S_k) < n - c$ there is a different l such that $P(C_l) > 2^{-n}$. Hence in view of Theorems 8(a)$'$ and 7(a)

$$\log_2 \#\{\text{consecutive } A : H(A) < n\}$$
$$\geq \log_2 \#\{\text{singleton } A : I'(A) < n - c\}$$
$$\geq n - c - I'(S_{n-c}) + O(1)$$
$$= n - I'(S_n) + O(1).$$

Theorem 16: There is a $\% = 0^g 1$ with the following property. Suppose the string p is an oracle program for the finite set A. For each $k \in A$, let ^1t_k be the time at which it is output. Also, let ^2t_k be the maximum time taken to halt by any program that the oracle is asked about during these ^1t_k time quanta. Finally, let $t_k = \max {}^1t_k, {}^2t_k$, and $l_k = 2^lk + k$. Then for all sequences x, $\%px$ is a program for the cofinite set $B = $ all natural numbers not of the form l_k ($k \in A$). By the lemma each k in A can be recovered from the corresponding l_k.

Proof: $\%$ instructs U to act as follows on px in order to produce B. U works in stages. At stage t ($t = 1, 2, 3, \ldots$) U simulates t time quanta of the computation $U'(px)$. U fakes the halting-problem oracle used by U' by answering that a program halts iff it takes $\leq t$ time quanta to do so. While simulating $U'(px)$, U notes the time at which each output k occurs. U also keeps track of the latest stage at which a change occurred in the chronological list of *yes/no* answers given by the fake oracle during the simulation before k is output. Thus at stage t there are current estimates for ^1t_k, for ^2t_k, and for $t_k = \max {}^1t_k, {}^2t_k$, for each k that currently seems to be in $U'(px)$. As t goes to infinity these estimates will attain the true values for $k \in A$, and will not exist or will go to infinity for $k \notin A$.

Meanwhile U enumerates B. That part of B output by stage t consists precisely of all natural numbers less than 2^{l+1} that are not of the form $2^lk + k$, for any k in the current approximation to $U'(px)$. Here t_k denotes the current estimate for the value of t_k.

It is not difficult to see that this $\%$ proves the theorem.

Theorem 17: $\log_2 \#\{\text{cofinite } A : H(A) < n\} \geq n - I'(C_n) + O(\log I'(C_n))$.

Proof: By Theorem 16, $c = |\%|$ has the property that for each consecutive set A such that $I'(A) < n - c$ there is a different cofinite set B such that $P(B) > 2^{-n}$. Hence in view of Theorems 13(a)$'$ and 7(a)

$$\log_2 \#\{\text{cofinite } B : H(B) < n\}$$
$$\geq \log_2 \#\{\text{consecutive } A : I'(A) < n - c\}$$
$$\geq n - c - I'(C_{n-c}) + O(\log I'(C_{n-c}))$$
$$\geq n - I'(C_n) + O(\log I'(C_n)).$$

Corollary 2: There is a $\% = 0^g 1$ with the property that for every sequence p, $\#U(\%p) = \#U'(p)$.

Proof: The $\%$ in the proof of Theorem 16 has this property.

7. The Remaining Upper Bounds

In this section we use several approximations to $P(A)$, and the notion of the canonical index of a finite set A [5, pp. 69-71]. This is defined to be $\Sigma 2^k$ ($k \in A$), and it establishes a one-to-one correspondence between the natural numbers and the finite sets of natural numbers. Let D_i be the finite set whose canonical index is i. We also need to use the concept of a recursive real number, which is a real x for which one can compute a convergent sequence of nested open intervals with rational endpoints that contain x [5, pp. 366, 371]. This is the formal definition corresponding to the intuitive notion that a computable real number is one whose decimal expansion can be calculated. The recursive reals constitute a field.

Definition:

Consider a sequence p produced by flipping an unbiased coin.

- $P(A)^t$ = the probability that (the output by time t of $U(p)$) = A.

Let s be an arbitrary string.

- $P(s)^t$ = the probability that $(\forall k < |s|)[k \in$ (the output by time t of $U(p)$) iff the k th bit of s is a $1]$.

- $P(s)$ = the probability that $(\forall k < |s|)[k \in U(p)$ iff the kth bit of s is a $1]$.

Note that $P(D_i)^t$ is a rational number that can be calculated from i and t, and $P(s)^t$ is a rational number that can be calculated from s and t.

Lemma 3:

(a) If A is finite, then $P(A) = \lim_{t \to \infty} P(A)^t$.

(b) $P(s) = \lim_{t \to \infty} P(s)^t$.

(c) $P(\Lambda) = 1$.

(d) $P(s) = P(s0) + P(s1)$.

(e) Consider a set A. Let a_n be the n -bit string whose kth bit is a 1 iff $k \in A$. Then $P(a_0) = 1$, $P(a_n) \geq P(a_{n+1})$, and $\lim_{n \to \infty} P(a_n) = P(A)$.

Theorem 18:

(a) $P(D_i)$ is a real recursive in the halting problem uniformly in i.

(b) $P(s)$ is a real recursive in the halting problem uniformly in s.

This means that given i and s one can use the oracle to obtain these real numbers as the limit of a convergent sequence of nested open intervals with rational end-points.

Proof: Note that $P(D_i) > n/m$ iff there is a k such that $P(D_i)^k > n/m$ and for all $t > k$, $P(D_i)^t \geq P(D_i)^k$. One can use the oracle to check whether or not a given i, n, m and k have this property, for there is a $\% = 0^g1$ such that $U(\%0^i10^n10^m10^k1)$ does not halt iff $P(D_i)^t \geq P(D_i)^k > n/m$ for all $t > k$. Thus if $P(D_i) > n/m$ one will eventually discover this by systematically checking all possible quadruples i, n, m and k. Similarly, one can use the oracle to discover that $P(D_i) < n/m$, that $P(s) > n/m$, and that $P(s) < n/m$. This is equivalent to the assertion that $P(D_i)$ and $P(s)$ are reals recursive in the halting problem uniformly in i and s.

Theorem 19: $P'(S_i) \gtrsim P(D_i)$.

Proof: It follows from Theorem 18(a) that there is a $\% = 0^g1$ with the following property. Consider a real number x in the interval between 0 and 1 and the sequence p_x that is its binary expansion. Then $U'(\%p_x) = S_i$ if x is in the open interval I_i of real numbers between $\sum_{k<i} P(D_k)$ and $\sum_{k\leq i} P(D_k)$. This shows that $c = |\%|$ has the property that $P'(S_i) \geq 2^{-c}$ (the length of the interval I_i) $= 2^{-c}P(D_i)$. (See [7, pp. 14-15] for a construction that is analogous.)

Theorem 20:

(a) $\log_2 \#\{$consecutive $A : H(A) < n\} = n - I'(S_n) + O(1)$.

(b) $\log_2 \#\{$finite $A : H(A) < n\} = n - I'(S_n) + O(1)$.

Proof: From Theorems 15, 19, 8(b)′, and 7(a), we see that

$$n - I'(S_n) + O(1)$$
$$\leq \log_2 \#\{\text{consecutive } A : H(A) < n\}$$
$$\leq \log_2 \#\{\text{finite } A : H(A) < n\}$$
$$\leq \log_2 \#\{\text{singleton } A : H'(A) < n + c\}$$
$$\leq n + c - I'(S_{n+c}) + O(1)$$
$$= n - I'(S_n) + O(1).$$

Theorem 21: $I'(\overline{A}, A) \leq H(A) + O(1)$.

Proof: Let us start by associating with each string s an interval I_s of length $P(s)$. First of all, I_Λ is the interval of reals between 0 and 1, which is okay because $P(\Lambda) = 1$. Then each I_s is partitioned into two parts: the subinterval I_{s0} of length $P(s0)$, followed by the subinterval I_{s1} of length $P(s1)$. This works because $P(s) = P(s0) + P(s1)$.

There is a $\% = 0^s1$ which makes U' behave as follows. After reading $\%$ U' expects to find the sequence p_x, the binary expansion of a real number x between 0 and 1. Initially U' sets $s = \Lambda$. U' then works in stages. At stage k ($k = 0, 1, 2, \dots$) U' initially knows that x is in the interval I_s, and contrives to decide whether it is in the subinterval I_{s0} or in the subinterval I_{s1}. To do this U' uses the oracle to calculate the end-points of these intervals with arbitrarily high precision, by means of the technique indicated in the proof of Theorem 18(b). And of course U' also has to read p_x to know the value of x, but it only reads the program tape when it is forced to do so in order to make a decision (this is the crux of the proof). If U' decides that x is in I_{s0} it outputs $2k$ and sets $s = s0$. If it decides that x is in I_{s1} it outputs $2k + 1$ and sets $s = s1$. Then U' proceeds to the next stage.

Why does this show that $I'(\overline{A}, A) \leq H(A) + O(1)$? From part (e) of the lemma it is not difficult to see that to each r.e. set A there corresponds an open interval I_A of length $P(A)$ consisting of reals x with the property that $U'(\%p_x) = \overline{A}$ join A. Moreover U' only reads as much of p_x as is necessary; in fact, if $P(A) > 2^{-n}$ there is an x in I_A for which this is at most $n + O(1)$ bits. Hence $I'(\overline{A}, A) \leq |\%| + H(A) + O(1) = H(A) + O(1)$.

Theorem 22:

(a) $\log_2 \#\{\text{cofinite } A : H(A) < n\} = n - I'(C_n) + O(\log I'(C_n))$.
(b) $\log_2 \#\{A : H(A) < n\} = n - I'(C_n) + O(\log I'(C_n))$.

Proof: From Theorems 17, 21, 5(a)′, 5(d)′, 13(b)′ and 7(a) we see that

$$n - I'(C_n) + O(\log I'(C_n))$$
$$\leq \log_2 \#\{\text{cofinite } A : H(A) < n\}$$
$$\leq \log_2 \#\{A : H(A) < n\}$$
$$\leq \log_2 \#\{A : I'(A) < n + c\}$$
$$\leq n + c - I'(C_{n+c}) + O(\log I'(C_{n+c}))$$
$$= n - I'(C_n) + O(\log I'(C_n)).$$

Corollary 3: $P'(\overline{A}) \gtrsim P(A)$.

Proof: By Theorems 2(b)′, 5(d)′ and 21, $H'(\overline{A}) \leq I'(\overline{A}) \leq I'(\overline{A}, A) + O(1) \leq H(A) + O(1)$. Hence $P'(\overline{A}) \gtrsim P(A)$.

8. The Probability of the Set of Natural Numbers Less than N

In the previous sections we established the results that were announced in Section 3. The techniques that were used to do this can also be applied to a topic of a somewhat different nature, $P(C_n)$.

$P(C_n)$ sheds light on two interesting quantities: $Q_1(n)$ the probability that a set has cardinality n, and $Q_0(n)$ the probability that the complement of a set has cardinality n. We also consider a gigantic function $G(n)$, which is the greatest natural number that can be obtained in the limit from below with probability greater than 2^{-n}.

Definition:

- $Q_1(n) = \Sigma \, P(A) \, (\#A = n)$.
- $Q_0(n) = \Sigma \, P(A) \, (\#\overline{A} = n)$.
- $G(n) = \max k \, (P(C_k) > 2^{-n})$.
- Let ρ be the defective probability measure on the sets of natural numbers that is defined as follows: $\rho A = \Sigma \, Q_1(n) \, (n \in A)$.
- Let μ be an arbitrary probability measure, possibly defective, on the sets of natural numbers. μ is said to be a C-measure if there is a function $u(n, t)$ such that $u(n, t) \geq u(n, t + 1)$ and $\mu C_n = \lim_{t \to \infty} u(n, t)$. Here it is required that $u(n, t)$ be a rational number that can be calculated from n and t. In other words, μ is a C-measure if μC_n can be obtained as a monotone limit from above uniformly in n.

Theorem 23:

(a) $Q_1(n) \approx P(C_n)$.
(b) $Q_0(n) \approx P'(C_n)$.
(c) ρ is a C-measure.
(d) If μ is a C-measure, then $\rho A \gtrsim \mu A$.
(e) If $H'(S_k) < n + O(1)$, then $k < G(n)$.
(f) $H'(S_{G(n)}) = n + O(1)$.

Proof:

(a) Note that $Q_1(n) \geq P(C_n)$. Also, there is a $\% = 0^g 1$ such that $U(\% p) = C_{\#U(p)}$ for all sequences p. Hence $P(C_n) \gtrsim Q_1(n)$.

(b) Keep part (a) in mind. By Corollary 2, $Q_0(n) \gtrsim Q_1'(n) \gtrsim P'(C_n)$. And since $P'(\overline{A}) \gtrsim P(A)$ (Corollary 3), $Q_0(n) \lesssim Q_1'(n) \lesssim P'(C_n)$.

(c) Lemma 1(c) states that the function $Q(n)^t$ defined in Section 5 plays the role of $u(n, t)$.

(d) A construction similar to the proof of Theorem 9 shows that there is a $\% = 0^g 1$ with the following property. Consider a real number x between 0 and 1 and the sequence p_x that is its binary expansion. $U(\% p_x) = C_n$ if x is in the open interval I_n of reals between μC_n and μC_{n+1}.

This proves part (d) because the length of the interval I_n is precisely μS_n, and hence $\rho S_n = Q_1(n) \geq 2^{-|\%|} \mu S_n$.

(e) By Theorem 2(c)', if $P'(S_k) > 2^{-n}$ then $I'(S_k) < n + O(1)$. Hence by Theorem 14, there is an $l > k$ such that $P(C_l) \gtrsim 2^{-n}$. Thus $k < l \leq G(n + O(1))$.

(f) Note that the canonical index of C_n is $2^k - 1$. It follows from Theorem 19 that if $P(C_k) > 2^{-n}$, then $P'(\{2^k - 1\}) \gtrsim 2^{-n}$. There is a $\% = 0^g 1$ such that $U'(\%p) = S_k$ if $U'(p) = \{2^k - 1\}$. Hence if $P(C_k) > 2^{-n}$, then $P'(S_k) \gtrsim P'(\{2^k - 1\}) \gtrsim 2^{-n}$. In other words, if $P(C_k) > 2^{-n}$ then $H'(S_k) \leq n + O(1)$. Note that by definition $P(C_{G(n)}) > 2^{-n}$. Hence $H'(S_{G(n)}) \leq n + O(1)$. Thus in view of (e), $H'(S_{G(n)}) = n + O(1)$.

Addendum

An important advance in the line of research proposed in this paper has been achieved by Solovay [8]; with the aid of a crucial lemma of D. A. Martin he shows that

$$I(A) \leq 3 H(A) + O(\log H(A)).$$

In [9] and [10] certain aspects of the questions treated in this paper are examined from a somewhat different point of view.

References

1. K. de Leeuw, E. F. Moore, C. E. Shannon and N. Shapiro, Computability by probabilistic machines, in *Automata Studies,* C. E. Shannon and J. McCarthy (Eds.), pp. 183-212. Princeton University Press, N.J. (1956).

2. G. J. Chaitin, Randomness and mathematical proof, *Scient. Am.* **232** (5), 47-52 (May 1975).

3. G. J. Chaitin, A theory of program size formally identical to information theory, *J. Ass. Comput. Mach.* **22** (3), 329-340 (July 1975).

4. C. E. Shannon and W. Weaver, *The Mathematical Theory of Communication.* University of Illinois, Urbana (1949).

5. H. Rogers, Jr., *Theory of Recursive Functions and Effective Computability.* McGraw-Hill, N.Y. (1967).

6. R. M. Solovay, unpublished manuscript on [3] dated May 1975.

7. S. K. Leung-Yan-Cheong and T. M. Cover, Some inequalities between Shannon entropy and Kolmogorov, Chaitin, and extension complexities, Technical Report 16, Dept. of Statistics, Stanford University, CA (October 1975).

8. R. M. Solovay, On random r.e. sets, *Proceedings of the Third Latin American Symposium on Mathematical Logic.* Campinas, Brazil, (July 1976), (to appear).

9. G. J. Chaitin, Information-theoretic characterizations of recursive infinite strings, *Theor. Comput. Sci.* **2**, 45-48 (1976).

10. G. J. Chaitin, Program size, oracles, and the jump operation, *Osaka J. Math.* (to appear).

Part V—Technical Papers on Blank-Endmarker Programs

INFORMATION-THEORETIC LIMITATIONS OF FORMAL SYSTEMS

Journal of the ACM 21 (1974), pp. 403-424.

GREGORY J. CHAITIN[43]
Buenos Aires, Argentina

Abstract

An attempt is made to apply information-theoretic computational complexity to metamathematics. The paper studies the number of bits of instructions that must be a given to a computer for it to perform finite and infinite tasks, and also the amount of time that it takes the computer to perform these tasks. This is applied to measuring the difficulty of proving a given set of theorems, in terms of the number of bits of axioms that are assumed, and the size of the proofs needed to deduce the theorems from the axioms.

Key Words and Phrases

complexity of sets, computational complexity, difficulty of theorem-proving, entropy of sets, formal systems, Gödel's incompleteness theorem, halting problem, information content of sets, information content of axioms, information theory, information time trade-offs, metamathematics, random strings, recursive functions, recursively enumerable sets, size of proofs, universal computers

CR Categories

5.21, 5.25, 5.27, 5.6

1. Introduction

This paper attempts to study information-theoretic aspects of computation in a very general setting. It is concerned with the information that must be supplied to a computer for it to carry out finite or infinite computational tasks, and also with the time it takes the computer to do this. These questions, which have come to be grouped under the heading of abstract computational complexity, are considered to be of interest in themselves. However, the motivation for this investigation is primarily its metamathematical applications.

Computational complexity differs from recursive function theory in that, instead of just asking whether it is possible to compute something, one asks exactly how much effort is needed to do this. Similarly, instead of the usual metamathematical approach, we propose to measure the difficulty of

---

[43]   Copyright © 1974, Association for Computing Machinery, Inc. General permission to republish, but not for profit, all or part of this material is granted provided that ACM's copyright notice is given and that reference is made to the publication, to its date of issue, and to the fact that reprinting privileges were granted by permission of the Association for Computing Machinery.

An early version of this paper was presented at the Courant Institute Computational Complexity Symposium, New York, October 1971. [28] includes a nontechnical exposition of some results of this paper. [1] and [2] announce related results.

Author's address: Rivadavia 3580, Dpto. 10A, Buenos Aires, Argentina.

proving something. How many bits of axioms are needed to be able to obtain a set of theorems? How long are the proofs needed to demonstrate them? What is the trade-off between how much is assumed and the size of the proofs?

We consider the axioms of a formal system to be a program for listing the set of theorems, and the time at which a theorem is written out to be the length of its proof.

We believe that this approach to metamathematics may yield valuable dividends. Mathematicians were at first greatly shocked at, and then ignored almost completely, Gödel's announcement that no set of axioms for number theory is complete. It wasn't clear what, in practice, was the significance of Gödel's theorem, how it should affect the everyday activities of mathematicians. Perhaps this was because the unprovable propositions appeared to be very pathological singular points.[44] [45]

The approach of this paper, in contrast, is to measure the power of a set of axioms, to measure the information that it contains. We shall see that there are circumstances in which one only gets out of a set of axioms what one puts in, and in which it is possible to reason in the following manner. If a set of theorems constitutes t bits of information, and a set of axioms contains less than t bits of information, then it is impossible to deduce these theorems from these axioms.

We consider that this paper is only a first step in the direction of such an approach to metamathematics;[46] a great deal of work remains to be done to clarify these matters. Nevertheless, we would like to sketch here the conclusions which we have tentatively drawn.[47]

After empirically exploring, in the tradition of Euler and Gauss, the properties of the natural numbers one may discover interesting regularities. One then has two options. The first is to accept the conjectures one has formulated on the basis of their empirical corroboration, as an experimental scientist might do. In this way one may have a great many laws to remember, but will not have to bother to deduce them from other principles. The other option is to try to find a theory for one's observations, or to see if they follow from existing theory. In this case it may be possible to reduce a great many observations into a few general principles from which they can be deduced. But there is a cost: one can now only arrive at the regularities one observed by means of long demonstrations.

Why use formal systems, instead of proceeding empirically? First of all, if the empirically derived conjectures aren't independent facts, reducing them to a few common principles allows one to have to remember less assumptions, and this is easier to do, and is much safer, as one is assuming less. The cost is, of course, the size of the proofs.

What attitude, then, does this suggest toward Gödel's theorem that any formalization of number theory is incomplete? It tends to provide theoretical justification for the attitude that number theorists have in fact adopted when they extensively utilize in their work hypotheses such as that of Riemann concerning the zeta function. Gödel's theorem does not mean that mathematicians must give up hope of understanding the properties of the natural numbers; it merely means that one may have to adopt new axioms as one seeks to order and interrelate, to organize and comprehend, ever

---

[44] In [3] and [4] von Neumann analyzes the effect of Gödel's theorem upon mathematicians. Weyl's reaction to Gödel's theorem is quoted by Bell [5]. The original source is [6]. See also Weyl's discussion [7] of Gödel's views regarding his incompleteness theorem.

[45] For nontechnical expositions of Gödel's incompleteness theorem, see [8, 9, 10, Sec. 1, pp. xv-xviii, 11, and 12]. [28] contains a nontechnical exposition of an incompleteness theorem analogous to Berry's paradox that is Theorem 4.1 of this paper.

[46] [13-16] are related in approach to this paper. [13, 15, and 16] are concerned with measuring the size of proofs and the effect of varying the axioms upon their size. In [14] Cohen "measures the strength of a [formal] system by the ordinals which can be handled in the system."

[47] The analysis that follows of the possible significance of the results of this paper has been influenced by [17 and 18], in addition to the references cited in Footnote [44]. Incidentally, it is interesting to examine [19, p. 112] in the light of this analysis.

INFORMATION-THEORETIC LIMITATIONS OF FORMAL SYSTEMS

more extensive mathematical observations. I.e. the mathematician shouldn't be more upset than the physicist when he needs to assume a new axiom; nor should he be too horrified when an axiom must be abandoned because it is found that it contradicts previously existing theory, or because it predicts properties of the natural numbers that are not corroborated empirically. In a word, we propose that there may be theoretical justification for regarding number theory somewhat more like a dynamic empirical science than as a closed static body of theory.

This paper grew out of work on the concept of an individual random, patternless, chaotic, unpredictable string of bits. This concept has been rigorously defined in several ways, and the properties of these random strings have been studied by several authors (see, for example, [20-28]). Most strings are random; they have no special distinguishing features; they are typical and hard to tell apart. But can it be proved that a particular string is random? The answer is that about n bits of axioms are needed to be able to prove that a particular n-bit string is random.

More precisely, the train of thought was as follows. The entropy, or information content, or complexity, of a string is defined to be the number of bits needed to specify it so effectively that it can be constructed. A random n-bit string is about n bits of information, i.e. has complexity/entropy/information content $\approx n$; there is essentially nothing better to do if one wishes to specify such a string than just show it directly. But the string consisting of 1,000,000 repetitions of the 6-bit pattern 000101 has far less than 6,000,000 bits of complexity. We have just specified it using far fewer bits.

What if one wishes to be able to determine each string of complexity $\leq n$ and its complexity? It turns out that this requires $n + O(1)$ bits of axioms; at least $n - c$ bits are necessary (Theorem 4.1), and $n + c$ bits are sufficient (Theorem 4.3). But the proofs will be enormously long unless one essentially directly takes as axioms all the theorems that one wishes to prove, and in that case there will be an enormously great number of bits of axioms (Theorem 7.6(c)).

Another theme of this paper arises from the following metamathematical considerations, which are well known (see, for example, [29]). In a formal system without a decision method, it is impossible to bound the size of a proof of a theorem by a recursive function of the number of characters in the statement of the theorem. For if there were such a function f, one could decide whether or not an arbitrary proposition p is a theorem, by merely checking if a proof for it appears among the finitely many possible proofs of size bounded by f of the number of characters in p.

Thus, in a formal system having no decision method, there are very profound theorems, theorems that have short statements, but need immensely long proofs. In Section 10 we study the function $e(n)$, necessarily nonrecursive, defined to be the least s such that all theorems of the formal system with $\leq n$ characters have proofs of size $\leq s$.

To close this introduction, we would like to mention without proof an example that shows particularly clearly the relationship between the number of bits of axioms that are assumed and what can be deduced. This example is based on the work of M. Davis, Ju. V. Matisjasevic, H. Putnam, and J. Robinson that settled Hilbert's tenth problem (cf. [30]). There is a polynomial P in $k + 2$ variables with integer coefficients that has the following property. Consider the infinite string whose ith bit is 1 or 0 depending on whether or not the set

$$S_i = \{ n \in N \mid \exists x_1, \ldots, x_k \in N \quad P(i, n, x_1, \ldots, x_k) = 0 \}$$

is infinite. Here N denotes the natural numbers. This infinite binary sequence is random, i.e. the complexity of an initial segment is asymptotic to its length. What is the number of bits of axioms that is needed to be able to prove for each natural number $i < n$ whether or not the set S_i is infinite? By using the methods of Section 4, it is easy to see that the number of bits of axioms that is needed is asymptotic to n.

2. Definitions Related to Computers and Complexity

This paper is concerned with measuring the difficulty of computing finite and infinite sets of binary strings. The binary strings are considered to be ordered in the following fashion: Λ, 0, 1, 00, 01, 10, 11, 000, 001, 010, 011, 100, 101, 110, 111, 0000, ... In order to be able to also study the difficulty of computing finite or infinite sets of natural numbers, we consider each binary string to simultaneously be a natural number: the nth binary string corresponds to the natural number n. Ordinal numbers are considered to start with 0, not 1. For example, we speak of the 0th string of length n.

In order to be able to study the difficulty of computing finite and infinite sets of mathematical propositions, we also consider that each binary string is simultaneously a proposition. Propositions use a finite alphabet of characters which we suppose includes all the usual mathematical symbols. We consider the nth binary string to correspond to the nth proposition, where the propositions are in lexicographical order defined by an arbitrary ordering of the symbols of their alphabet.

Henceforth, we say "string" instead of "binary string," it being understood that this refers to a binary string. It should be clear from the context whether we are considering something to be a string, a natural number, or a proposition.

Operations with strings include exponentiation: 0^k and 1^k denote the string of k 0's and k 1's, respectively. $\lg(s)$ denotes the length of a string s. Note that the length $\lg(n)$ of a natural number n is therefore $\lfloor \log_2(n + 1) \rfloor$. The maximum element of a finite set of strings S is denoted by max S, and we stipulate that max $\emptyset = 0$. $\#(S)$ denotes the number of elements in a finite set S.

We use these notational conventions in a somewhat tricky way to indicate how to compactly code several pieces of information into a single string. Two coding techniques are used.

(a) Consider two natural numbers n and k such that $0 \leq k < 2^n$. We code n and k into the string $s = 0^n + k$, i.e. the kth string of length n. Given the string s, one recovers n and k as follows: $n = \lg(s)$, $k = s - 0^{\lg(s)}$. This technique is used in the proofs of Theorems 4.3, 6.1, 7.4, and 10.1. In three of these proofs k is $\#(S)$, where S is a subset of the strings having length $< n$; n and $\#(S)$ are coded into the string $s = 0^n + \#(S)$. In the case of Theorem 6.1, k is the number that corresponds to a string s of length $< n$ (thus $0 \leq k < 2^n - 1$); n and s are coded into the string $s' = 0^n + s$.

(b) Consider a string p and a natural number k. We code p and k into the string $s = 0^{\lg(k)}1kp$, i.e. the string consisting of $\lg(k)$ 0's followed by a 1 followed by the kth string followed by the string p. The length of the initial run of 0's is the same as the length of the k th string and is used to separate kp in two and recover k and p from s. Note that $\lg(s) = \lg(p) + 2\lg(k) + 1$. This technique is used in the proof of Theorem 10.4. The proof of Theorem 4.1 uses a simpler technique: p and k are coded into the string $s = 0^k1p$. But this coding is less economical, for $\lg(s) = \lg(p) + k + 1$.

We use an extremely general definition of computer; this has the advantage that if one can show that something is difficult to compute using any such computer, this will be a very strong result. A computer is defined by indicating whether it has halted and what it has output, as a function of its program and the time. The formal definition of a computer C is an ordered pair $\langle C, H_C \rangle$ consisting of two total recursive functions

$$C:X^* \times N \to \{S \in 2^{X^*} \mid S \text{ is finite}\}$$

and $H_C:X^* \times N \to X$. Here $X = \{0, 1\}$, X^* is the set of all strings, and N is the set of all natural numbers. It is assumed that the functions C and H_C have the following two properties:

(a) $C(p,t) \subset C(p,t + 1)$, and
(b) if $H_C(p,t) = 1$, then $H_C(p,t + 1) = 1$ and $C(p,t) = C(p,t + 1)$.

$C(p,t)$ is the finite set of strings output by the computer C up to time t when its program is p. If $H_C(p,t) = 1$ the computer C is halted at time t when its program is p. If $H_C(p,t) = 0$ the computer C isn't halted at time t when its program is p. Henceforth whether $H_C(p,t) = 1$ or 0 will be indicated by stating that "$C(p,t)$ is halted" or that "$C(p,t)$ isn't halted." Property (a) states that $C(p,t)$ is the cumulative output, and property (b) states that a computer that is halted remains halted and never outputs anything else.

$C(p)$, the output of the computation that C performs when it is given the program p, is defined to be $\bigcup_t C(p,t)$. It is said that "$C(p)$ halts" iff there is a t such that $C(p,t)$ is halted. Furthermore, if $C(p)$ halts, the time at which it halts is defined to be the least t such that $C(p,t)$ is halted. We say that the program p calculates the finite set S when run on C if $C(p) = S$ and halts. We say that the program p enumerates the finite or infinite set S when run on C if $C(p) = S$.

We now define a class of computers that are especially suitable to use for measuring the information needed to specify a computation. A computer U is said to be universal if it has the following property. For any computer C, there must be a natural number, denoted $\text{sim}(C)$ (the cost of simulating C), such that the following holds. For any program p, there exists a program p' such that: $\lg(p') \leq \lg(p) + \text{sim}(C)$, $U(p')$ halts iff $C(p)$ halts, and $U(p') = C(p)$.

The idea of this definition is as follows. The universal computers are information-theoretically the most economical ones; their programs are shortest. More precisely, a universal computer U is able to simulate any other computer, and the program p' for U that simulates the program p for C need not be much longer than p. If there are instructions of length n for computing something using the computer C, then there are instructions of length $\leq n + \text{sim}(C)$ for carrying out the same computation using U; i.e. at most $\text{sim}(C)$ bits must be added to the length of the instructions, to indicate the computer that is to be simulated. Note that we do not assume that there is an effective procedure for obtaining p' given C and p. We have no need for the concept of an effectively universal computer in this paper. Nevertheless, the most natural examples of universal computers are effectively universal. See the Appendix for examples of universal computers.

We shall suppose that a particular universal computer U has somehow been chosen, and shall use it as our standard computer for measuring the information needed to specify a computation. The choice of U corresponds to the choice of the standard of measurement.

We now define $I(S)$, the information needed to calculate the finite set S, and $I_e(S)$, the information needed to enumerate the finite or infinite set S.

$$I(S) = \min \lg(p) \ (U(p) = S \text{ and halts});$$

$$I_e(S) = \begin{cases} \min \lg(p) \ (U(p) = S), \\ \infty \text{ if there are no such } p. \end{cases}$$

We say that $I(S)$ is the complexity of the finite set S, and that $I_e(S)$ is the e-complexity (enumeration complexity) of the finite or infinite set S. Note that $I(S)$ is the number of bits in the shortest program for U that calculates S, and $I_e(S)$ is the number of bits in the shortest program for U that enumerates S. Also, $I_e(S)$, the e-complexity of a set S, is ∞ if S isn't r.e. (recursively enumerable).

We say that a program p such that $U(p) = S$ and halts is a description of S, and a program p such that $U(p) = S$ is an e-description (enumeration description) of S. Moreover, if $U(p) = S$ and halts and $\lg(p) = I(S)$, then we say that p is a minimal description of S. Likewise, if $U(p) = S$ and $\lg(p) = I_e(S)$, then we say that p is a minimal e-description of S.

Finally, we define $I_e(f)$, the e-complexity of a partial function f. This is defined to be the e-complexity of the graph of f, i.e. the set of all ordered pairs of the form $(n, f(n))$. Here the ordered pair (i,j) is defined to be the natural number $(2i + 1)2^j - 1$; this is an effective 1-1 correspondence between the ordered pairs of natural numbers and the natural numbers. Note that $I_e(f) = \infty$ if f isn't partial recursive.

Before considering basic properties of these concepts, we introduce an abbreviated notation. Instead of $I(\{s\})$ and $I_e(\{s\})$ we shall write $I(s)$ and $I_e(s)$; i.e. the complexity or e-complexity of a string is defined to be complexity or e-complexity of its singleton set.

We now present basic properties of these concepts. First of all, note that there are precisely 2^n programs of length n, and $2^{n+1} - 1$ programs of length $\leq n$. It follows that the number of different sets of complexity n and the number of different sets of e-complexity n are both $\leq 2^n$. Also, the number of different sets of complexity $\leq n$ and the number of different sets of e-complexity $\leq n$ are both $\leq 2^{n+1} - 1$; i.e. the number of different objects of complexity or e-complexity $\leq n$ is bounded by the number of different descriptions or e-descriptions of length $\leq n$, which is $2^{n+1} - 1$. Thus it might be said that almost all sets are arbitrarily complex.

It is immediate from the definition of complexity and of a universal computer that $I_e(C(p)) \leq \lg(p) + \text{sim}(C)$, and $I(C(p)) \leq \lg(p) + \text{sim}(C)$ if $C(p)$ halts. This is used often, and without explicit mention. The following theorem lists for reference other basic properties of complexity and e-complexity that are used in this paper.

Theorem 2.1:

(a) There is a c such that for all strings s, $I(s) \leq \lg(s) + c$.
(b) There is a c such that for all finite sets S, $I(S) \leq \max S + c$.
(c) For any computer C, there is a c such that for all programs p, $I_e(C(p)) \leq I(p) + c$, and $I(C(p)) \leq I(p) + c$ if $C(p)$ halts.

Proof: (a) There is a computer C such that $C(s) = \{s\}$ and halts for all programs s. Thus $I(s) \leq \lg(s) + \text{sim}(C)$.

(b) There is a computer C such that $C(p)$ halts for all programs p, and $n \in C(p)$ iff $n < \lg(p)$ and the nth bit of p is a 1. Thus $I(S) \leq \max S + 1 + \text{sim}(C)$.

(c) There is a computer C' that does the following when it is given the program s. First C' simulates running s on U, i.e. it simulates $U(s)$. If and when $U(s)$ halts, C' has determined the set calculated by U when it is given the program s. If this isn't a singleton set, C' halts. If it is a singleton set $\{p\}$, C' then simulates running p on C. As C' determines the strings output by $C(p)$, it also outputs them. And C' halts if C halts during the simulated run.

In summary, $C'(s) = C(p)$ and halts iff $C(p)$ does, if s is a description of the program p. Thus, if s is a minimal description of the string p, then

$$\begin{cases} I_e(C(p)) = I_e(C'(s)) \leq \lg(s) + \text{sim}(C') = I(p) + \text{sim}(C'), \text{ and} \\ I(C(p)) = I(C'(s)) \leq \lg(s) + \text{sim}(C') = I(p) + \text{sim}(C') \text{ if } C(p) \text{ halts. Q.E.D.} \end{cases}$$

It follows from Theorem 2.1(a) that all strings of length n have complexity $\leq n + c$. In conjunction with the fact that $< 2^{n-k}$ strings are of complexity $< n - k$, this shows that the great majority of the strings of length n are of complexity $\approx n$. These are the random strings of length n. By taking $C = U$ in Theorem 2.1(c), it follows that there is a c such that for any minimal description p, $I(p) + c \geq I(U(p)) = \lg(p)$. Thus minimal descriptions are highly random strings. Likewise, minimal e-descriptions are highly random. This corresponds in information theory to the fact that the most informative messages are the most unexpected ones, the ones with least regularities and redundancies, and appear to be noise, not meaningful messages.

3. Definitions Related to Formal Systems

This paper deals with the information and time needed to carry out computations. However, we wish to apply these results to formal systems. This section explains how this is done.

The abstract definition used by Post that a formal system is an r.e. set of propositions is close to the viewpoint of this paper (see [31]).[48] However, we are not quite this unconcerned with the internal details of formal systems.

The historical motivation for formal systems was of course to construct deductive theories with completely objective, formal criteria for the validity of a demonstration. Thus, a fundamental characteristic of a formal system is an algorithm for checking the validity of proofs. From the existence of this proof verification algorithm, it follows that the set of all theorems that can be deduced from the axioms p by means of the rules of inference by proofs $\leq t$ characters in length is given by a total recursive function C of p and t. To calculate $C(p,t)$ one applies the proof verification algorithm to each of the finitely many possible demonstrations having $\leq t$ characters.

These considerations motivate the following definition. The rules of inference of a class of formal systems is a total recursive function $C:X^* \times N \rightarrow \{S \in 2^{X^*} \mid S \text{ is finite}\}$ with the property that $C(p,t) \subset C(p,t+1)$. The value of $C(p,t)$ is the finite (possibly empty) set of the theorems that can be proven from the axioms p by means of proofs $\leq t$ in size. Here p is a string and t is a natural number. $C(p) = \cup_t C(p,t)$ is the set of theorems that are consequences of the axioms p. The ordered pair $\langle C, p \rangle$, which implies both the choice of rules of inference and axioms, is a particular formal system.

Note that this definition is the same as the definition of a computer with the notion of "halting" omitted. Thus given any rules of inference, there is a computer that never halts whose output up to time t consists precisely of those propositions that can be deduced by proofs of size $\leq t$ from the axioms the computer is given as its program. And given any computer, there are rules of inference such that the set of theorems that can be deduced by proofs of size $\leq t$ from the program, is precisely the set of strings output by the computer up to time t. For this reason we consider the following notions to be synonymous: "computer" and "rules of inference," "program" and "axioms," and "output up to time t" and "theorems with proofs of size $\leq t$."

The rules of inference that correspond to the universal computer U are especially interesting, because they permit axioms to be very economical. When using the rules of inference U, the number of bits of axioms needed to deduce a given set of propositions is precisely the e-complexity of the set of propositions. If n bit of axioms are needed to obtain a set T of theorems using the rules of inference U, then at least $n - \text{sim}(C)$ bits of axioms are needed to obtain them using the rules of inference C; i.e. if $C(a) = T$, then $\lg(a) \geq I_e(T) - \text{sim}(C)$. Thus it could be said that U is among the rules of inference that permit axioms to be most economical. In Section 4 we are interested exclusively in the number of bits needed to deduce certain sets of propositions, not in the size of the proofs. We shall therefore only consider the rules of inference U in Section 4, i.e. formal systems of the form $\langle U, p \rangle$.

As a final comment regarding the rules of inference U, we would like to point out the interesting fact that if these rules of inference are used, then a minimal set of axioms for obtaining a given set of theorems must necessarily be random. This is just another way of saying that a minimal e-description is a highly random string, which was mentioned at the end of Section 2.

The following theorem also plays a role in the interpretation of our results in terms of formal systems.

Theorem 3.1:

Let f be a recursive function, and g be a recursive predicate.

---

[48]   For standard definitions of formal systems, see, for example, [32-34] and [10, p. 117].

(a) Let C be a computer. There is a computer C' that never halts such that $C'(p,t) = \{f(s) \mid s \in C(p,t) \ \& \ g(s)\}$ for all p and t.

(b) There is a c such that $I_e(\{f(s) \mid s \in S \ \& \ g(s)\}) \leq I_e(S) + c$ for all r.e. sets S.

Proof: (a) is immediate; (b) follows by taking $C = U$ in part (a). Q.E.D.

The following is an example of the use of Theorem 3.1. Suppose we wish to study the size of the proofs that "$n \in H$" in a formal system $\langle C, p \rangle$, where n is a numeral for a natural number. If we have a result concerning the speed with which any computer can enumerate the set H, we apply this result to the computer C' that has the property that $n \in C'(p,t)$ iff "$n \in H$" $\in C(p,t)$ for all n, p, and t. In this case the predicate g selects those strings that are propositions of the form "$n \in H$," and the function f transforms "$n \in H$" to n.

Here is another kind of example. Suppose there is a computer C that enumerates a set H very quickly. Then there is a computer C' that enumerates propositions of the form "$n \in H$" just as quickly. In this case the predicate g is taken to be always true, and the function f transforms n to "$n \in H$."

4. The Number of Bits of Axioms Needed to Determine the Complexity of Specific Strings

The set of all programs that halt when run on U is r.e. Similarly, the set of all true propositions of the form "$I(s) \leq n$" where s is a string and n is a natural number, is an r.e. set. In other words, if a program halts, or if a string is of complexity less than or equal to n, one will eventually find this out. To do this one need only try on U longer and longer test runs of more and more programs, and do this in a systematic way.

The problem is proving that a program doesn't halt, or that a string is of complexity greater than n. In this section we study how many bits of axioms are needed, and, as was pointed out in Section 3, it is sufficient to consider only the rules of inference U. We shall see that with n bits of axioms it is impossible to prove that a particular string is of complexity greater than $n + c$, where c doesn't depend on the particular axioms chosen (Theorem 4.1). It follows from this that if a formal system has n bits of axioms, then there is a program of length $\leq n + c$ that doesn't halt, but the fact that this program doesn't halt can't be proven in this formal system (Theorem 4.2).

Afterward, we show that $n + c$ bits of axioms suffice to be able to determine each program of length not greater than n that halts (Theorem 4.4), and thus to determine each string of complexity less than or equal to n, and its complexity (Theorem 4.3). Furthermore, the remaining strings must be of complexity greater than n.

Next, we construct an r.e. set of strings P that has the property that infinitely many strings aren't in it, but a formal system with n bits of axioms can't prove that a particular string isn't an element of P if the length of this string is greater than $n + c$ (Theorem 4.5). It follows that P is what Post called a simple set; that is, P is r.e., and its complement is infinite, but contains no infinite r.e. subset (see [31], Sec. 5, pp. 319-320). Moreover, $n + c$ bits suffice to determine each string of length not greater than n that isn't an element of P.

Finally, we show that not only are n bits of axioms insufficient to exhibit a string of complexity $> n + c$, they are also insufficient to exhibit (by means of an e-description) an r.e. set of e-complexity greater than $n + c$ (Theorem 4.6). This is because no set can be of e-complexity much greater than the complexity of one of its e-descriptions, and thus $I_e(U(s)) > k$ implies $I(s) > k - c$, where c is a constant that doesn't depend on s.

Although these results clarify how many bits of axioms are needed to determine the complexity of individual strings, they raise several questions regarding the size of proofs.

$n + c$ bits of axioms suffice to determine each string of complexity $\leq n$ and its complexity, but the method used here to do this appears to be extremely slow, that is, the proofs appear to be extremely long. Is this necessarily the case? The answer is "yes," as is shown in Section 7.

We have pointed out that there is a formal system having as theorems all true propositions of the form "$U(p)$ halts." The size of the proof that $U(p)$ halts must grow faster than any recursive function f of $\lg(p)$. For suppose that such a recursive bound on the length of proofs existed. Then all true propositions of the form "$U(p)$ doesn't halt" could be enumerated by checking to see if there is no proof that $U(p)$ halts of size $< f(\lg(p))$. This is impossible, by Theorem 4.2. The size of these proofs is studied in Section 10.

Theorem 4.1: (a) There is a c such that for all programs p, if a proposition of the form "$I(s) > n$" (s a string, n a natural number) is in $U(p)$ only if $I(s) > n$, then "$I(s) > n$" is in $U(p)$ only if $n < \lg(p) + c$.

In other words: (b) There is a c such that for all formal systems $\langle U, p \rangle$, if "$I(s) > n$" is a theorem of $\langle U, p \rangle$ only if it is true, then "$I(s) > n$" is a theorem of $\langle U, p \rangle$ only if $n < \lg(p) + c$.

For any r.e. set of propositions T, one obtains the following from (a) by taking p to be a minimal e-description of T: (c) If T has the property that "$I(s) > n$" is in T only if $I(s) > n$, then T has the property that "$I(s) > n$" is in T only if $n < I_e(T) + c$.

Idea of Proof: The following is essentially the classical Berry paradox.[49] For each natural number n greater than 1, consider "the least natural number that can't be defined in less than N characters." Here N denotes the numeral for the number n. This is a $\lfloor \log_{10} n \rfloor + c$ character phrase defining a number that supposedly needs at least n characters to be defined. This is a paradox if $\lfloor \log_{10} n \rfloor + c < n$, which holds for all sufficiently great values of n.

The following is a sharper version of Berry's paradox. Consider: "the least natural number whose definition requires more characters than there are in this phrase." This c-character phrase defines a number that supposedly needs more than c characters to be defined.

The following version is analogous to our proof. Consider this program: "Calculate the first string that can be proven in $\langle U, p \rangle$ to be of complexity greater than the number of bits in this program, where p is the following string: ..." Here "first" refers to first in the recursive enumeration $U(p,t)$ ($t = 0,1,2, \ldots$) of the theorems of the formal system $\langle U, p \rangle$. This program is only a constant number c of bits longer than the number of bits in p. It is no longer a paradox; it shows that in $\langle U, p \rangle$ no string can be proven to be of complexity greater than $c + \lg(p) = c + $ the number of bits of axioms of $\langle U, p \rangle$.

Proof: Consider the computer C that does the following when it is given the program p'. First, it solves the equation $p' = 0^k 1 p$. If this isn't possible (i.e. $p' = 0^k$), then C halts without outputting anything. If this is possible, C continues by simulating running the program p on U. It generates $U(p)$ searching for a proposition of the form "$I(s) > n$" in which s is a string, n is a natural number, and $n \geq \lg(p') + k$. If and when it finds such a proposition "$I(s) > n$" in $U(p)$, it outputs s and halts.

Suppose that p satisfies the hypothesis of the theorem, i.e. "$I(s) > n$" is in $U(p)$ only if $I(s) > n$. Consider $C(0^{\text{sim}(C)} 1 p)$. If $C(0^{\text{sim}(C)} 1 p) = \{s\}$, then $I(s) \leq \lg(0^{\text{sim}(C)} 1 p) + \text{sim}(C)$

---

[49] Although due to Berry, its importance was recognized by Russell and it was published by him [35, p. 153].

$= \lg(p) + 2\text{sim}(C) + 1$. But C outputs s and halts because it found the proposition "$I(s) > n$" in $U(p)$ and

$$n \geq \lg(p') + k = \lg(0^{\text{sim}(C)}1p) + \text{sim}(C) = \lg(p) + 2\text{sim}(C) + 1.$$

Thus, by the hypothesis of the theorem, $I(s) > n \geq \lg(p) + 2\text{sim}(C) + 1$, which contradicts the upper bound on $I(s)$. Consequently, $C(0^{\text{sim}(C)}1p)$ doesn't output anything (i.e. equals $\emptyset$), for there is no proposition "$I(s) > n$" in $U(p)$ with $n \geq \lg(p) + 2\text{sim}(C) + 1$. The theorem is proved with $c = 2\text{sim}(C) + 1$. Q.E.D.

Definition 4.1: $H = \{p \mid U(p) \text{ halts}\}$. ($H$ is r.e., as was pointed out in the first paragraph of this section.)

Theorem 4.2: There is a c such that for all formal systems $\langle U, p \rangle$, if a proposition of the form "$s \in H$" or "$s \notin H$" (s a string) is a theorem of $\langle U, p \rangle$ only if it is true, then there is a string s of length $\leq \lg(p) + c$ such that neither "$s \in H$" nor "$s \notin H$" is a theorem of $\langle U, p \rangle$.

Proof: Consider the computer C that does the following when it is given the program p. It simulates running p on U, and as it generates $U(p)$, it checks each string in it to see if it is a proposition of the form "$s \in H$" or "$s \notin H$," where s is a string. As soon as C has determined in this way for each string s of length less than or equal to some natural number n whether or not "$s \in H$" or "$s \notin H$" is in $U(p)$, it does the following.

C supposes that these propositions are true, and thus that it has determined the set $\{s \in H \mid \lg(s) \leq n\}$. Then it simulates running each of the programs in this set on U until U halts, and thus determines the set $S = \cup \, U(s) \, (s \in H \ \& \ \lg(s) \leq n)$. C then outputs the proposition "$I(f) > n$," where f is the first string not in S, and then continues generating $U(p)$ as was indicated in the first paragraph of this proof. Inasmuch as f isn't output by any program of length $\leq n$ that halts, it must in fact be of complexity $> n$.

Thus, $C(p)$ enumerates true propositions of the form "$I(f) > n$" if p satisfies the hypothesis of the theorem. Hence, by Theorem 4.1, "$I(f) > n$" is in $C(p)$ only if $n < I_e(C(p)) + c' \leq \lg(p) + \text{sim}(C) + c'$. It is easy to see that the theorem is proved with $c = \text{sim}(C) + c'$. Q.E.D.

Theorem 4.3: Consider the set T_n consisting of all true propositions of the form "$I(s) = k$" (s a string, k a natural number $\leq n$) and all true propositions of the form "$I(s) > n$." $I_e(T_n) = n + O(1)$.

In other words, a formal system $\langle U, p' \rangle$ whose theorems consist precisely of all true propositions of the form "$I(s) = k$" with $k \leq n$, and all true propositions of the form "$I(s) > n$," requires $n + O(1)$ bits of axioms; i.e. $n - c$ bits are necessary and $n + c$ bits are sufficient to obtain this set of theorems.

Idea of Proof: If one knows n and how many programs of length $\leq n$ halt when run of U, then one can find them all, and see what they calculate. n and this number h can be coded into an $(n + 1)$-bit string. In other words, the axiom of this formal system with theorem set T_n is essentially "the number of programs of length $\leq n$ that halt when run on U is h," where n and h are particular natural numbers. This axiom is $n + O(1)$ bits of information.

Proof: By Theorem 4.1, $I_e(T_n) \geq n - c$. It remains to show that $I_e(T_n) \leq n + c$.

Consider the computer C that does the following when it is given the program p of length ≥ 1. It generates the r.e. set H until it has found $p - 0^{\lg(p)}$ programs of length $\leq \lg(p) - 1$ that halt when run on U. If and when it has found this set S of programs, it simulates running each program in S on U

INFORMATION-THEORETIC LIMITATIONS OF FORMAL SYSTEMS

until it halts. C then examines each string that is calculated by a program in S, and determines the length of the shortest program in S that calculates it. If $p - 0^{\lg(p)} =$ the number h of programs of length $\leq \lg(p) - 1$ that halt when run of U, then C has determined each string of complexity $\leq \lg(p) - 1$ and its complexity. If $p - 0^{\lg(p)} < h$, then C's estimates of the complexity of strings are too high. And if $p - 0^{\lg(p)} > h$, then C never finishes generating H. Finally, C outputs its estimates as propositions of the form "$I(s) = k$" with $k \leq \lg(p) - 1$, and as propositions of the form "$I(s) > k$" with $k = \lg(p) - 1$ indicating that all other strings are of complexity $> \lg(p) - 1$.

We now show how C can be used to enumerate T_n economically. Consider $h = \#(\{s \in H \mid \lg(s) \leq n\})$. As there are precisely $2^{n+1} - 1$ strings of length $\leq n$, $0 \leq h \leq 2^{n+1} - 1$. Let p be $0^{n+1} + h$, that is, the hth string of length $n + 1$. Then $C(p) = T_n$, and thus $I_e(T_n) \leq \lg(p) + \text{sim}(C) = n + 1 + \text{sim}(C)$. Q.E.D.

Theorem 4.4: Let T_n be the set of all true propositions of the form "$s \in H$" or "$s \notin H$" with s a string of length $\leq n$. $I_e(T_n) = n + O(1)$.

In other words, a formal system $\langle U, p \rangle$ whose theorems consist precisely of all true propositions of the form "$s \in H$" or "$s \notin H$" with $\lg(s) \leq n$, requires $n + O(1)$ bits of axioms; i.e. $n - c$ bits are necessary and $n + c$ bits are sufficient to obtain this set of theorems.

Proof: Theorem 4.2 shows that $I_e(T_n) \geq n - c$. The proof that $I_e(T_n) \leq n + c$ is obtained from the proof of Theorem 4.3 by simplifying the definition of the computer C so that it outputs T_n, instead of, in effect, using T_n to determine each string of complexity $\leq n$ and its complexity. Q.E.D.

Definition 4.2: $P = \{s \mid I(s) < \lg(s)\}$; i.e. P contains each string s whose complexity $I(s)$ is less than its length $\lg(s)$.

Theorem 4.5: (a) P is r.e., i.e. there is a formal system with the property that "$s \in P$" is a theorem iff $s \in P$.

(b) $\bar{P}$ is infinite, because for each n there is a string of length n that isn't an element of P.

(c) There is a c such that for all formal systems $\langle U, p \rangle$, if "$s \notin P$" is a theorem only if it is true, then "$s \notin P$" is a theorem only if $\lg(s) < \lg(p) + c$. Thus, by the definition of e-complexity, if an r.e. set T of propositions has the property that "$s \notin P$" is in T only if it is true, then "$s \notin P$" is in T only if $\lg(s) < I_e(T) + c$.

(d) There is a c such that for all r.e. sets of strings S, if S contains no string in P (i.e. $S \subset \bar{P}$), then $\lg(\max S) < I_e(S) + c$. Thus $\max S < 0^{I_e(S)+c} = 2^{I_e(S)+c} - 1$, and $\#(S) < 2^{I_e(S)+c}$.

(e) Let T_n be the set of all true propositions of the form "$s \notin P$" with $\lg(s) \leq n$. $I_e(T_n) = n + O(1)$. In other words, a formal system $\langle U, p \rangle$ whose theorems consist precisely of all true propositions of the form "$s \notin P$" with $\lg(s) \leq n$, requires $n + O(1)$ bits of axioms; i.e. $n - c$ bits are necessary and $n + c$ bits are sufficient to obtain this set of theorems.

Proof: (a) This is an immediate consequence of the fact that the set of all true propositions of the form "$I(s) \leq n$" is r.e.

(b) We must show that for each n there is a string of length n whose complexity is greater than or equal to its length. There are 2^n strings of length n. As there are exactly $2^n - 1$ program of length $< n$, there are $< 2^n$ strings of complexity $< n$. Thus at least one string of length n must be of complexity $\geq n$.

(c) Consider the computer C that does the following when it is given the program p. It simulates running p on U. As C generates $U(p)$, it examines each string in it to see if it is a proposition of the form "$s \notin P$," where s is a string of length ≥ 1. If it is, C outputs the proposition "$I(s) > n$" where $n = \lg(s) - 1$.

If p satisfies the hypothesis, i.e. "$s \notin P$" is in $U(p)$ only if it is true, then $C(p)$ enumerates true propositions of the form "$I(s) > n$" with $n = \lg(s) - 1$. It follows by Theorem 4.1 that n must be $< I_e(C(p)) + c' \leq \lg(p) + \text{sim}(C) + c'$. Thus $\lg(s) - 1 < \lg(p) + \text{sim}(C) + c'$, and part (c) of the theorem is proved with $c = \text{sim}(C) + c' + 1$.

(d) Consider the computer C that does the following when it is given the program p. It simulates running p on U. As C generates $U(p)$, it takes each string s in $U(p)$, and outputs the proposition "$s \notin P$."

Suppose S contains no string in P. Let p be a minimal e-description of S, i.e. $U(p) = S$ and $\lg(p) = I_e(S)$. Then $C(p)$ enumerates true propositions of the form "$s \notin P$" with $s \in S$. By part (c) of this theorem,

$$\lg(s) < I_e(C(p)) + c' \leq \lg(p) + \text{sim}(C) + c' = I_e(S) + \text{sim}(C) + c'.$$

Part (d) of the theorem is proved with $c = \text{sim}(C) + c'$.

(e) That $I_e(T_n) \geq n - c$ follows from part (c) of this theorem. The proof that $I_e(T_n) \leq n + c$ is obtained by changing the definition of the computer C in the proof of Theorem 4.3 in the following manner. After C has determined each string of complexity $\leq n$ and its complexity, C determines each string s of complexity $\leq n$ whose complexity is greater than or equal to its length, and then C outputs each such s in a proposition of the form "$s \notin P$." Q.E.D.

Theorem 4.6: (a) There is a c such that for all programs p, if a proposition of the form "$I_e(U(s)) > n$" (s a string, n a natural number) is in $U(p)$ only if $I_e(U(s)) > n$, then "$I_e(U(s)) > n$" is in $U(p)$ only if $n < \lg(p) + c$.

In other words: (b) There is a c such that for all formal systems $\langle U, p \rangle$, if "$I_e(U(s)) > n$" is a theorem of $\langle U, p \rangle$ only if it is true, then "$I_e(U(s)) > n$" is a theorem of $\langle U, p \rangle$ only if $n < \lg(p) + c$.

For any r.e. set of propositions T, one obtains the following from (a) by taking p to be a minimal e-description of T: (c) If T has the property that "$I_e(U(s)) > n$" is in T only if $I_e(U(s)) > n$, then T has the property that "$I_e(U(s)) > n$" is in T only if $n < I_e(T) + c$.

Proof: By Theorem 2.1(c), there is a c' such that $I_e(U(s)) > n$ implies $I(s) > n - c'$.

Consider the computer C that does the following when it is given the program p. It simulates running p on U. As C generates $U(p)$, it checks each string in it to see if it is a proposition of the form "$I_e(U(s)) > n$" with s a string and n a natural number. Each time it finds such a proposition in which $n \geq c'$, C outputs the proposition "$I(s) > m$" where $m = n - c' \geq 0$.

If p satisfies the hypothesis of the theorem, then $C(p)$ enumerates true propositions of the form "$I(s) > m$." "$I(s) > m$" ($m = n - c' \geq 0$) is in $C(p)$ iff "$I_e(U(s)) > n$" ($n \geq c'$) is in $U(p)$. By Theorem 4.1, "$I(s) > m$" is in $C(p)$ only if

$$m < I_e(C(p)) + c'' \leq \lg(p) + \text{sim}(C) + c''.$$

Thus "$I_e(U(s)) > n$" ($n \geq c'$) is in $U(p)$ only if $n - c' < \lg(p) + \text{sim}(C) + c''$. The theorem is proved with $c = \text{sim}(C) + c'' + c'$. Q.E.D.

5. The Greatest Natural Number of Complexity $\leq$ N

The growth of $a(n)$, the greatest natural number of complexity $\leq n$ as a partial function of n, serves as a benchmark for measuring a number of computational phenomena. The general approach in

Sections 6 to 10 will be to use a partial function of n to measure some quantity of computational interest, and to compare the growth of this partial function as n increases with that of $a(n)$.

We compare rates of growth in the following fashion.

Definition 5.1: We say that a partial function f grows at least as quickly as another partial function g, written $f \geq {}'g$ or $g \leq {}'f$, when a shift of f overbounds g. That is, when there is a c such that for all n, if $g(n)$ is defined, then $f(n + c)$ is defined and $f(n + c) \geq g(n)$. Note that $f \leq {}'g$ and $g \leq {}'h$ implies $f \leq {}'h$.

Definition 5.2: We say that the partial functions f and g grow equally quickly, written $f = {}'g$, iff $f \leq {}'g$ and $g \leq {}'f$.

We now formally define $a(n)$, and list its basic properties for future reference.

Definition 5.3: $a(n) = \max k \, (I(k) \leq n)$. The maximum is taken over all natural numbers k of complexity $\leq n$. If there are no such k, then $a(n)$ is undefined.

Theorem 5.1:

(a) If $a(n)$ is defined, then $I(a(n)) \leq n$.
(b) If $a(n)$ is defined, then $a(n + 1)$ is defined and $a(n) \leq a(n + 1)$.
(c) If $I(n) \leq m$, then $n \leq a(m)$.
(d) $n \leq a(I(n))$.
(e) If $a(m)$ is defined, then $n > a(m)$ implies $I(n) > m$.
(f) If $I(n) > i$ for all $n \geq m$, then $m > a(i)$ if $a(i)$ is defined.
(g) $I(a(n)) = n + O(1)$.
(h) There is a c such that for all finite sets S of strings, $\max S \leq a(I(S) + c)$.
(i) There is a c such that for all finite sets S of strings and all n, if $a(n)$ is defined and $a(n) \in S$, then $I(S) > n - c$.

Proof: (a) to (f) follow immediately from the definition of $a(n)$.

(g) Consider the two computers C and C' that always halt and such that $C(n) = \{n + 1\}$ and $C'(n + 1) = \{n\}$. It follows by Theorem 2.1(c) that $I(n) = I(n + 1) + O(1)$. By part (e) of this theorem, if $a(n)$ is defined then $I(a(n) + 1) > n$. By part (a) of this theorem, $I(a(n)) \leq n$. Hence if $a(n)$ is defined we have $I(a(n)) = I(a(n) + 1) + O(1)$, $I(a(n)) \leq n$, and $I(a(n) + 1) > n$. It follows that $I(a(n)) = n + O(1)$.

(h) Consider the computer C such that $C(p) = \{ \max S \}$ and halts if p is a description of S, i.e. if $U(p) = S$ and halts. It follows that $I(\max S) \leq I(S) + \mathrm{sim}(C)$. Thus by part (c) of this theorem, $\max S \leq a(I(S) + c)$, where $c = \mathrm{sim}(C)$.

(i) Consider the computer C such that $C(p) = \{1 + \max S\}$ and halts if p is a description of S, i.e. if $U(p) = S$ and halts. It follows that $I(1 + \max S) \leq I(S) + \mathrm{sim}(C)$. If $a(n) \in S$, then $1 + \max S > a(n)$, and thus by part (e) of this theorem $I(1 + \max S) > n$. Hence $n < I(1 + \max S) \leq I(S) + \mathrm{sim}(C)$, and thus $I(S) > n - c$, where $c = \mathrm{sim}(C)$. Q.E.D.

6. How Fast Does the Greatest Natural Number of Complexity $\leq$ N Grow with Increasing N?

In Theorem 6.2 we show that an equivalent definition of $a(n)$ is the greatest value at n of any partial recursive function of complexity $\leq n$. In Theorem 6.3 we use this to show that any partial function

$\geq$ $'a$ eventually overtakes any partial recursive function. This will apply directly to all the functions that will be shown in succeeding sections to be $='$ to a.

In Theorem 6.4 it is shown that for any partial recursive function f, $f(a(.)) \leq$ $'a$. Thus there is a c such that for all n, if $a(n)$ is defined, then $a(n) < a(n + c)$ (Theorem 6.5).

Theorem 6.1: There is a c such that if $f{:}N \to N$ is a partial recursive function defined at $'n$ and $n \geq I_e(f)$, then $I(f(n)) \leq n + c$.

Proof: Given a minimal e-description s of the graph of f, we add it to 0^{n+1}. As $n \geq I_e(f) = \lg(s)$, the resulting string $p = 0^{n+1} + s$ has both n ($= \lg(p) - 1$) and the graph of f ($= U(s) = U(p - 0^{\lg(p)})$) coded into it. Given this string p as its program, a computer C generates the graph of f searching for the pair $(n, f(n))$. If and when it is found, the computer outputs $f(n)$ and halts. Thus, $f(n)$, if defined, is of complexity $\leq \lg(p) + \text{sim}(C) = n + 1 + \text{sim}(C)$. Q.E.D.

Definition 6.1: $b(n) = \max f(n)$ $(I_e(f) \leq n)$. The maximum is taken over all partial recursive functions $f{:}N \to N$ that are defined at n and are of e-complexity $\leq n$. If there are no such functions, then $b(n)$ is undefined.

Theorem 6.2: $a =$ $'b$.

Proof: First we show that $b \leq$ $'a$. If $b(n)$ is defined, then there is a partial recursive function $f{:}N \to N$ defined at n with $I_e(f) \leq n$, such that $f(n) = b(n)$. By Theorem 6.1, $I(f(n)) \leq n + c$, and thus $f(n) \leq a(n + c)$ by Theorem 5.1(c). Hence if $b(n)$ is defined, $b(n) = f(n) \leq a(n + c)$, and thus $b \leq$ $'a$.

Now we show that $a \leq$ $'b$. Suppose that $a(n)$ is defined, and consider the constant function $f_n{:}N \to N$ whose value is always $a(n)$, and the computer C such that $C(n) = \{(0,n),(1,n),(2,n), \dots \}$. It follows by Theorem 2.1(c) that $I_e(f_n) \leq I(a(n)) + c$, which by Theorem 5.1(a) is $\leq n + c$. Thus if $a(n)$ is defined, $a(n) = f_n(n + c) \leq \max f(n + c)$ $(I_e(f) \leq n + c) = b(n + c)$. Hence $a \leq$ $'b$. Q.E.D.

Theorem 6.3: Let the partial function $x{:}N \to N$ have the property that $x \geq$ $'a$. There is a constant c' such that the following holds for all partial recursive functions $f{:}N \to N$. If $f(n)$ is defined and $n \geq I_e(f) + c'$, then $x(n)$ is defined and $x(n) \geq f(n)$.

Proof: By Theorem 6.2 and the transitivity of $\geq$ $'$, $x \geq$ $'a \geq$ $'b$. Thus there is a c such that $x(n + c)$ is defined and $x(n + c) \geq b(n)$ if $b(n)$ is defined. Consider the shifted function $f'(n) = f(n + c)$. The existence of a computer C such that $(i,j) \in C(p)$ iff $(i + c,j) \in U(p)$ shows that $I_e(f') \leq I_e(f) + \text{sim}(C)$. By the definition of b, $x(n + c) \geq b(n) \geq f'(n)$ if f' is defined at n and $I_e(f') \leq n$. Thus $x(n + c) \geq f(n + c)$ if f is defined at $n + c$ and $I_e(f') \leq I_e(f) + \text{sim}(C) \leq n$. In other words, $x(n) \geq f(n)$ if f is defined at n and $I_e(f) + \text{sim}(C) + c \leq n$. The theorem is proved with $c' = \text{sim}(C) + c$. Q.E.D.

Theorem 6.4: Let $f{:}N \to N$ be a partial recursive function. $f(a(.)) \leq$ $'a$.

Proof: There is a computer C such that $C(n) = \{f(n)\}$ and halts if $f(n)$ is defined. Thus by Theorem 2.1(c), if $f(n)$ is defined, $I(f(n)) \leq I(n) + c$. Substituting $a(n)$ for n, we obtain $I(f(a(n))) \leq I(a(n)) + c \leq n + c$, for by Theorem 5.1(a), $I(a(n)) \leq n$. Thus if $f(a(n))$ is defined, $f(a(n)) \leq a(n + c)$, by Theorem 5.1(c). Q.E.D.

INFORMATION-THEORETIC LIMITATIONS OF FORMAL SYSTEMS

Theorem 6.5: There is a c such that for all n, if $a(n)$ is defined, then $a(n) < a(n + c)$.

Proof: Taking $f(n) = n + 1$ in Theorem 6.4, we obtain $a(.) + 1 \leq {}'a$. Q.E.D.

7. The Resources Needed to Calculate/Enumerate the Set of All Strings of Complexity $\leq$ N

7.1: We first discuss the metamathematical implications of the material in this section.

The basic fact used in this section (see the proof of Theorem 7.3) is that for any computer C there is a c such that for all n, if $a(n)$ is defined then max $\cup C(p,a(n))$ $(\lg(p) \leq a(n))$ is less than $a(n + c)$. Thus $a(n + c)$ cannot be output by programs of length $\leq a(n)$ in time $\leq a(n)$. If we use Theorem 3.1(a) to take C to be such that $s \in C(p,t)$ iff "$I(s) = k$" $\in C^*(p,t)$, and we recall that $a(n + c)$ is a string of complexity $\leq n + c$, we obtain the following result. Any formal system $\langle C^*, p \rangle$ whose theorems include all true propositions of the form "$I(s) = k$" with $k \leq n + c$, must either have more than $a(n)$ bits of axioms, or need proofs of size greater than $a(n)$ to be able to demonstrate these propositions. Here c depends only on the rules of inference C^*. This is a strong result, in view of the fact that $a(n)$ is greater than or equal to any partial recursive function $f(n)$ for $n \geq I_e(f) + c'$ (Theorem 6.3).

The idea of Section 9 is to show that both extremes are possible and there is a drastic trade-off. We can deduce these results from a few bits of axioms ($\leq n + c$ bits by Theorem 4.3) by means of enormous proofs, or we can directly take as axioms all that we wish to prove. This gives short proofs, but we are assuming an enormous number of bits of axioms.

From the fact that $a(n + c) > \max \cup C(p,a(n))$ $(\lg(p) \leq a(n))$, it also follows that if one wishes to prove a numerical upper bound on $a(n + c)$, one faces the same drastic alternatives. Lin and Rado, in trying to determine particular values of $\Sigma(n)$ and $SH(n)$, have, in fact, essentially been trying to do this (see [36]). In their paper they explain the difficulties they encountered and overcame for $n = 3$, and expect to be insurmountable for greater values of n.

7.2: Now we begin the formal exposition, which is couched exclusively in terms of computers.

In this section we study the set $K(n)$ consisting of all strings of complexity $\leq n$. This set turns out to be extremely difficult to calculate, or even to enumerate a superset of — either the program or the time needed must be extremely large. In order to measure this difficulty, we will first measure the resources needed to output $a(n)$.

Definition 7.1: $K(n) = \{s \mid I(s) \leq n\}$. Note that this set may be empty, and $\#(K(n))$ isn't greater than $2^{n+1} - 1$, inasmuch as there are exactly $2^{n+1} - 1$ programs of length $\leq n$.

We shall show that $a(n)$ and the resources required to calculate/enumerate $K(n)$ grow equally quickly. What do we mean by the resources required to calculate a finite set, or to enumerate a superset of it? It is assumed that the computer C is being used to do this.

Definition 7.2: Let S be a finite set of strings. $r(S)$, the resources required to calculate S, is the least r such that there is a program p of length $\leq r$ having the property that $C(p,r) = S$ and is halted. If there is no such r, $r(S)$ is undefined. $r_e(S)$, the resources required to enumerate a superset of S, is the least r such that there is a program p of length $\leq r$ with the property that $S \subset C(p,r)$. If there is no such r, $r_e(S)$ is undefined. We abbreviate $r(\{s\})$ and $r_e(\{s\})$ as $r(s)$ and $r_e(s)$.

We shall find very useful the notion of the set of all output produced by the computer C with information and time resources limited to r. We denote this by C_r.

Definition 7.3: $C_r = \cup\ C(p,r)$ $(\lg(p) \leq r)$.

We now list for future reference basic properties of these concepts.

Theorem 7.0:

(a)
$$a(n) = \begin{cases} \max K(n) \text{ if } K(n) \neq \emptyset, \\ \text{undefined if } K(n) = \emptyset. \end{cases}$$

(b) $K(n) \neq \emptyset$, and $a(n)$ is defined, iff $n \geq n^*$. Here $n^* = \min I(s)$, where the minimum is taken over all strings s.
(c) For all r, $C_r \subset C_{r+1}$.
•   In (d) to (k), S and S' are arbitrary finite sets of strings.
(d) $S \subset C_{r_e(S)}$ if $r_e(S)$ is defined.
(e) $r_e(S) \leq r(S)$ if $r(S)$ is defined.
(f) If $r_e(S')$ is defined, then $S \subset S'$ implies $r_e(S) \leq r_e(S')$.
(g) If $S \subset C(p,t)$, then either $\lg(p) \geq r_e(S)$ or $t \geq r_e(S)$.
(h) If $C(p) = S$ and halts, then either $\lg(p) \geq r(S)$, or the time at which $C(p)$ halts is $\geq r(S)$.
(i) If $C(p) = S$ and halts, and $\lg(p) < r(S)$, then $C(p)$ halts at time $\geq r(S)$.
•   In (j) and (k) it is assumed that C is U. Thus $r(S)$ and $r_e(S)$ are always defined.
(j) If $r(S) > I(S)$, then there is a program p of length $I(S)$ such that $U(p) = S$ and halts at time $\geq r(S)$.
(k) $r(S) \geq I(S)$.

Proof: These results follow immediately from the definitions. Q.E.D.

Theorem 7.1:

There is a c such that for all finite sets S of strings,
(a) $I(r(S)) \leq I(S) + c$ if $r(S)$ is defined, and
(b) $I(r_e(S)) \leq I(S) + c$ if $r_e(S)$ is defined.

Proof: (a) Consider the computer C' that does the following when it is given the program p. First, it simulates running p on U. If and when U halts during the simulated run, C' has determined the finite set $S = U(p)$ of strings. Then C' repeats the following operations for $r = 0,1,2, \ldots$

C' determines $C(p',r)$ for each program p' of length $\leq r$. It checks those $C(p', r)$ $(\lg(p') \leq r)$ that are halted to see if one of them is equal to S. If none of them are, C' adds 1 to r and repeats this operation. If one of them is equal to S, C' outputs r and halts.

Let p be a minimal description of a finite set S of strings, i.e. $U(p) = S$ and halts, and $\lg(p) = I(S)$. Then if $r(S)$ is defined, $C'(p) = \{r(S)\}$ and halts, and thus $I(r(S)) \leq \lg(p) + \text{sim}(C') = I(S) + \text{sim}(C')$. This proves part (a) of the theorem.

(b) The proof of part (b) of the theorem is obtained from the proof of part (a) by changing the definition of the computer C' so that it checks all $C(p', r)$ $(\lg(p') \leq r)$ to see if one of them includes S, instead of checking all those $C(p', r)$ $(\lg(p') \leq r)$ that are halted to see if one of them is equal to S. Q.E.D.

Theorem 7.2: max $C_{a(.)} \leq {}'a$.

Proof: Theorem 2.1(c) and the existence of a computer C' such that $C'(r) = C_r$ and halts, shows that there is a c such that for all r, $I(C_r) \leq I(r) + c$. Thus by Theorem 5.1(h) and (b) there is a c' such that for all r, max $C_r \leq a(I(r) + c')$. Hence if $a(n)$ is defined, max $C_{a(n)} \leq a(I(a(n)) + c') \leq a(n + c')$ by Theorem 5.1(a) and (b). Q.E.D.

Theorem 7.3:

(a) If $r_e(a(n))$ is defined when $a(n)$ is, then $r_e(a(.)) = {}'a$.
(b) If $r(a(n))$ is defined when $a(n)$ is, then $r(a(.)) = {}'a$.

Proof: By Theorem 7.1, if $r_e(a(n))$ and $r(a(n))$ are defined, $I(r_e(a(n))) \leq I(a(n)) + c$ and $I(r(a(n))) \leq I(a(n)) + c$. By Theorem 5.1(a), $I(a(n)) \leq n$. Thus $I(r_e(a(n))) \leq n + c$ and $I(r(a(n))) \leq n + c$. Applying Theorem 5.1(c), we obtain $r_e(a(n)) \leq a(n + c)$ and $r(a(n)) \leq a(n + c)$. Thus we have shown that $r_e(a(.)) \leq {}'a$ and $r(a(.)) \leq {}'a$, no matter what C is.

$r(S)$, if defined, is $\geq r_e(S)$ (Theorem 7.0(e)), and thus to finish the proof it is sufficient to show that $a \leq {}'r_e(a(.))$ if $r_e(a(n))$ is defined when $a(n)$ is. By Theorems 7.2 and 6.5 there is a c such that for all n, if $a(n)$ is defined, then max $C_{a(n)} < a(n + c)$, and thus $a(n + c) \notin C_{a(n)}$. And inasmuch as for all finite sets S, $S \subset C_{r_e(S)}$ (Theorem 7.0(d)), it follows that $a(n + c) \in C_{r_e(a(n+c))}$.

In summary, there is a c such that for all n, if $a(n)$ is defined, then $a(n + c) \notin C_{a(n)}$, and $a(n + c) \in C_{r_e(a(n+c))}$.

As for all r, $C_r \subset C_{r+1}$ (Theorem 7.0(c)), it follows that if $a(n)$ is defined then $a(n) < r_e(a(n + c))$. Thus $a \leq {}'r_e(a(.))$. Q.E.D.

Theorem 7.4: $I(K(n)) = n + O(1)$.

Proof: As was essentially shown in the proof of Theorem 4.3, there is a computer C' such that $C'(0^{n+1} + \#(\{p \in H \mid \lg(p) \leq n\})) = K(n)$ and halts, for all n. Thus $I(K(n)) \leq n + 1 + \text{sim}(C')$ for all n.

$K(n) = \emptyset$ can hold for only finitely many values of n, by Theorem 7.0(b). By Theorem 7.0(a), for all other values of n, $a(n) \in K(n)$, and thus, by Theorem 5.1(i), there is a c such that $I(K(n)) \geq n - c$ for all n. Q.E.D.

Theorem 7.5:

(a) If $r_e(K(n))$ is defined for all n, then $r_e(K(.)) = {}'a$.
(b) If $r(K(n))$ is defined for all n, then $r(K(.)) = {}'a$.

Proof: By Theorem 7.1, if $r_e(K(n))$ and $r(K(n))$ are defined, $I(r_e(K(n))) \leq I(K(n)) + c$, and $I(r(K(n))) \leq I(K(n)) + c$. $I(K(n)) = n + O(1)$ (Theorem 7.4), and thus there is a c' that doesn't depend on n such that $I(r_e(K(n))) \leq n + c'$, and $I(r(K(n))) \leq n + c'$. Applying Theorem 5.1(c), we obtain $r_e(K(n)) \leq a(n + c')$, and $r(K(n)) \leq a(n + c')$. Thus we have shown that $r_e(K(.)) \leq {}'a$ and $r(K(.)) \leq {}'a$, no matter what C is.

For all finite sets S and S' of strings, if $S \subset S'$, then $r_e(S) \leq r_e(S') \leq r(S')$ if these are defined (Theorem 7.0(f), (e)). As $a(n) \in K(n)$ if $a(n)$ is defined (Theorem 7.0(a)), we have $r_e(a(n)) \leq r_e(K(n)) \leq r(K(n))$ if these are defined. By Theorem 7.3(a), $r_e(a(.)) \geq {}'a$ if $r_e(a(n))$ is defined when $a(n)$ is, and thus $r_e(K(.)) \geq {}'a$ and $r(K(.)) \geq {}'a$ if these are defined for all n. Q.E.D.

Theorem 7.6: Suppose $r_e(a(n))$ is defined when $a(n)$ is. There is a c such that the following holds for all partial recursive functions $f:N \to N$. If $n \geq I_e(f) + c$ and $f(n)$ is defined, then

(a) $a(n)$ is defined,
(b) if $a(n) \in C(p,t)$, then either $\lg(p) \geq f(n)$ or $t \geq f(n)$, and
(c) if $K(n) \subset C(p,t)$, then either $\lg(p) \geq f(n)$, or $t \geq f(n)$.

Proof: (a) and (b) By Theorem 7.3(a), $r_e(a(.)) \geq {}'a$. Taking $r_e(a(.))$ to be the partial function $x(.)$ in the hypothesis of Theorem 6.3, we deduce that if $n \geq I_e(f) + c$ and $f(n)$ is defined, then $a(n)$ is defined and $r_e(a(n)) \geq f(n)$. Here c doesn't depend on f.

Thus if $a(n) \in C(p,t)$, then by Theorem 7.0(g) it follows that either $\lg(p) \geq r_e(a(n)) \geq f(n)$ or $t \geq r_e(a(n)) \geq f(n)$.

(c) Part (c) of this theorem is an immediate consequence of parts (a) and (b) and the fact that if $a(n)$ is defined then $a(n) \in K(n)$ (see Theorem 7.0(a)). Q.E.D.

8. The Minimum Time Such That All Programs of Length $\leq$ N That Halt Have Done So

In this section we show that for any computer, $a \geq {}'$ the minimum time such that all programs of length $\leq n$ that halt have done so (Theorem 8.1). Moreover, in the case of U this is true with " $= {}'$" instead of " $\geq {}'$" (Theorem 8.2).

The situation revealed in the proof of Theorem 8.2 can be stated in the following vague but suggestive manner. Suppose that one wishes to calculate $a(n)$ or $K(n)$ using the standard computer U. To do this one only needs about n bits of information. But a program of length $n + O(1)$ for calculating $a(n)$ is among the programs of length $\leq n + O(1)$ that take the most time to halt. Likewise, an $(n + O(1))$-bit program for calculating $K(n)$ is among the programs of length $\leq n + O(1)$ that take the most time to halt. These are among the most difficult calculations that can be accomplished by program having not more than n bits.

Definition 8.1: $d_C(n) = $ the least t such that for all p of length $\leq n$, if $C(p)$ halts, then $C(p,t)$ is halted. This is the minimum time at which all programs of length $\leq n$ that halt have done so. Although it is 0 if no program of length $\leq n$ halts, we stipulate that $d_C(n)$ is undefined in this case.

Theorem 8.1: $d_C \leq {}'a$.

Proof: Consider the computer C' that does the following when it is given the program p. C' simulates $C(p,t)$ for $t = 0,1,2, \ldots$ until $C(p,t)$ is halted. If and when this occurs, C' outputs the final value of t, which is the time at which $C(p)$ halts. Finally, C' halts.

If $d_C(n)$ is defined, then there is a program p of length $\leq n$ that halts when run on C and does this at time $d_C(n)$. Then $C'(p) = \{d_C(n)\}$ and halts. Thus $I(d_C(n)) \leq \lg(p) + \text{sim}(C') \leq n + \text{sim}(C')$. By Theorem 5.1(c), we conclude that $d_C(n)$ is, if defined, $\leq a(n + \text{sim}(C'))$. Q.E.D.

Theorem 8.2: $d_U = {}'a$.

Proof: In view of Theorem 8.1, $d_U \leq {}'a$. Thus we need only show that $d_U \geq {}'a$.

Recall that $a(n)$ is defined iff $n \geq n^*$ (Theorem 7.0(b)). As $C = U$ is a universal computer, $r(a(n))$ is defined if $a(n)$ is defined. Thus Theorem 7.3(b) applies to this choice of C, and $r(a(.)) \geq {}'a$. That is to say, there is a c such that for all $n \geq n^*$, $r(a(n + c)) \geq a(n)$.

INFORMATION-THEORETIC LIMITATIONS OF FORMAL SYSTEMS

As $a \geq {}'a$, taking $x = a$ and $f(n) = n + c + 1$ in Theorem 6.3, we obtain the following. There is a c' such that for all $n \geq n^* + c'$, $a(n) \geq f(n) = n + c + 1$. We conclude that for all $n \geq n^* + c'$, $a(n) > n + c$.

By Theorem 5.1(a), $n + c \geq I(a(n + c))$ for all $n \geq n^*$.

The preceding results may be summarized in the following chain of inequalities. For all $n \geq n^* + c'$, $r(a(n + c)) \geq a(n) > n + c \geq I(a(n + c))$.

As $r(a(n + c)) \geq I(a(n + c))$, the hypothesis of Theorem 7.0(j) is satisfied, and we conclude the following. There is a program p of length $I(a(n + c)) \leq n + c$ such that $U(p) = \{a(n + c)\}$ and halts at time $\geq r(a(n + c)) \geq a(n)$. Thus for all $n \geq n^* + c'$, $d_U(n + c) \geq a(n)$.

Applying Theorem 5.1(b) to this lower bound on $d_U(n + c)$, we conclude that for all $n \geq n^*$, $d_U(n + c' + c) \geq a(n + c') \geq a(n)$. Q.E.D.

9. Examples of Trade-Offs Between Information and Time

Consider calculating $a(n)$ using the computer U and the computer C defined as follows. For all programs p, $C(p,0) = \{p\}$ and is halted.

Since $I(a(n)) \leq n$ (Theorem 5.1(a)), there is a program $\leq n$ bits long for calculating $a(n)$ using U. But inasmuch as $r(a(.)) \geq {}'a$ (Theorem 7.3(b)) and $d_U \leq {}'a$ (Theorem 8.1), this program takes "about" $a(n)$ units of time to halt (see the proof of Theorem 8.2). More precisely, with finitely many exceptions, this program takes between $a(n - c)$ and $a(n + c)$ units of time to halt.

What happens if one uses C to calculate $a(n)$? Inasmuch as $C(a(n)) = \{a(n)\}$ and halts at time 0, C can calculate $a(n)$ immediately. But this program, although fast, is $\lg(a(n)) = \lfloor \log_2(a(n) + 1) \rfloor$ bits long. Thus $r(a(n))$ is precisely $\lg(a(n))$ if one uses this computer.

Now for our second example. Suppose one wishes to enumerate a superset of $K(n)$, and is using the following two computers, which never halt: $C(p,t) = \{s \mid \lg(s) \leq t\}$ and $C'(p,t) = \{s \mid \lg(s) \leq \lg(p)\}$. These two computers have the property that $K(n)$, if not empty, is included in $C(p,t)$, or is included in $C'(p,t)$, iff $t \geq \lg(a(n))$, or iff $\lg(p) \geq \lg(a(n))$, respectively. Thus for these two computers, $r_e(K(n))$, which we know by Theorem 7.5(a) must be $= {}'$ to a, is precisely given by the following: $r_e(K(n)) = 0$ if $a(n)$ is undefined, and $\lg(a(n))$ otherwise.

It is also interesting to slow down or speed up the computer C by changing its time scale recursively. Let $f:N \to N$ be an arbitrary unbounded total recursive function with the property that for all n, $f(n) \leq f(n + 1)$. C^f, the f speed-up/slowdown of C, is defined as follows: $C^f(p,t) = \{s \mid f(\lg(s)) \leq t\}$. For the computer C^f, $r_e(K(n))$ is precisely given by the following: $r_e(K(n)) = 0$ if $a(n)$ is undefined, and $f(\lg(a(n)))$ otherwise. The fact that by Theorem 7.5(a) this must be $= {}'$ to a, is related to Theorem 6.4 that for any partial recursive function f, $f(a(.)) \leq {}'a$.

Now, for our third example, we consider trade-offs in calculating $K(n)$. We use U and the computer C defined as follows. For all p and n, $C(p,0)$ is halted, and $n \in C(p,0)$ iff $\lg(p) > n$ and the nth bit of the string p is a 1.

Inasmuch as $I(K(n)) = n + O(1)$ (Theorem 7.4), if we use U there is a program about n bits long for calculating $K(n)$. But this program takes "about" $a(n)$ units of time to halt, in view of the fact that $r(K(.)) \geq {}'a$ (Theorem 7.5(b)) and $d_U \leq {}'a$ (Theorem 8.1) (see the proof of Theorem 8.2). More precisely, with finitely many exceptions, this program takes between $a(n - c)$ and $a(n + c)$ units of time to halt.

On the other hand, using the computer C we can calculate $K(n)$ immediately. But the shortest program for doing this has length precisely $1 + \max K(n) = a(n) + 1$ if $a(n)$ is defined, and has length

0 otherwise. In other words, for this computer $r(K(n)) = 0$ if $a(n)$ is undefined, and $a(n) + 1$ otherwise.

We have thus seen three examples of a drastic trade-off between information and time resources. In this setting information and time play symmetrical roles, especially in the case of the resources needed to enumerate a superset.

10. The Speed of Recursive Enumerations

10.1: We first discuss the metamathematical implications of the material in this section.

Consider a particular formal system, and a particular r.e. set of strings R. Suppose that a proposition of the form "$s \in R$" is a theorem of this formal system iff it is true, i.e. iff the string s is an element of R. Define $e(n)$ to be the least m such that all theorems of the formal system of the form "$s \in R$" with $\lg(s) \leq n$ have proofs of size $\leq m$. By using Theorem 3.1(a) we can draw the following conclusions from the results of this section. First, $e \leq {}'a$ for any R. Second, $e = {}'a$ iff

$$I(\{s \in R \mid \lg(s) \leq n\}) = n + O(1). \tag{*}$$

Thus r.e. sets R for which $e = {}'a$ are the ones that require the longest proofs to show that "$s \in R$", and this is the case iff R satisfies (*). It is shown in this section that the r.e. set of strings $\{p \mid p \in U(p)\}$ has property (*), and the reader can show without difficulty that H and P are also r.e. sets of strings that have property (*). Thus we have three examples of R for which $e = {}'a$.

10.2: Now we begin the formal exposition, which is couched exclusively in terms of computers.

Consider an r.e. set of strings R and a particular computer C^* and p^* such that $C^*(p^*) = R$. How quickly is R enumerated? This is, what is the time $e(n)$ that it takes to output all elements of R of length $\leq n$?

Definition 10.1: $R_n = \{s \in R \mid \lg(s) \leq n\}$. $e(n) =$ the least t such that $R_n \subset C^*(p^*,t)$.

We shall see that the rate of growth of the total function $e(n)$ can be related to the growth of the complexity of R_n. In this way we shall show that some r.e. sets R are the most difficult to enumerate, i.e. take the most time.

Theorem 10.1[50]: There is a c such that for all n, $I(R_n) \leq n + c$.

Proof: $0 \leq \#(R_n) \leq 2^{n+1} - 1$, for there are precisely $2^{n+1} - 1$ strings of length $\leq n$. Consider p, the $\#(R_n)$-th string of length $n + 1$; i.e. $p = 0^{n+1} + \#(R_n)$. This string has both n ($ = \lg(p) - 1$) and $\#(R_n)$ ($ = p - 0^{\lg(p)}$) coded into it. When this string p is its program, the computer C generates the r.e. set R by simulating $C^*(p^*)$, until it has found $\#(R_n)$ strings of length $\leq n$ in R. C then outputs this set of strings, which is R_n, and halts. Thus $I(R_n) \leq \lg(p) + \text{sim}(C) = n + 1 + \text{sim}(C)$. Q.E.D.

Theorem 10.2:

(a) There is a c such that for all n, $e(n) \leq a(I(R_n) + c)$.

(b) $e \leq {}'a$.

Proof: (a) Consider the computer C that does the following. Given a description p of R_n as its program, the computer C first simulates running p on U in order to determine R_n. Then it simulates

---

[50]   This theorem, with a different proof, is due to Loveland [37, p. 64].

$C^*(p^*,t)$ for $t = 0,1,2, \ldots$ until $R_n \subseteq C^*(p^*,t)$. C then outputs the final value of t, which is $e(n)$, and halts.

This shows that $\leq \text{sim}(C)$ bits need be added to the length of a description of R_n to bound the length of a description of $e(n)$; i.e. if $U(p) = R_n$ and halts, then $C(p) = \{e(n)\}$ and halts, and thus $I(e(n)) \leq \lg(p) + \text{sim}(C)$. Taking p to be a minimal description of R_n, we have $\lg(p) = I(R_n)$, and thus $I(e(n)) \leq I(R_n) + \text{sim}(C)$. By Theorem 5.1(c), this gives us $e(n) \leq a(I(R_n) + \text{sim}(C))$. Part (a) of the theorem is proved with $c = \text{sim}(C)$.

(b) By part (a) of this theorem, $e(n) \leq a(I(R_n) + c)$. And by Theorem 10.1, $I(R_n) \leq n + c'$ for all n. Applying Theorem 5.1(b), we obtain $e(n) \leq a(I(R_n) + c) \leq a(n + c' + c)$ for all n. Thus $e \leq {}'a$. Q.E.D.

Theorem 10.3: If $a \leq {}'e$, then there is a c such that $I(R_n) \geq n - c$ for all n.

Proof: By Theorem 7.0(b) and the definition of $\leq {}'$, if $a \leq {}'e$, then there is a c_0 such that for all $n \geq n^*$, $a(n) \leq e(n + c_0)$. And by Theorem 10.2(a), there is a c_1 such that $e(n + c_0) \leq a(I(R_{n+c_0}) + c_1)$ for all n. We conclude that for all $n \geq n^*$, $a(n) \leq a(I(R_{n+c_0}) + c_1)$.

By Theorems 6.5 and 5.1(b), there is a c_2 such that if $a(m)$ is defined and $m \leq n - c_2$, then $a(m) < a(n)$. As we have shown in the first paragraph of this proof that for all $n \geq n^*$, $a(n) \leq a(I(R_{n+c_0}) + c_1)$, it follows that $I(R_{n+c_0}) > n - c_2$.

In other words, for all $n \geq n^*$, $I(R_{n+c_0}) > (n + c_0) - c_0 - c_1 - c_2$. And thus for all n, $I(R_n) \geq n - c_0 - c_1 - c_2 - M$, where $M = \max_{n<n^*+c_0} n - c_0 - c_1 - c_2$ if this is positive, and 0 otherwise. The theorem is proved with $c = c_0 + c_1 + c_2 + M$. Q.E.D.

Theorem 10.4:

If there is a c such that $I(R_n) \geq n - c$ for all n, then
(a) there is a c' such that if $t \geq e(n)$, then $I(t) > n - c'$, and
(b) $e \geq {}'a$.

Proof: By Theorem 5.1(f) it follows from (a) that $e(n) > a(n - c')$ if $a(n - c')$ is defined. Hence $e(n + c') \geq a(n)$ if $a(n)$ is defined, i.e. $e \geq {}'a$. Thus to complete the proof we need only show that (a) follows from the hypothesis.

We consider the case in which $t \geq e(n)$ and $n \geq I(t) = n - k$, for if $I(t) > n$ then any c' will do.

There is a computer C that does the following when it is given the program $0^{\lg(k)}1kp$, where p is a minimal description of t. First, C determines $\lg(p) + k = I(t) + k = (n - k) + k = n$. Second, C simulates running p on U in order to determine $U(p) = \{t\}$. C now uses its knowledge of n and t in order to calculate R_n. To do this C first simulates running p^* on C^* in order to determine $C^*(p^*,t)$, and finally C outputs all strings in $C^*(p^*,t)$ that are of length $\leq n$, which is R_n, and halts.

In summary, C has the property that if $t \geq e(n)$, $I(t) = n - k$, and p is a minimal description of t, then $C(0^{\lg(k)}1kp) = R_n$ and halts, and thus

$$I(R_n)$$
$$\leq \lg(0^{\lg(k)}1kp) + \text{sim}(C)$$
$$= \lg(p) + 2\lg(k) + \text{sim}(C) + 1$$
$$= I(t) + 2\lg(k) + \text{sim}(C) + 1$$
$$= n - k + 2\lg(k) + \text{sim}(C) + 1.$$

Taking into account the hypothesis of this theorem, we obtain the following for all n: if $t \geq e(n)$ and $I(t) = n - k$, then $n - c \leq I(R_n) \leq n - k + 2\lg(k) + \text{sim}(C) + 1$, and thus $c + \text{sim}(C) + 1 \geq k - 2\lg(k)$. As $\lg(k) = \lfloor \log_2(k + 1) \rfloor$, this implies that there is a c' such that for all n, if $t \geq e(n)$ and $I(t) = n - k$, then $k < c'$. We conclude that for all n, if $t \geq e(n)$, then either $I(t) > n$ or $I(t) = n - k > n - c'$. Thus in either case $I(t) > n - c'$. Q.E.D.

Theorem 10.5: If $R = \{p \mid p \in U(p)\}$, then there is a c such that $I(R_n) > n - c$ for all n.

Proof: Consider the following computer C. When given the program p, C first simulates running p on U until U halts. If and when it finishes doing this, C then outputs each string $s \notin U(p)$, and never halts.

If the program p is a minimal description of R_n, then C enumerates a set that cannot be enumerated by any program p' run on U having $\leq n$ bits. The reason is that if $\lg(p') \leq n$, then $p' \in C(p)$ iff $p' \notin R_n$ iff $p' \notin U(p')$. Thus $\leq \text{sim}(C)$ bits need be added to the length $I(R_n)$ of a minimal description p of R_n to bound the length of an e-description of the set $C(p)$ of e-complexity $> n$; i.e. $n < I_e(C(p)) \leq \lg(p) + \text{sim}(C) = I(R_n) + \text{sim}(C)$. Hence $n < I(R_n) + c$, where $c = \text{sim}(C)$. Q.E.D.

Theorem 10.6:

(a) $e \leq {}'a$ and $\exists c \, \forall n \, I(R_n) \leq n + c$.

(b) $e \geq {}'a$ iff $\exists c \, \forall n \, I(R_n) \geq n - c$.

(c) If $R = \{p \mid p \in U(p)\}$, then $e = {}'a$ and $I(R_n) = n + O(1)$.

Proof: (a) is Theorems 10.2(b) and 10.1. (b) is Theorem 10.4(b) and 10.3. And (c) follows immediately from parts (a) and (b) and Theorem 10.5. Q.E.D.

Appendix. Examples of Universal Computers

In this Appendix we use the formalism of Rogers.[51] In particular, P_x denotes the xth Turing machine, $\varphi_x^{(2)}$ denotes the partial recursive function $N \times N \to N$ that P_x calculates, and D_x denotes the xth finite set of natural numbers. Here the index x is an arbitrary natural number.

First we give a more formal definition of computer than in Section 2.

A partial recursive function $c : N \times N \to N$ is said to be adequate (as a defining function for a computer C) iff it has the following three properties:

(a) it is a total function;

(b) $D_{c(p,t)} \subset D_{c(p,t+1)}$;

(c) if the natural number 0 is an element of $D_{c(p,t)}$, then $D_{c(p,t)} = D_{c(p,t+1)}$.

A computer C is defined by means of an adequate function $c : N \times N \to N$ as follows.

(a) $C(p,t)$ is halted iff the natural number 0 is an element of $D_{c(p,t)}$.

(b) $C(p,t)$ is the set of strings $\{n \mid n + 1 \in D_{c(p,t)}\}$; i.e. the nth string is in $C(p,t)$ iff the natural number $n + 1$ is an element of $D_{c(p,t)}$.

We now give an name to each computer. The natural number i is said to be an adequate index iff $\varphi_i^{(2)}$ is an adequate function. If i is an adequate index, C^i denotes the computer whose defining function is $\varphi_i^{(2)}$. If i isn't an adequate index, then "C^i" isn't the name of a computer.

We now define a universal computer U in such a way that it has the property that if i is an adequate index, then $U(0^i1p) = C^i(p)$ and halts iff $C^i(p)$ halts. In what follows i and t denote arbitrary natural

---

[51] See [38, pp. 13-15, 21, 70].

numbers, and p denotes an arbitrary string. $U(0^i, t)$ is defined to be equal to $\emptyset$ and to be halted. $U(0^i1p, t)$ is defined recursively. If $t \geq 1$ and $U(0^i1p, t - 1)$ is halted, then $U(0^i1p, t) = U(0^i1p, t - 1)$ and is halted. Otherwise $U(0^i1p, t)$ is the set of strings $\{n \mid n + 1 \in W\}$ and is halted iff $0 \in W$. Here

$$W = \bigcup_{i < t_0} D_{\varphi_i^{(2)}(p,t)},$$

and t_0 is the greatest natural number $\leq t$ such that if $t' < t_0$ then P_i applied to $\langle p, t' \rangle$ yields an output in $\leq t$ steps.

The universal computer U that we have just defined is, in fact, effectively universal: to simulate the computation that C^i performs when it is given the program p, one gives U the program $p' = 0^i1p$, and thus p' can be obtained from p in an effective manner. Our second example of a universal computer, U', is not effectively universal, i.e. there is no effective procedure for obtaining p' from p.[52]

U' is defined as follows:

$$\begin{cases} U'(\Lambda, t) = \emptyset \text{ and is halted,} \\ U'(0p, t) = U(p, t) - \{1\} \text{ and is halted iff } U(p,t) \text{ is, and} \\ U'(1p, t) = U(p, t) \cup \{1\} \text{ and is halted iff } U(p,t) \text{ is.} \end{cases}$$

I.e. U' is almost identical to U, except that it eliminates the string 1 from the output, or forces the string 1 to be included in the output, depending on whether the first bit of its program is 0 or not. It is easy to see that U' cannot be effectively universal. If it were, given any program p for U, by examining the first bit of the program p' for U' that simulates it, one could decide whether or not the string 1 is in $U(p)$. But there cannot be an effective procedure for deciding, given any p, whether or not the string 1 is in $U(p)$.

Added in Proof

The following additional references have come to our attention.

Part of Gödel's analysis of Cantor's continuum problem [39] is highly relevant to the philosophical considerations of Section 1. Cf. especially [39, pp. 265, 272].

Schwartz [40, pp. 26-28] first reformulates our Theorem 4.1 using the hypothesis that the formal system in question is a consistent extension of arithmetic. He then considerably extends Theorem 4.1 [40, pp. 32-34]. The following is a paraphrase of these pages.

Consider a recursive function $f:N \to N$ that grows very quickly, say $f(n) = n!!!!!!!!!!!$. A string s is said to have *property f* if the fact that p is a description of $\{s\}$ either implies that $\lg(p) \geq \lg(s)$ or that $U(p)$ halts at time $> f(\lg(s))$. Clearly a 1000-bit string with property f is very difficult to calculate. Nevertheless, a counting argument shows that there are strings of all lengths with property f, and they can be found in an effective manner [40, Lemma 7, p. 32]. In fact, the first string of length n with property f is given by a recursive function of n, and is therefore of complexity $\leq \log_2 n + c$. This is thus an example of an extreme trade-off between program size and the length of computation. Furthermore, an argument analogous to the demonstration of Theorem 4.1 shows that proofs that specific strings have property f must necessarily be extremely tedious (if some natural hypotheses concerning U and the formal system in question are satisfied) [40, Theorem 8, pp. 33-34].

---

[52]   The definition of U' is an adaptation of [38, p. 42, Exercise 2-11].

[41, Item 2, pp. 12-20] sheds light on the significance of these results. Cf. especially the first unital-icized paragraphs of answers numbers 4 and 8 to the question "What is programming?" [41, pp. 13, 15-16]. Cf. also [40, Appendix, pp. 63-69].

Index of Symbols

References

1. CHAITIN, G. J. Information-theoretic aspects of Post's construction of a simple set. On the difficulty of generating all binary strings of complexity less than n. (Abstracts.) *AMS Notices 19* (1972), pp. A-712, A-764.

2. CHAITIN, G. J. On the greatest natural number of definitional or information complexity $\leq n$. There are few minimal descriptions. (Abstracts.) *Recursive Function Theory: Newsletter,* no. 4 (1973), pp. 11-14, Dep. of Math., U. of California, Berkeley.

3. VON NEUMANN, J. Method in the physical sciences. *J. von Neumann – Collected Works, Vol. VI,* A. H. Taub, Ed., MacMillan, New York, 1963, No. 36, pp. 491-498.

4. VON NEUMANN, J. The mathematician. In *The World of Mathematics, Vol. 4,* J. R. Newman, Ed., Simon and Schuster, New York, 1956, pp. 2053-2063.

5. BELL, E. T. *Mathematics: Queen and Servant of Science.* McGraw-Hill, New York, 1951, pp. 414-415.

6. WEYL, H. Mathematics and logic. *Amer. Math. Mon. 53* (1946), 1-13.

7. WEYL, H. *Philosophy of Mathematics and Natural Science.* Princeton U. Press, Princeton, N.J., 1949, pp. 234-235.

8. TURING, A. M. Solvable and unsolvable problems. In *Science News,* no. 31 (1954), A. W. Heaslett, Ed., Penguin Books, Harmondsworth, Middlesex, England, pp. 7-23.

9. NAGEL, E., AND NEWMAN, J. R. *Gödel's Proof.* Routledge & Kegan Paul, London, 1959.

10. DAVIS, M. *Computability and Unsolvability.* McGraw-Hill, New York, 1958.

11. QUINE, W. V. Paradox. *Scientific American 206,* 4 (April 1962), 84-96.

12. KLEENE, S. C. *Mathematical Logic.* Wiley, New York, 1968, Ch. V, pp. 223-282.

13. GÖDEL, K. On the length of proofs. In *The Undecidable,* M. Davis, Ed., Raven Press, Hewlett, N.Y., 1965, pp. 82-83.

14. COHEN, P. J. *Set Theory and the Continuum Hypothesis.* Benjamin, New York, 1966, p. 45.

15. ARBIB, M. A. Speed-up theorems and incompleteness theorems. In *Automata Theory,* E. R. Cainiello, Ed., Academic Press, New York, 1966, pp. 6-24.

16. EHRENFEUCHT, A., AND MYCIELSKI, J. Abbreviating proofs by adding new axioms. *AMS Bull. 77* (1971), 366-367.

17. POLYA, G. Heuristic reasoning in the theory of numbers. *Amer. Math. Mon. 66* (1959), 375-384.

18. EINSTEIN, A. Remarks on Bertrand Russell's theory of knowledge. In *The Philosophy of Bertrand Russell*, P. A. Schilpp, Ed., Northwestern U., Evanston, Ill., 1944, pp. 277-291.

19. HAWKINS, D. Mathematical sieves. *Scientific American 199*, 6 (Dec. 1958), 105-112.

20. KOLMOGOROV, A. N. Logical basis for information theory and probability theory. *IEEE Trans. IT-14* (1968), 662-664.

21. MARTIN-LÖF, P. Algorithms and randomness. *Rev. of Internat. Statist. Inst. 37* (1969), 265-272.

22. LOVELAND, D. W. A variant of the Kolmogorov concept of complexity. *Inform. and Contr. 15* (1969), 510-526.

23. CHAITIN, G. J. On the difficulty of computations. *IEEE Trans. IT-16* (1970), 5-9.

24. WILLIS, D. G. Computational complexity and probability constructions. *J. ACM 17*, 2 (April 1970), 241-259.

25. ZVONKIN, A. K., AND LEVIN, L. A. The complexity of finite objects and the development of the concepts of information and randomness by means of the theory of algorithms. *Russian Math. Surveys 25*, 6 (Nov.-Dec. 1970), 83-124.

26. SCHNORR, C. P. *Zufälligkeit und Wahrscheinlichkeit – Eine algorithmische Begründung der Wahrscheinlichkeitstheorie.* Springer, Berlin, 1971.

27. FINE, T. L. *Theories of Probability – An Examination of Foundations.* Academic Press, New York, 1973.

28. CHAITIN, G. J. Information-theoretic computational complexity. *IEEE Trans. IT-20* (1974), 10-15.

29. DELONG, H. *A Profile of Mathematical Logic.* Addison-Wesley, Reading, Mass., 1970, Sec. 28.2, pp. 208-209.

30. DAVIS, M. Hilbert's tenth problem is unsolvable. *Amer. Math. Mon. 80* (1973), 233-269.

31. POST, E. Recursively enumerable sets of positive integers and their decision problems. In *The Undecidable*, M. Davis, Ed., Raven Press, Hewlett, N.Y., 1965, pp. 305-307.

32. MINSKY, M. L. *Computation: Finite and Infinite Machines.* Prentice-Hall, Englewood Cliffs, N.J., 1967, Sec. 12.2-12.5, pp. 222-232.

33. SHOENFIELD, J. R. *Mathematical Logic.* Addison-Wesley, Reading, Mass., 1967, Sec. 1.2, pp. 2-6.

34. MENDELSON, E. *Introduction to Mathematical Logic.* Van Nostrand Reinhold, New York, 1964, pp. 29-30.

35. RUSSELL, B. Mathematical logic as based on the theory of types. In *From Frege to Gödel*, J. van Heijenoort, Ed., Harvard U. Press, Cambridge, Mass., 1967, pp. 150-182.

36. LIN, S., AND RADO, T. Computer studies of Turing machine problems. *J. ACM 12*, 2 (April 1965), 196-212.

37. LOVELAND, D. W. On minimal-program complexity measures. Conf. Rec. of the ACM Symposium on Theory of Computing, Marina del Rey, California, May 1969, pp. 61-65.

38. ROGERS, H. *Theory of Recursive Functions and Effective Computability.* McGraw-Hill, New York, 1967.

39. GÖDEL, K. What is Cantor's continuum problem? In *Philosophy of Mathematics,* Benacerraf, P., and Putnam, H., Eds., Prentice-Hall, Englewood Cliffs, N.J., 1964, pp. 258-273.

40. SCHWARTZ, J. T. A short survey of computational complexity theory. Notes, Courant Institute of Mathematical Sciences, NYU, New York, 1972.

41. SCHWARTZ, J. T. *On Programming: An Interim Report on the SETL Project. Installment I: Generalities.* Lecture Notes, Courant Institute of Mathematical Sciences, NYU, New York, 1973.

42. CHAITIN, G. J. A theory of program size formally identical to information theory. Res. Rep. RC4805, IBM Res. Center, Yorktown Heights, N.Y., 1974.

RECEIVED OCTOBER 1971; REVISED JULY 1973

A NOTE ON MONTE CARLO PRIMALITY TESTS AND ALGORITHMIC INFORMATION THEORY

Communications on Pure and Applied Mathematics 31 (1978), pp. 521-527.

GREGORY J. CHAITIN
IBM Thomas J. Watson Research Center

JACOB T. SCHWARTZ[53]
Courant Institute of Mathematical Sciences

Abstract

Solovay and Strassen, and Miller and Rabin have discovered fast algorithms for testing primality which use coin-flipping and whose conclusions are only probably correct. On the other hand, algorithmic information theory provides a precise mathematical definition of the notion of random or patternless sequence. In this paper we shall describe conditions under which if the sequence of coin tosses in the Solovay-Strassen and Miller-Rabin algorithms is replaced by a sequence of heads and tails that is of maximal algorithmic information content, i.e., has maximal algorithmic randomness, then one obtains an error-free test for primality. These results are only of theoretical interest, since it is a manifestation of the Gödel incompleteness phenomenon that it is impossible to "certify" a sequence to be random by means of a proof, even though most sequences have this property. Thus by using certified random sequences one can in principle, but not in practice, convert probabilistic tests for primality into deterministic ones.

Algorithmic Information Theory

To prepare for discussion of the Solovay-Strassen and Miller-Rabin algorithms, we first summarize some of the basic concepts of algorithmic information theory [1]-[4].[54]

Consider a universal Turing machine U whose programs are in binary. By "universal" we mean that for any other Turing machine M whose programs p are in binary there is a prefix μ such that $U(\mu p)$ always carries out the same computation as $M(p)$.

$I(X)$, the algorithmic information content of X, is defined to be the size in bits of the smallest programs for U to compute X. There is absolutely no restriction on the running time or storage space used by these programs. If X is a finite object such as a natural number or bit string, this includes the proviso that U halt after printing X. If X is an infinite object such as a set of natural numbers or of bit strings, then of course U does not halt. Sets, as opposed to sequences, may have their members printed in arbitrary order. X can also be an r.e. function f; in that case U prints the graph of f, i.e., the set of all ordered pairs $\langle x, f(x) \rangle$. Note that variations in the definition of U give rise to at most $O(1)$ differences in the resulting I, by the definition of universality.

---

[53] The second author has been supported by US DOE, Contract EY-76-C-02-3077*000. We wish to thank John Gill III and Charles Bennett for helpful discussions. Reproduction in whole or in part is permitted for any purpose of the United States Government.
[54] We could equally well have used in this paper the newer formalism of [7], in which programs are "self-delimiting."

It is easy to show (cf. [1]-[4]) that the maximum value of $I(s)$ taken over all n-bit strings s is equal to $n + O(1)$, and that an overwhelming majority of the s of length n have $I(s)$ very close to n. Such s have maximum information content or "entropy" and are highly random, patternless, incompressible, and typical. They are said to be "algorithmically random." The greater the difference between $I(s)$ and the length of s, the less random s is, the more atypical it is, and the more pattern it has. It is convenient to say that "s is c-random" if $I(s) \geq n - c$, where n is the length of s. Less than 2^{n-c} n-bit strings are not c-random. As for natural numbers, $I(n) \leq \log_2 n + O(1)$ and most n have $I(n)$ very close to $\log_2 n$. Strangely enough, though most strings are random, it is impossible to prove that specific strings have this property! For an explanation of this paradox see [1]-[6].

The Solovay-Strassen and Miller-Rabin Algorithms [8]-[10]

The general form of these algorithms is as follows: To test whether n is prime, take k natural numbers uniformly distributed between 1 and $n - 1$, inclusive, and for each one i check whether the predicate $W(i, n)$ holds. (Read "i is a witness of n's compositeness.") If so, n is composite. If not, n is prime with probability $1 - 2^{-k}$. This is because, as proved in [8]-[10], at least half the i's from 1 to $n - 1$ are in fact witnesses of n's compositeness, if n is indeed composite, and none of them are if n is prime. The definition of W is different in the Solovay-Strassen and the Miller-Rabin algorithms, but both algorithms are of this form, where $W(i, n)$ can be computed quickly, i.e., the running time of a program which computes $W(i, n)$ is bounded by a polynomial in the size of n, in other words, by a polynomial in $\log n$.

We shall now show that if sufficiently long random sequences are supplied, the probabilistic reasoning of [8]-[10] can be converted into a rigorous proof of primality. To state our precise results, we need to make the following definition:

Definition 1: Let s be an m-bit sequence, and let J be an integer. Find the smallest integer k such that $(J - 1)^{k+1} > 2^m - 1$, and (by converting s to base $J - 1$ representation) find the unique sequence $d_k d_{k-1} \ldots d_0$ of base $J - 1$ digits such that

$$\sum_{0 \leq i \leq k} d_i (J - 1)^i = s.$$

Calculate

$$Z(s, J) \;=\; \neg W(1 + d_0, J) \; \& \; \ldots \; \& \; \neg W(1 + d_{k-1}, J),$$

where $W(i, n)$ is as above. Then we say that J passes the s-test for primality if and only if $Z(s, J)$ is true.

Lemma 1: Let m, J, k, and Z be as in Definition 1. Then the number of m-bit sequences s for which $Z(s, J)$ is true is 2^m if J is prime, but is not more than 2^{m+1-k} if J is not a prime.

Proof: If J is prime, then $W(i, J)$ is always false, so our assertion is trivial. Now suppose J is composite, so that $W(i, J)$ is true for at least $(J - 1)/2$ values of i. Since $d_k (J - 1)^k \leq 2^m$, i.e.,

$$0 \leq d_k \leq 2^m (J - 1)^{-k},$$

it follows that the number of s satisfying $Z(s, J)$ is at most

$$\left[2^m (J - 1)^{-k} + 1 \right] \left[(J - 1)/2 \right]^k \;=\; \left[2^m + (J - 1)^k \right] 2^{-k},$$

or 2^{m+1-k} since $(J - 1)^k \leq 2^m$.

A NOTE ON MONTE CARLO PRIMALITY TESTS AND ALGORITHMIC INFORMATION THEORY

It is now easy to prove our results:

Theorem 1: For all sufficiently large c, if s is any c-random $j(j + 2c)$-bit sequence and J any integer whose binary representation is j bits long, then $Z(s, J)$ if and only if J is a prime.

Theorem 2: For all sufficiently large c, if s is any c-random $2j(i + c)$-bit sequence and J any integer whose binary representation is j bits long and whose information content $I(J)$ is not more than i, then $Z(s, J)$ if and only if J is a prime.

Proof of Theorem 1: Denote the cardinality of a set e by writing $|e|$, and let $\sigma(m)$ be the set of all m-bit sequences. Let J be a non-prime integer j bits long. By Lemma 1,

$$\left|\{s \in \sigma(j(j + 2c)) : Z(s, J)\}\right| \leq 2^{j(j+2c)+1-(j+2c)}.$$

Hence

(1) $$\left|\{s \in \sigma(j(j + 2c)) : \exists J \in \sigma(j)[J \text{ is composite } \& Z(s, J)]\}\right| \leq 2^{j(j+2c)+1-2c}.$$

Since any member s of the set S appearing in (1) can be calculated uniquely if we are given c and the ordinal number of the position of s in S expressed as a $j(j + 2c) + 1 - 2c$ bit string, it follows that

$$I(s) \leq j(j + 2c) + 1 - 2c + 2I(c) + O(1) \leq j(j + 2c) - 2c + O(\log c).$$

(The coefficient 2 in the term $2I(c)$ is present because when two strings are encoded into a single one by concatenating them, it is necessary to add information indicating where to separate them. The most straight-forward technique for providing punctuation doubles the length of the shorter string.) Hence if c is sufficiently large, no c-random $j(j + 2c)$ bit string can belong to S.

Proof of Theorem 2: Arguing as in the proof of Theorem 1, let J be a non-prime integer j bits long such that $I(J) \leq i$. By Lemma 1,

(1′) $$\left|\{s \in \sigma(2j(i + c)) : Z(s, J)\}\right| \leq 2^{2j(i+c)+1-2(i+c)}.$$

Since any member s of the set S' appearing in (1′) can be calculated uniquely if we are given J and the ordinal number of the position of s in S' expressed as a $2j(i + c) + 1 - 2(i + c)$ bit string, it follows that

$$I(s) \leq 2j(i + c) + 1 - 2(i + c) + 2I(J) + O(1) \leq 2j(i + c) - 2c + O(1).$$

(The coefficient 2 in the term $2I(J)$ is present for the same reason as in the proof of Theorem 1.) Hence if c is sufficiently large, no c-random $2j(i + c)$ bit sequence can belong to S'.

Applications of the Foregoing Results

Let s be a probabilistically determined sequence in which 0's and 1's appear independently with probabilities α, $1 - \alpha$, where $0 < \alpha < 1$. Group s into successive pairs of bits, and then drop all 00 and 11 pairs and convert each 01 (respectively 10) pair into a 0 (respectively, a 1). This gives a sequence s' in which 0's and 1's appear independently with exactly equal probabilities. If s' is n bits long, then the probability that $I(s') < n - c$ is less than 2^{-c}; thus c-random sequences can be derived easily from probabilistic experiments. Theorem 2 gives the number of potential witnesses of compositeness which must be checked to ensure that primality for numbers of special form is determined correctly with high probability (or with certainty, if some oracle gave us a long bit string known to satisfy the randomness criterion of algorithmic information theory). Mersenne numbers $N = 2^n - 1$ only require checking $O(\log n) = O(\log\log N)$ potential witnesses. Fermat numbers

$$N = 2^{2^n} + 1$$

only require checking $O(\log n) = O(\text{logloglog } N)$ potential witnesses. Eisenstein-Bell[55] numbers

$$N = 2^{2^{2^{\cdots}} \ (n \ 2\text{'s altogether})} + 1$$

only require checking $O(\log n) = O(\log^k N)$ (for any k) potential witnesses. A number of the form $10^n + k$ only requires checking $O(\log n) + O(\log k)$ potential witnesses.

Concerning Theorem 1 it is worthwhile to remark the following: Using the extended Riemann Hypothesis, Miller was able to show that if n is composite, then the first natural number that is a witness of n's compositeness (under the Miller-Rabin version of the predicate W) is always less than $O((\log n)^2)$. In contrast, we only need to check a "certified" random sample of $\log_2 n + O(1)$ potential witnesses.

Additional Remarks

The central idea of the Solovay-Strassen and Miller-Rabin algorithms and of the preceding discussion can be expressed as follows: Consider a specific propositional formula F in n variables for which we somehow know that the percentage of satisfiability is greater than 75% or less than 25%. We wish to decide which of these two possibilities is in fact the case. The obvious way of deciding is to evaluate F at all 2^n possible n-tuples of the variables. But only $O(I(F))$ data points are necessary to decide which case holds by sampling, if one posses an algorithmically random sequence $O(n \ I(F))$ bits long. Thus one need only evaluate F for $O(I(F))$ n-tuples of its variables, if the random sample is "certified."

These algorithms would be even more interesting if it were possible to show that they are faster than any deterministic algorithms which accomplish the same task. Gill [12], [13] in fact attacked the problem of showing that there are tasks which can be accomplished faster by a Monte Carlo algorithm than deterministically, before the current surge of interest in these matters caused by the discovery of several probabilistic algorithms which are much better than any known deterministic ones for the same task.

The discussion of extensible formal systems given in [14] raises the question of how to find systematic sources of new axioms, likely to be consistent with the existing axioms of logic and set theory, which can shorten the proofs of interesting theorems. From the metamathematical results of [1]-[3], we know that no statement of the form "s is c-random" can be proved if s has a length significantly greater than c. This raises the question of whether statements of the form "s is c-random" are generally useful new axioms. (Note that Ehrenfeucht and Mycielski [15] show that by adding any previously unprovable statement X to a formal system, one always shortens very greatly the lengths of infinitely many proofs. Their argument is roughly as follows: Consider a proposition of the form "either X or algorithm A halts," where A in fact halts but takes a very long time to do so. Previously the proof of this assertion was very long; one had to simulate A's computation until it halted. Now the proof is immediate, for X is an axiom. See also Gödel [16].)

---

55   Quotation from Bell [11]: "F. M. G. Eisenstein (1823-1852), a first-rate arithmetician, stated (1844) as a problem that there are an infinity of primes in the sequence

$$2^2 + 1, \ 2^{2^2} + 1, \ 2^{2^{2^2}} + 1, \ \ldots .$$

Doubtless he had a proof. This looks like the sort of thing an ingenious amateur might settle. If anyone asks why I have not done it myself – I am neither an amateur nor ingenious."

Hence it is reasonable to ask whether the addition of axioms "*s* is *c*-random" is likely either to allow interesting new theorems to be proved, or to shorten the proof of interesting theorems which could have been proved anyhow (but perhaps by unreachably long proofs). The following discussion of this issue is very informal and is intended to be merely suggestive. On the one hand, it is easy to see that interesting new theorems are probably not obtained in this manner. The argument is as follows. It it were highly probable that a particular theorem *T* can be deduced from axioms of the form "*s* is *c*-random," then *T* could in fact be proved without extending the axiom system. For even without extending the axiom system one could show that "if *s* is random, then *T*" holds for many *s*, and thus *T* would follow from the fact that most *s* are indeed random. In other words, we would have before us a proof by cases in which we do not know which case holds, but can show that most do. Hence it seems that interesting new theorems will probably not be obtained by extending a formal system in this way.

As to the possibility of interesting proof-shortenings, we can note that Ehrenfeucht-Mycielski theorems are not very interesting ones. Quick Monte Carlo algorithms for primality suggest another possibility. Perhaps adding axioms of the form "*s* is random" makes it possible to obtain shorter proofs of primality? Pratt's work [17] suggests caution, but the following more general conjecture seems reasonable. If it is in fact the case that for some tasks Monte Carlo algorithms are much better than deterministic ones, then it may also be the case that some interesting theorems have much shorter proofs when a formal system is extended by adding axioms of the form "*s* is random."

Bibliography

1. Chaitin, G. J., *Information-theoretic computational complexity,* IEEE Trans. Info. Theor. IT-20, 1974, pp. 10-15.

2. Chaitin, G. J., *Information-theoretic limitations of formal systems,* J. ACM 21, 1974, pp. 403-424.

3. Chaitin, G. J., *Randomness and mathematical proof,* Sci. Amer. 232, 5, May 1975, pp. 47-52.

4. Schwartz, J. T., *Complexity of statement, computation and proof,* AMS Audio Recordings of Mathematical Lectures 67, 1972.

5. Levin, M., *Mathematical logic for computer scientists,* MIT Project MAC TR-131, June 1974, pp. 145-147, 153.

6. Davis, M., *What is a computation?* in *Mathematics Today – Twelve Informal Essays,* Springer-Verlag, New York, to appear in 1978.

7. Chaitin, G. J., *Algorithmic information theory,* IBM J. Res. Develop. 21, 1977, pp. 350-359, 496.

8. Solovay, R., and Strassen, V., *A fast Monte-Carlo test for primality,* SIAM J. Comput. 6, 1977, pp. 84-85.

9. Miller, G. L., *Riemann's hypothesis and tests for primality,* J. Comput. Syst. Sci. 13, 1976, pp. 300-317.

10. Rabin, M. O., *Probabilistic algorithms* in *Algorithms and Complexity – New Directions and Recent Results,* J. F. Traub (ed.), Academic Press, New York, 1976, pp. 21-39.

11. Bell, E. T., *Mathematics – Queen and Servant of Science,* McGraw-Hill, New York, 1951, pp. 225-226.

12. Gill, J. T. III, *Computational complexity of probabilistic Turing machines,* Proc. 6th Annual ACM Symp. Theory of Computing, Seattle, Washington, April 1974, pp. 91-95.

13. Gill, J. T. III, *Computational complexity of probabilistic Turing machines*, SIAM J. Comput. 6, 1977, pp. 675-695.

14. Davis, M., and Schwartz, J. T., *Correct-Program Technology/Extensibility of Verifiers — Two Papers on Program Verification*, Courant Computer Science Report #12, Courant Institute of Mathematical Sciences, New York University, September 1977.

15. Ehrenfeucht, A., and Mycielski, J., *Abbreviating proofs by adding new axioms*, AMS Bull. 77, 1971, pp. 366-367.

16. Gödel, K., *On the length of proofs* in *The Undecidable — Basic Papers on Undecidable Propositions, Unsolvable Problems and Computable Functions*, M. Davis (ed.), Raven Press, Hewlett, New York, 1965, pp. 82-83.

17. Pratt, V. R., *Every prime has a succinct certificate*, SIAM J. Comput. 4, 1975, pp. 214-220.

Received January, 1978.

INFORMATION-THEORETIC CHARACTERIZATIONS OF RECURSIVE INFINITE STRINGS

Theoretical Computer Science 2 (1976), pp. 45-48.

Gregory J. CHAITIN
IBM Thomas J. Watson Research Center,
Yorktown Heights, N.Y. 10598, USA

Communicated by A. Meyer
Received November 1974
Revised March 1975

Abstract

Loveland and Meyer have studied necessary and sufficient conditions for an infinite binary string x to be recursive in terms of the program-size complexity relative to n of its n-bit prefixes x_n. Meyer has shown that x is recursive iff $\exists c$, $\forall n$, $K(x_n/n) \leq c$, and Loveland has shown that this is false if one merely stipulates that $K(x_n/n) \leq c$ for infinitely many n. We strengthen Meyer's theorem. From the fact that there are few minimal-size programs for calculating a given result, we obtain a necessary and sufficient condition for x to be recursive in terms of the absolute program-size complexity of its prefixes: x is recursive iff $\exists c$, $\forall n$, $K(x_n) \leq K(n) + c$. Again Loveland's method shows that this is no longer a sufficient condition for x to be recursive if one merely stipulates that $K(x_n) \leq K(n) + c$ for infinitely many n.

$N = \{0, 1, 2, \ldots \}$ is the set of natural numbers, $S = \{\Lambda, 0, 1, 00, 01, 10, 11, 000, \ldots \}$ is the set of strings, and X is the set of infinite strings. All strings and infinite strings are binary. The variables c, i, m and n range over N; the variables p, q, s and t range over S; and the variable x ranges over X.

$|s|$ is the length of a string s, and s_n and x_n are the prefixes of length n of s and x. x is recursive iff there is a recursive function $f:N \to S$ such that $x_n = f(n)$ for all n. $B(n)$ is the nth element of S; the function $B:N \to S$ is a recursive bijection. The quantity $|B(n)| = \lfloor \log_2 (n + 1) \rfloor$ plays an important role in this paper.

A *computer* C is a partial recursive function $C : S \times S \to S$. $C(p, q)$ is the output resulting from giving C the program p and the data q. The *relative complexity* $K_C : S \times S \to N$ is defined as follows: $K_C(s/t) = \min |p| \; (C(p, t) = s)$. The complexity $K_C : S \to N$ is defined as follows: $K_C(s) = K_C(s/\Lambda)$. $K_C(s)$ is the length of the shortest program for calculating s on C without any data. A computer U is *universal* iff for each computer C there is a constant c such that $K_U(s/t) \leq K_C(s/t) + c$ for all s and t.

Pick a standard Gödel numbering of the partial recursive functions $C : S \times S \to S$, and define the computer U as follows. $U(0^i, q) = \Lambda$, and $U(0^i 1p, q) = C_i(p, q)$, where C_i is the ith computer. U is universal, and is our standard computer for measuring complexities. The "U" in "K_U" is henceforth omitted.

The following situation occurs in the proofs of our main theorems, Theorems 3 and 6. There is an algorithm A for enumerating a set of strings. There are certain inputs to A, e.g. n and m. $A(n, m)$ denotes the enumeration (ordered set) produced by A from the inputs n and m. And $\mathrm{ind}(s, A(n, m))$

denotes the index of s in the enumeration $A(n, m)$; $\text{ind}(., A(., .)) : S \times N \times N \rightarrow N$ is a partial recursive function. A key step in the proofs of Theorems 3 and 6 is that if $s \in A(n, m)$ and one knows A, n, m, and $\text{ind}(s, A(n, m))$, then one can calculate s.

Theorem 1:

(a) $\exists c,\ \forall s, t\ K(s/t) \leq |s| + c$.

(b) $\forall n, t\ \exists s,\ |s| = n$ and $K(s/t) \geq n$.

(c) There is a recursive function $f : S \times S \times N \rightarrow N$ such that $f(s, t, n) \geq f(s, t, n + 1)$ and $K(s/t) = \lim_{n \to \infty} f(s, t, n)$.

(d) $\exists c,\ \forall s,\ K(B(|s|)) - c \leq K(s) \leq |s| + c$.

Proof:

(a) There is a computer C such that $C(p, q) = p$ for all p and q. Thus $K_C(s/t) = |s|$, and $K(s/t) \leq K_C(s/t) + c = |s| + c$.

(b) There are 2^n strings s of length n, but only $2^n - 1$ programs for $U(., t)$ of length $< n$. Thus at least one s of length n needs a program of length $\geq n$.

(c) Since U is a partial recursive function, its graph $\{\langle p, q, U(p, q)\rangle\}$ is an r.e. set. Let U_n be the first n triples in a fixed recursive enumeration of the graph of U. Recall the upper bound $|s| + c$ on $K(s/t)$. Take

$$f(s, t, n) = \min \{|s| + c\} \cup \{|p| : \langle p, t, s \rangle \in U_n\}.$$

(d) Theorem 1(a) yields $K(s) = K(s/\Lambda) \leq |s| + c$. And there is a computer C such that $C(p, q) = B(|U(p, q)|)$. Thus $K_C(B(|s|)) \leq K(s)$ and $K(B(|s|)) \leq K(s) + c$. $\square$

Theorem 2:

(a) If x is recursive, then there is a c such that $K(x_n/B(n)) \leq c$ and $K(x_n) \leq K(B(n)) + c$ for all n.

(b) For each c there are only finitely many x such that $\forall n,\ K(x_n/B(n)) \leq c$, and each of these x is recursive (Meyer).

(c) There is a c such that nondenumerably many x have the property that $K(x_n/B(n)) \leq c$ and $K(x_n) \leq K(B(n)) + c$ for infinitely many n (Loveland).

Proof:

(a) By definition, if x is recursive there is a recursive function $f : N \rightarrow S$ such that $f(n) = x_n$. There is a computer C such that $C(p, q) = f(B^{-1}(q))$. Thus $K_C(x_n/B(n)) = 0$ and $K(x_n/B(n)) \leq c$. There is also a computer C such that $C(p, q) = f(B^{-1}(U(p, q)))$. Thus $K_C(x_n) = K(B(n))$ and $K(x_n) \leq K(B(n)) + c$.

(b) See [4, pp. 525-526].

(c) See [4, pp. 515-516]. $\square$

Theorem 3: Consider a computer D. A t-description of s is a string p such that $D(p, t) = s$. There is a recursive function $f_D : N \rightarrow N$ with the property that no string s has more than $f_D(n)$ t-descriptions of length $< K(s/t) + n$.

Proof: There are $< 2^n$ t-descriptions of length $< n$. Thus there are $< 2^{n-m}$ strings s with $\geq 2^m$ t-descriptions of length $< n$. Since the graph of D is an r.e. set, given n, m and t, one can recursively enumerate the $< 2^{n-m}$ strings having $\geq 2^m$ t-descriptions of length $< n$. Pick an algorithm $A(n, m, t)$ for doing this.

There is a computer C with the following property. Suppose s has $\geq 2^m$ t-descriptions of length $< n$. Then $s = C(0^{|p|}1pq, t)$, where $p = B(m)$ and q is the $\text{ind}(s, A(n, m, t))$ th string of length $n - m$. C recovers information from this program in the following order: $|B(m)|$, m, $n - m$, n, $\text{ind}(s, A(n, m, t))$, and s. Thus

$$K_C(s/t) \leq |0^{|p|}1pq| = |q| + 2|p| + 1 = n - m + 2|B(m)| + 1,$$

and $K(s/t) \leq n - m + 2|B(m)| + c$. Note that A, C and c all depend on D.

We restate this. Suppose s has $\geq 2^m$ t-descriptions of length $< K(s/t) + n$. Then $K(s/t) \leq K(s/t) + n - m + 2|B(m)| + c$. I.e., $m - 2|B(m)| - c \leq n$. Let $f_D(n)$ be two raised to the least m such that $m - 2|B(m)| - c > n$. s cannot have $\geq f_D(n)$ t-descriptions of length $< K(s/t) + n$. $\square$

Theorem 4: There is a recursive function $f_4 : N \to N$ with the property that for no n and m are there more than $f_4(m)$ strings s of length n with $K(s) < K(B(n)) + m$.

Proof: In Theorem 3 consider only Λ-descriptions and take D to be the computer defined as follows: $D(p, q) = B(|U(p, q)|)$. It follows that there is a recursive function $f_D : N \to N$ with the property that for no n and m are there more than $f_D(m)$ programs p of length $< K(B(n)) + m$ such that $|U(p, \Lambda)| = n$. $f_4 = f_D$ is the function whose existence we wished to prove. $\square$

Theorem 5:

(a) There is an algorithm $A_5(n, m)$ that enumerates the prefixes of length n of strings s of length $2n$ such that $K(s_i) \leq |B(i)| + m$ for all $i \in [n, 2n]$.

(b) There is a recursive function $f_5 : N \to N$ with the property that $A_5(n, m)$ never has more than $f_5(m)$ elements.

Proof:

(a) The existence of A_5 is an immediate consequence of the fact that $K(s)$ can be recursively approximated arbitrarily closely from above (Theorem 1(c)).

(b) By using the counting argument that established Theorem 1(b), it also follows that $\exists c$, $\forall n$, $\exists i \in [n, 2n]$ such that $K(B(i)) > |B(i)| - c$. For such i the condition that $K(s_i) \leq |B(i)| + m$ implies $K(s_i) < K(B(i)) + m + c$, which by Theorem 4 can hold for at most $f_4(m + c)$ different strings s_i of length i. Thus there are at most $f_5(m)$ elements in $A_5(n, m)$, where $f_5(m) = f_4(m + c)$. $\square$

Theorem 6: For each c there are only finitely many x such that $\forall n$, $K(x_n) \leq |B(n)| + c$, and each of these x is recursive.

Proof: By hypothesis x has the property that $\forall n$, $K(x_n) \leq |B(n)| + c$. Thus $\forall n$, $x_n \in A_5(n, c)$, where A_5 is the enumeration algorithm of Theorem 5(a). There is a computer C (depending on c) such that

$$x_n = C(B(\text{ind}(x_n, A_5(n, c))), B(n)) \text{ for all } n.$$

Thus

$$K_C(x_n/B(n)) \leq |B(\mathrm{ind}(x_n, A_5(n, c)))| \leq |B(f_5(c))| \quad \text{by Theorem 5(b).}$$

As $K_C(x_n/B(n)) \leq |B(f_5(c))|$ for all n, it follows that there is a c' such that $K(x_n/B(n)) \leq c'$ for all n. Applying Theorem 2(b) we conclude that x is recursive and there can only be finitely many such x. $\square$

Theorem 7: x is recursive iff $\exists c, \forall n, K(x_n) \leq K(B(n)) + c$.

Proof: The "only if" is Theorem 2(a). The "if" follows from Theorem 6 and the fact that $\exists c, \forall n, K(x_n) \leq K(B(n)) + c$ implies $\exists c, \forall n, K(x_n) \leq |B(n)| + c$, which is an immediate consequence of Theorem 1(a). $\square$

Can this information-theoretic characterization of recursive infinite strings be reformulated in terms of other definitions of program-size complexity? It is easy to see that Theorem 7 also holds for Schnorr's process complexity [5]. This is not the case for the algorithmic entropy H (see [3]). Although recursive x satisfy $\exists c, \forall n, H(x_n) \leq H(B(n)) + c$, Solovay (private communication) has announced there is a nonrecursive x that also has this property.

Theorems 6 and 7 reveal a complexity gap, because $K(B(n))$ is sometimes much smaller than $|B(n)|$.

References

1. G. J. Chaitin, Information-theoretic aspects of the Turing degrees, Abstract 72T-E77, *AMS Notices* **19** (1972) A-601, A-602.

2. G. J. Chaitin, There are few minimal descriptions, A necessary and sufficient condition for an infinite binary string to be recursive, (abstracts), *Recursive Function Theory Newsletter* (January 1973) 13-14.

3. G. J. Chaitin, A theory of program size formally identical to information theory, *J. ACM* **22** (1975) 329-340.

4. D. W. Loveland, A variant of the Kolmogorov concept of complexity, *Information and Control* **15** (1969) 510-526.

5. C. P. Schnorr, Process complexity and effective random tests, *J. Comput. System Sci.* **7** (1973) 376-388.

PROGRAM SIZE, ORACLES, AND THE JUMP OPERATION

Osaka Journal of Mathematics 14 (1977), pp. 139-149.

Gregory J. CHAITIN

(Received January 17, 1976)

Abstract

There are a number of questions regarding the size of programs for calculating natural numbers, sequences, sets, and functions, which are best answered by considering computations in which one is allowed to consult an oracle for the halting problem. Questions of this kind suggested by work of T. Kamae and D. W. Loveland are treated.

1. Computer Programs, Oracles, Information Measures, and Codings

In this paper we use as much as possible Rogers' terminology and notation [1, pp. xv-xix]. Thus $N = \{0, 1, 2, \ldots \}$ is the set of (natural) numbers; $i, j, k, n, v, w, x, y, z$ are elements of N; A, B, X are subsets of N; f, g, h are functions from N into N; φ, ψ are partial functions from N into N; $\langle x_1, \ldots, x_k \rangle$ denotes the ordered k-tuple consisting of the numbers $x_1, \ldots, x_k$; the *lambda* notation $\lambda x[\ldots x \ldots]$ is used to denote the partial function of x whose value is $\ldots x \ldots$; and the *mu* notation $\mu x[\ldots x \ldots]$ is used to denote the least x such that $\ldots x \ldots$ is true.

The size of the number x, denoted $lg(x)$, is defined to be the number of bits in the xth binary string. The binary strings are $\Lambda, 0, 1, 00, 01, 10, 11, 000, \ldots$ Thus $lg(x)$ is the integer part of $\log_2 (x + 1)$. Note that there are 2^n numbers x of size n, and $2^n - 1$ numbers x of size less than n.

We are interested in the size of programs for a certain class of computers. The zth computer in this class is defined in terms of $\varphi_z^{(2)X}$ [1, pp. 128-134], which is the two-variable partial X-recursive function with Gödel number z. These computers use an oracle for deciding membership in the set X, and the zth computer produces the output $\varphi_z^{(2)X}(x, y)$ when given the program x and the data y. Thus the output depends on the set X as well as the numbers x and y.

We now choose the standard universal computer U that can simulate any other computer. U is defined as follows:

$$U^X((2x + 1)2^z - 1, y) = \varphi_z^{(2)X}(x, y).$$

Thus for each computer C there is a constant c such that any program of size n for C can be simulated by a program of size $\leq n + c$ for U.

Having picked the standard computer U, we can now define the program size measures that will be used throughout this paper.

The fundamental concept we shall deal with is $I(\psi/X)$, which is the number of bits of information needed to specify an algorithm relative to X for the partial function ψ, or, more briefly, the information in ψ relative to X. This is defined to be the size of the smallest program for ψ:

$$I(\psi/X) = \min lg(x) \ (\psi = \lambda y[U^X(x, y)]).$$

Here it is understood that $I(\psi/X) = \infty$ if ψ is not partial X-recursive.

$I(x \rightarrow y/X)$, which is the information relative to X to go from the number x to the number y, is defined as follows:

$$I(x \rightarrow y/X) = \min I(\psi/X) \ (\psi(x) = y).$$

And $I(x/X)$, which is the information in the number x relative to the set X, is defined as follows:

$$I(x/X) = I(0 \rightarrow x/X).$$

Finally $I(\psi/X)$ is used to define three versions $I(A/X)$, $I_r(A/X)$, and $I_f(A/X)$ of the information relative to X of a set A. These correspond to the three ways of naming a set [1, pp. 69-71]: by r.e. indices, by characteristic indices, and by canonical indices. The first definition is as follows:

$$I(A/X) = I(\lambda x[\text{if } x \in A \text{ then } 1 \text{ else undefined}]/X).$$

Thus $I(A/X) < \infty$ iff A is r.e. in X. The second definition is as follows:

$$I_r(A/X) = I(\lambda x[\text{if } x \in A \text{ then } 1 \text{ else } 0]/X).$$

Thus $I_r(A/X) < \infty$ iff A is recursive in X. And the third definition, which applies only to finite sets, is as follows:

$$I_f(A/X) = I(\sum_{x \in A} 2^x/X).$$

The following notational convention is used: $I(\psi)$, $I(x \rightarrow y)$, $I(x)$, $I(A)$, $I_r(A)$, and $I_f(A)$ are abbreviations for $I(\psi/\emptyset)$, $I(x \rightarrow y/\emptyset)$, $I(x/\emptyset)$, $I(A/\emptyset)$, $I_r(A/\emptyset)$, and $I_f(A/\emptyset)$, respectively.

We use the coding τ^* of finite sequences of numbers into individual numbers [1, p. 71]; τ^* is an effective one-one mapping from $\bigcup_{k=0}^{\infty} N^k$ onto N. And we also use the notation $\bar{f}(x)$ for τ^* of the sequence $\langle f(0), f(1), \ldots, f(x-1) \rangle$ [1, p. 377]; for any function f, $\bar{f}(x)$ is the code number for the finite initial segment of f of length x.

The following theorems, whose straight-forward proofs are omitted, give some basic properties of these concepts.

Theorem 1:

(a) $I(x/X) \leq lg(x) + c$
(b) There are less than 2^n numbers x with $I(x/X) < n$.
(c) $|I(x/X) - I(y/X)| \leq 2lg(|x - y|) + c$
(d) The set of all true propositions of the form "$I(x \rightarrow y/X) \leq z$" is r.e. in X.
(e) $I(x \rightarrow y/X) \leq I(y/X) + c$

Recall that there are 2^n numbers x of size n, that is, there are 2^n numbers x with $lg(x) = n$. In view of (a) and (b) most x of size n have $I(x/X) \approx n$. Such x are said to be X-random. In other words, x is said to be X-random if $I(x/X)$ is approximately equal to $lg(x)$; most x have this property.

Theorem 2:

(a) $I(\tau^*(\langle x, y \rangle)) \leq I(\tau^*(\langle y, x \rangle)) + c$
(b) $I(\tau^*(\langle x, y \rangle) \rightarrow \tau^*(\langle y, x \rangle)) \leq c$
(c) $I(x) \leq I(\tau^*(\langle x, y \rangle)) + c$
(d) $I(\tau^*(\langle x, y \rangle) \rightarrow x) \leq c$

Theorem 3:

(a) $I(x \rightarrow \psi(x)/X) \leq I(\psi/X)$

(b) For each ψ that is partial X-recursive there is a c such that $I(\psi(x)/X) \leq I(x/X) + c$.

(c) $I(x \rightarrow \bar{f}(x)/X) \leq I(f/X) + c$

(d) $I(\bar{f}(x) \rightarrow x/X) \leq c$ and $I(x/X) \leq I(\bar{f}(x)/X) + c$

Theorem 4:

(a) $I(x/X) \leq I(\lambda y[x]/X)$ and $I(\lambda y[x]/X) \leq I(x/X) + c$

(b) $I(x/X) \leq I_f(\{x\}/X) + c$ and $I_f(\{x\}/X) \leq I(x/X) + c$

(c) $I(x/X) \leq I_r(\{x\}/X) + c$ and $I_r(\{x\}/X) \leq I(x/X) + c$

(d) $I(x/X) \leq I(\{x\}/X) + c$ and $I(\{x\}/X) \leq I(x/X) + c$

(e) $I_r(A/X) \leq I_f(A/X) + c$ and $I(A/X) \leq I_r(A/X) + c$

See [2] for a different approach to defining program size measures for functions, numbers, and sets.

2. The Jump and Limit Operations

The jump X' of a set X is defined in such a manner that having an oracle for deciding membership in X' is equivalent to being able to solve the halting problem for algorithms relative to X [1, pp. 254-265].

In this paper we study a number of questions regarding the information $I(\psi)$ in ψ relative to the empty set, that are best answered by considering $I(\psi/\emptyset')$ and $I(\psi/\emptyset'')$, which are the information in ψ relative to the halting problem and relative to the jump of the halting problem. The thesis of this paper is that in order to understand $I(\psi/X)$ with $X = \emptyset$, which is the case of practical significance, it is sometimes necessary to jump higher in the arithmetical hierarchy to $X = \emptyset'$ or $X = \emptyset''$.

The following theorem, whose straight-forward proof is omitted, gives some facts about how the jump operation affects program size measures.

Theorem 5:

(a) $\lambda xy[I(x \rightarrow y/X)]$ and $\lambda x[I(x/X)]$ are X'-recursive.

(b) $I(\psi/X') \leq I(\psi/X) + c$

(c) For each n consider the least x such that $lg(x) \geq n$ and $I(x/X) \geq n$. This x has the property that $lg(x) = n$, $I(x/X) \leq n + c_1$, and $I(x/X') \leq I(n/X') + c_2 \leq lg(n) + c_3$.

(d) $I(\bar{A}/X') \leq I(A/X) + c$

(e) $I_r(A/X') \leq I(A/X) + c$

(f) If A is finite $I_f(A/X') \leq I(A/X) + c$

(g) $I(X'/X) \leq c$ and $I_r(X'/X) = \infty$

It follows from (b) that X'-randomness implies X-randomness. However (c) shows that the converse is false: there are X-random numbers that are not at all X'-random.

Having examined the jump operation, we now introduce the limit operation. The following theorem shows that the limit operation is in a certain sense analogous to the jump operation. This theorem is the tool we shall use to study work of Kamae and Loveland in the following sections.

Definition: Consider a function f. $\lim_x f(x)$ denotes a number z having the property that there is an x_0 such that $f(x) = z$ if $x \geq x_0$. If no such z exists, $\lim_x f(x)$ is undefined. In other words $\lim_x f(x)$ is the value that $f(x)$ assumes for almost all x (if there is such a value).

Theorem 6:

(a) If $I(z/X') < n$, then there is a function f such that $z = \lim_x f(x)$ and $I(f/X) < n + c$.

(b) If $I(f/X) < n$ and $\lim_x f(x) = z$, then $I(z/X') < n + c$.

Proof:

(a) By hypothesis there is a program w of size less than n such that $U^{X'}(w, 0) = z$. Given w and an arbitrary number x, one calculates $f(x)$ using the oracle for membership in X as follows. Choose a fixed algorithm relative to X for enumerating X'.

One performs x steps of the computation $U^{X'}(w, 0)$. This is done using a fake oracle for X' that answers that v is in X' iff v is obtained during the first x steps of the algorithm relative to X for enumerating X'. If a result is obtained by performing x steps of $U^{X'}(w, 0)$ in this manner, that is the value of $f(x)$. If not, $f(x)$ is 0.

It is easy to see that $\lim_x f(x) = U^{X'}(w, 0) = z$ and $I(f/X) \leq lg(w) + c < n + c$.

(b) By hypothesis there is a program w of size less than n such that $\lim_x U^X(w, x) = z$. Given w one can use the oracle for X' to calculate z. At stage i one asks the oracle whether there is a $j > i$ such that $U^X(w, j) \neq U^X(w, i)$. If so, one goes to stage $i + 1$ and tries again. If not, one is finished because $U^X(w, i) = z$. This shows that $I(z/X') \leq lg(w) + c < n + c$. Q.E.D.

See [3] for applications of oracles and the jump operation in the context of self-delimiting programs for sets and probability constructs; in this paper we are only interested in programs with endmarkers.

3. The Kamae Information Measure

In this section we study an information measure $K(x)$ suggested by work of Kamae [4] (see also [5]).

$I(y \to x)$ is less than or equal to $I(x) + c$, and it is natural to call $I(x) - I(y \to x)$ the degree to which y is helpful to x. Let us look at some examples. By definition $I(x) = I(0 \to x)$, and so 0 is no help at all. On the other hand some y are very helpful: $I(y \to x) < c$ for all those y whose prime factorization has 2 raised to the power x. Thus every x has infinitely many y that are extremely helpful to it.

Kamae proves in [4] that for each c there is an x such that $I(y \to x) < I(x) - c$ holds for almost all y. In other words, for each c there is an x such that almost all y are helpful to x more than c. This is surprising; one would have expected there to be a c with the property that every x has infinitely many y that are helpful to it less than c, that is, infinitely many y with $I(y \to x) > I(x) - c$.

We shall now study how $I(y \to x/X)$ varies when x is held fixed and y goes to infinity. Note that $I(y \to x/X)$ is bounded (in fact, by $I(x/X) + c$). This suggests the following definition: $K(x/X)$ is the greatest w such that $I(y \to x/X) = w$ holds for infinitely many y. In other words, $K(x/X)$ is the least v such that $I(y \to x/X) \leq v$ holds for almost all y.

Note that there are less than 2^n numbers x with $K(x/X) < n$, so that $K(x/X)$ clearly measures bits of information in some sense. The trivial inequality $K(x/X) \leq I(x/X) + c$ relates $K(x/X)$ and $I(x/X)$, but the following theorem shows that $K(x/X)$ is actually much more intimately related to the information measures $I(x/X')$ and $I(x/X'')$ than to $I(x/X)$.

Theorem 7:

(a) $K(x/X) \leq I(x/X') + c$

PROGRAM SIZE, ORACLES AND THE JUMP OPERATION

(b) $I(x/X'') \le K(x/X) + c$

Proof:

(a) Consider a number x_0. By Theorem 6a there is a function f such that $\lim_y f(y) = x_0$ and $I(f/X) \le I(x_0/X') + c$. Hence $I(y \to f(y)/X) \le I(f/X) \le I(x_0/X') + c$. In as much as $f(y) = x_0$ for almost all y, it follows that $I(y \to x_0/X) \le I(x_0/X') + c$ for almost all y. Hence $K(x_0/X) \le I(x_0/X') + c$.

(b) By using an oracle for membership in X' one can decide whether or not $I(y \to x/X) < n$. Thus by using an oracle for membership in X'' one can decide whether or not y_0 has the property that $I(y \to x/X) < n$ for all $y \ge y_0$. It follows that the set $A_n = \{x \mid K(x/X) < n\}$ is r.e. in X'' uniformly in n.

Suppose that $x_0 \in A_n$. Consider $j = 2^n + k$, where $k=$(the position of x_0 in a fixed X''-recursive enumeration of A_n uniform in n). Since there are less than 2^n numbers in A_n, $k < 2^n$ and $2^n \le j < 2^{n+1}$. Thus one can recover from j the values of n and k. And if one is given j one can calculate x_0 using an oracle for membership in X''. Thus if $K(x_0/X) < n$, then $I(x_0/X'') \le lg(j) + c_1 \le n + c_2$. Q.E.D.

What is the significance of this theorem? First of all, note that most x are $\emptyset''$-random and thus have $lg(x) \approx I(x/\emptyset'') \approx I(x/\emptyset') \approx I(x) \approx K(x)$. In other words, there is a c such that every n has the property that at least 99% of the x of size n have all four quantities $I(x/\emptyset'')$, $I(x/\emptyset')$, $I(x)$, and $K(x)$ inside the interval between $n - c$ and $n + c$. These x are "normal" because there are infinitely many y that do not help x at all, that is, there are infinitely many y with $I(y \to x) > I(x) - c$.

Now let us look at the other extreme, at the "abnormal" x discovered by Kamae.

Consider the first $\emptyset$-random number of size n, where n itself is $\emptyset''$-random. More precisely, let x_n be the first x such that $lg(x) = n$ and $I(x) \ge n$. (There is such an x_n, because there are 2^n numbers x of size n, and at most $2^n - 1$ of these x have $I(x) < n$.) Moreover, we stipulate that n itself be $\emptyset''$-random, so that $I(n/\emptyset'') \approx lg(n)$.

It is easy to see that these x_n have the property that $lg(n) \approx I(x_n/\emptyset'') \approx I(x_n/\emptyset') \approx K(x_n)$ and $I(x_n) \approx lg(x_n) = n$. Thus most y help these x_n a great deal, because $I(x_n) \approx n$ and for almost all y, $I(y \to x_n) \lesssim log_2 n$.

Theorem 7 enables us to make very precise statements about $K(x)$ when $I(x/\emptyset'') \approx I(x/\emptyset')$. But where is $K(x)$ in the interval between $I(x/\emptyset'')$ and $I(x/\emptyset')$ when this interval is wide? The following theorem shows that if $I(x/\emptyset'')$ and $I(x/\emptyset')$ are many orders of magnitude apart, then $K(x)$ will be of the same order of magnitude as $I(x/\emptyset')$. To be more precise, Theorems 7a and 8 show that

$$\frac{1}{2}I(x/\emptyset') - c \le K(x) \le I(x/\emptyset') + c.$$

Theorem 8: If $K(x_0/X) < n$, then $I(x_0/X') < 2n + c$.

Proof: Consider a fixed number n and a fixed set X. Let $\#_x$ be the cardinality of the set $B_x = \{z \mid I(x \to z/X) < n\}$. Note that $\#_x$ is bounded (in fact, by $2^n - 1$). Let i be the greatest w such that $\#_x = w$ holds for infinitely many x, which is also the least v such that $\#_x \le v$ holds for almost all x. Let $j = \mu z[\#_x \le i \text{ if } x \ge z]$, and let A be the infinite set of x greater than or equal to j such that $\#_x = i$. Thus B_x has exactly i elements if $x \in A$.

It is not difficult to see that if one knows n and i, then one can calculate j by using an oracle for membership in X'. And if one knows n, i, and j, by using an oracle for membership in X one can enumerate A and simultaneously calculate for each $x \in A$ the canonical index $\Sigma 2^z$ ($z \in B_x$) of the i-element set B_x.

Define $J(x)$ as follows: $J(x) =$ (the greatest w such that $I(y \to x/X) = w$ holds for infinitely many $y \in A$) = (the least v such that $I(y \to x/X) \leq v$ holds for almost all $y \in A$). It is not difficult to see from the previous paragraph that if one is given n and i and uses an oracle for membership in X', one can enumerate the set of all x such that $J(x) < n$.

Note that there are less than 2^n numbers x with $J(x) < n$, and that if $K(x/X) < n$, then $J(x) < n$. Suppose that x_0 has the property that $K(x_0/X) < n$. Consider the number $k = (2^n + i)2^n + i_2$, where $i_2 =$ (the position of x_0 in the above-mentioned X'-recursive enumeration of $\{x \mid J(x) < n\}$). Since $i < 2^n$ and $i_2 < 2^n$, one can recover n, i, and i_2 from k.

It is not difficult to see that if one is given k, then one can calculate x_0 using an oracle for membership in X'. Thus if $K(x_0/X) < n$, then $I(x_0/X') \leq lg(k) + c_1 < 2n + c_2$. Q.E.D.

4. The Loveland Information Measure

Define $L(f/X)$ to be $\max_x I(x \to \bar{f}(x)/X)$, and to be ∞ if $I(x \to \bar{f}(x)/X)$ is unbounded. This concept is suggested by work of Loveland [6]. Since there are less than 2^n functions f with $L(f/X) < n$, it is clear that in some sense $L(f/X)$ measures bits of information. $I(x \to \bar{f}(x)/X)$ is bounded if f is X-recursive, and conversely A. R. Meyer [6, pp. 525-526] has shown that if $I(x \to \bar{f}(x)/X)$ is bounded then f is X-recursive. Thus $L(f/X) < \infty$ iff $I(f/X) < \infty$.

But can something more precise be said about the relationship between $L(f)$ and $I(f)$? $L(f) \leq I(f) + c$, but as is pointed out in [6, p. 515], the proof that $I(f) < \infty$ if $L(f) < \infty$ is nonconstructive and does not give an upper bound on $I(f)$ in terms of $L(f)$. We shall show that in fact $I(f)$ can be enormous for reasonable values of $L(f)$. The proof that $I(f) < \infty$ if $L(f) < \infty$ may therefore be said to be extremely nonconstructive.

In [7] it is shown that $I(f) < \infty$ iff there is a c such that $I(\bar{f}(x)) - I(x) \leq c$ for all x. This result it now also seen to be extremely nonconstructive, because $I(f)$ may be enormous for reasonable c.

Furthermore, R. M. Solovay has studied in [8] what is the situation if the endmarker program size measure I used here is replaced by the self-delimiting program size measure H of [9]. He shows that there is a nonrecursive function f such that $H(\bar{f}(x)) - H(x)$ is bounded. This result previously seemed to contrast sharply with the fact that f is recursive if $I(x \to \bar{f}(x))$ is bounded [6] or if $I(\bar{f}(x)) - I(x)$ is bounded [7]. But now a harmonious whole is perceived since the sufficient conditions for f to be recursive just barely manage to keep $I(f)$ from being ∞.

Theorem 9: If $I(k/X') \leq n$, then there is a function f such that $L(f/X) \leq n + c$ and $I(f/X) \geq k - c$.

Proof: First we define the function g as follows: $g(x)$ is the first non-zero y such that $I(y/X) \geq x$. Note that g is X'-recursive.

By hypothesis $I(k/X') \leq n$. Hence $I(g(k)/X') \leq I(k/X') + c_1 \leq n + c_1$. By Theorem 6a, there is a function h such that $I(h/X) \leq n + c_2$ and $\lim_x h(x) = g(k)$. Let $x_0 = \mu z[h(x) = g(k)$ if $x \geq z]$. Thus $h(x) = g(k)$ if $x \geq x_0$.

The function f whose existence is claimed is defined as follows:

$$f(x) = \begin{cases} 0 & \text{if } x < x_0, \text{ and} \\ h(x) & \text{if } x \geq x_0. \end{cases}$$

Thus $f(x) = g(k)$ if $x \geq x_0$.

First we obtain a lower bound for $I(f/X)$. The following inequality holds for any function f:

$$I(f(\mu x[f(x) \neq 0])/X) \leq I(f/X) + c_3.$$

Hence for this particular f we see that $I(f/X) + c_3 \geq I(f(x_0)/X) = I(g(k)/X)$. Thus, by the definition of g, $I(f/X) + c_3 \geq I(g(k)/X) \geq k$.

Now to obtain an upper bound for $I(x \rightarrow \bar{f}(x)/X)$. There are two cases: $x \leq x_0$ and $x > x_0$. If $x \leq x_0$, then $\bar{f}(x)$ is the code number for a sequence of x 0's and thus $I(x \rightarrow \bar{f}(x)/X) \leq I(\lambda x[\tau^*(\langle 0 \rangle^x)]/X) = c_4$, where $\langle 0 \rangle^x$ denotes a sequence of x 0's. If $x > x_0$, then

$$
\begin{aligned}
I(x &\rightarrow \bar{f}(x)/X) \\
&\leq I(x \rightarrow \tau^*(\langle h(x), x_0 \rangle)/X) + c_5 \\
&= I(x \rightarrow \tau^*(\langle h(x), \mu z[h(x) = h(y) \text{ if } x \geq y \geq z]\rangle)/X) + c_5 \\
&\leq I(h/X) + c_6 \\
&\leq n + c_2 + c_6.
\end{aligned}
$$

Thus $I(x \rightarrow \bar{f}(x)/X)$ is either bounded by c_4 or by $n + c_2 + c_6$. Hence $I(x \rightarrow \bar{f}(x)/X) \leq n + c_7$ and $L(f/X) \leq n + c_7$.

To recapitulate, we have shown that this f has the property that $I(f/X) \geq k - c_3$ and $L(f/X) \leq n + c_7$. Taking $c = \max c_3, c_7$, we see that $I(f/X) \geq k - c$ and $L(f/X) \leq n + c$. Q.E.D.

Why does Theorem 9 show that $I(f)$ can be enormous even though $L(f)$ has a reasonable value? Consider the function $g(x)$ defined to be $(\ldots ((x!)!)! \ldots !)$ in which there are x !'s. $g(x)$ quickly becomes astronomical as x increases. However, $I(g(n)/\emptyset') \leq I(g(n)) + c_1 \leq I(n) + c_2 \leq \lg(n) + c_3$, and $\lg(n) + c_3$ is less than n for almost all n. Hence almost all n have the property that there is a function f with $L(f) \leq n + c$ and $I(f) \geq g(n) - c$.

In fact the situation is much worse. It is easy to define a function h that is $\emptyset'$-recursive and grows more quickly than any recursive function. In other words, h is recursive in the halting problem and for any recursive function g, $h(x) > g(x)$ for almost all x. As before we see that $I(h(n)/\emptyset') < n$ for almost all n. Hence almost all n have the property that there is a function f with $L(f) \leq n + c$ and $I(f) \geq h(n) - c$.

5. Other Applications

In this section some other applications of oracles and the jump operation are presented without proof.

First of all, we would like to examine a question raised by C. P. Schnorr [10, p. 189] concerning the relationship between $I(x)$ and the limiting relative frequency of programs for x. However, it is more appropriate to ask what is the relationship between the self-delimiting program size measure $H(x)$ [9] and the limiting relative frequency of programs for x (with endmarkers). Define $F(x, n)$ to be $-\log_2$ of (the number of programs w less than or equal to n such that $U^\emptyset(w, 0) = x)/(n + 1)$. Then Theorem 12 of [10] is analogous to the following:

Theorem 10: There is a c such that every x satisfies $F(x, n) \leq H(x) + c$ for almost all n.

This shows that if $H(x)$ is small, then x has many programs. Schnorr asks whether the converse is true. In fact it is not:

Theorem 11: There is a c such that every x satisfies $F(x, n) \leq H(x/\emptyset') + c$ for almost all n.

Thus even though $H(x)$ is large, x will have many programs if $H(x/\emptyset')$ is small.

We would like to end by examining the maximum finite cardinality $\#A$ and co-cardinality $\#\overline{A}$ attainable by a set A of bounded program size. First we define the partial function G:

$$G(x/X) = \max z \, (I(z/X) \leq x).$$

The following easily established results show how gigantic G is:

(a) If ψ is partial X-recursive and $x > I(\psi/X) + c$, then $\psi(x)$, if defined, is less than $G(x/X)$.

(b) If ψ is partial X-recursive, then there is a c such that $\psi(G(x/X))$, if defined, is less than $G(x + c/X)$.

Theorem 12:

(a) $G(x - c) < \max \#A \, (I_f(A) \leq x) < G(x + c)$

(b) $G(x - c/\emptyset') < \max \#A \, (I_r(A) \leq x) < G(x + c/\emptyset')$

(c) $G(x - c/\emptyset') < \max \#\overline{A} \, (I_r(A) \leq x) < G(x + c/\emptyset')$

(d) $G(x - c/\emptyset') < \max \#A \, (I(A) \leq x) < G(x + c/\emptyset')$

(e) $G(x - c/\emptyset'') < \max \#\overline{A} \, (I(A) \leq x) < G(x + c/\emptyset'')$

Here it is understood that the maximizations are only taken over those cardinalities which are finite.

The proof of (e) is beyond the scope of the method used in this paper; (e) is closely related to the fact that $\{x \mid W_x$ is co-finite$\}$ is Σ_3-complete [1, p. 328] and to Theorem 16 of [3].

Appendix

Theorem 3b can be strengthened to the following:

$$I(\psi(x)/X) \leq I(x/X) + I(\psi/X) + lg(I(\psi/X)) + lg(lg(I(\psi/X))) + 2lg(lg(lg(I(\psi/X)))) + c.$$

There are many other similar inequalities.

To formulate sharp results of this kind it is necessary to abandon the formalism of this paper, in which programs have endmarkers. Instead one must use the self-delimiting program formalism of [9] and [3] in which programs can be concatenated and merged. In that setting the following inequalities are immediate:

$$H(\psi(x)/X) \leq H(x/X) + H(\psi/X) + c,$$
$$H(\lambda x[\psi(\varphi(x))]/X) \leq H(\varphi/X) + H(\psi/X) + c.$$

IBM THOMAS J. WATSON RESEARCH CENTER

References

1. H. Rogers, Jr.: Theory of recursive functions and effective computability, McGraw-Hill, New York, 1967.

2. G. J. Chaitin: *Information-theoretic limitations of formal systems,* J. ACM. **21** (1974), 403-424.

3. G. J. Chaitin: *Algorithmic entropy of sets,* Comput. Math. Appl. **2** (1976), 233-245.

4. T. Kamae: *On Kolmogorov's complexity and information,* Osaka J. Math. **10** (1973), 305-307.

5. R. P. Daley: *A note on a result of Kamae,* Osaka J. Math **12** (1975), 283-284.

6. D. W. Loveland: *A variant of the Kolmogorov concept of complexity,* Information and Control **15** (1969), 510-526.

PROGRAM SIZE, ORACLES AND THE JUMP OPERATION

7. G. J. Chaitin: *Information-theoretic characterizations of recursive infinite strings*, Theoretical Comput. Sci. **2** (1976), 45-48.

8. R. M. Solovay: unpublished manuscript on [9] dated May 1975.

9. G. J. Chaitin: *A theory of program size formally identical to information theory*, J. ACM **22** (1975), 329-340.

10. C. P. Schnorr: *Optimal enumerations and optimal Gödel numberings*, Math. Systems Theory **8** (1975), 182-191.

11. G. J. Chaitin: *Algorithmic information theory*, IBM J. Res. Develop. **21** (1977), in press.

Part VI—Technical Papers on Turing Machines

ON THE LENGTH OF PROGRAMS FOR COMPUTING FINITE BINARY SEQUENCES

Journal of the ACM 13 (1966), pp. 547-569.

GREGORY J. CHAITIN[56]
The City College of the City University of New York, New York, N.Y.

Abstract

The use of Turing machines for calculating finite binary sequences is studied from the point of view of information theory and the theory of recursive functions. Various results are obtained concerning the number of instructions in programs. A modified form of Turing machine is studied from the same point of view. An application to the problem of defining a patternless sequence is proposed in terms of the concepts here developed.

Introduction

In this paper the Turing machine is regarded as a general purpose computer and some practical questions are asked about programming it. Given an arbitrary finite binary sequence, what is the length of the shortest program for calculating it? What are the properties of those binary sequences of a given length which require the longest programs? Do most of the binary sequences of a given length require programs of about the same length?

The questions posed above are answered in Part 1. In the course of answering them, the logical design of the Turing machine is examined as to redundancies, and it is found that it is possible to increase the efficiency of the Turing machine as a computing instrument without a major alteration in the philosophy of its logical design. Also, the following question raised by C. E. Shannon [1] is partially answered: What effect does the number of different symbols that a Turing machine can write on its tape have on the length of the program required for a given calculation?

In Part 2 a major alteration in the logical design of the Turing machine is introduced, and then all the questions about the lengths of programs which had previously been asked about the first computer are asked again. The change in the logical design may be described in the following terms: Programs for Turing machines may have transfers from any part of the program to any other part, but in the programs for the Turing machines which are considered in Part 2 there is a fixed upper bound on the length of transfers.

Part 3 deals with the somewhat philosophical problem of defining a random or patternless binary sequence. The following definition is proposed: Patternless finite binary sequences of a given length are sequences which in order to be computed require programs of approximately the same length as the longest programs required to compute any binary sequences of that given length. Previous work along these lines and its relationship to the present proposal are discussed briefly.

---

[56] This paper was written in part with the help of NSF Undergraduate Research Participation Grant GY-161.

Part 1

1.1. We define an *N*-state *M*-tape-symbol Turing machine by an *N*-row by *M*-column table. Each of the *NM* places in this table must have an entry consisting of an ordered pair (i, j) of natural numbers, where i goes from 0 to *N* and *j* goes from 1 to $M + 2$. These entries constitute, when specified, the program of the *N* -state *M*-tape-symbol Turing machine. They are to be interpreted as follows: An entry (i, j) in the *k* th row and the *p*th column of the table means that when the machine is in its *k*th state and the square of its one-way infinite tape which is being scanned is marked with the *p*th symbol, then the machine is to go to its *i*th state if $i \neq 0$ (the machine is to halt if $i = 0$) after performing the operation of

1.   moving the tape one square to the right if $j = M + 2$,
2.   moving the tape one square to the left if $j = M + 1$, and
3.   marking (overprinting) the square of the tape being scanned with the *j*th symbol if $1 \leq j \leq M$.

Special names are given to the first, second and third symbols. They are, respectively, the blank (for unmarked square), 0 and 1.

A Turing machine may be represented schematically as follows:

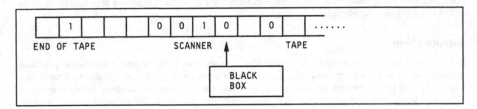

It is stipulated that

(1.1A)   Initially the machine is in its first state and scanning the first square of the tape.

(1.1B)   No Turing machine may in the course of a calculation scan the end square of the tape and then move the tape one square to the right.

(1.1C)   Initially all squares of the tape are blank.

Since throughout this paper we shall be concerned with computing finite binary sequences, when we say that a Turing machine calculates a particular finite binary sequence (say, 01111000), we shall mean that the machine stops with the sequence written at the end of the tape, with all other squares of the tape blank and with its scanner on the first blank square of the tape. For example, the following Turing machine has just calculated the sequence mentioned above:

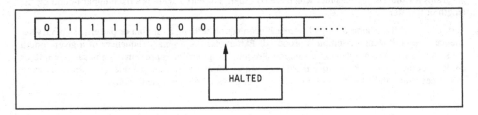

1.2. There are exactly

ON THE LENGTH OF PROGRAMS FOR COMPUTING FINITE BINARY SEQUENCES

$$((N + 1)(M + 2))^{NM}$$

possible programs for an N-state M-tape-symbol Turing machine. Thus to specify a single one of these programs requires

$$\log_2 (((N + 1)(M + 2))^{NM})$$

bits of information, which is asymptotic to $NM \log_2 N$ bits for M fixed and N large. Therefore a program for an N-state M-tape-symbol Turing machine (considering M to be fixed and N to be large) can be regarded as consisting of about $NM \log_2 N$ bits of information. It may be, however, that in view of the fact that different programs may cause the machine to behave in exactly the same way, a substantial portion of the information necessary to specify a program is redundant in its specification of the behavior of the machine. This in fact turns out to be the case. It will be shown in what follows that for M fixed and N large at least $1/M$ of the bits of information of a program are redundant. Later we shall be in a position to ask to what extent the remaining portion of $(1 - 1/M)$ of the bits is redundant.

The basic reason for this redundancy is that any renumbering of the rows of the table (this amounts to a renaming of the states of the machine) in no way changes the behavior that a given program will cause the machine to have. Thus the states can be named in a manner determined by the sequencing of the program, and this makes possible the omission of state numbers from the program. This idea is by no means new. It may be seen in most computers with random access memories. In these computers the address of the next instruction to be executed is usually 1 more than the address of the current instruction, and this makes it generally unnecessary to use memory space in order to give the address of the next instruction to be executed. Since we are not concerned with the practical engineering feasibility of a logical design, we can take this idea a step farther.

1.3. In the presentation of the redesigned Turing machine let us begin with an example of the manner in which one can take a program for a Turing machine and reorder its rows (rename its states) until it is in the format of the redesigned machine. In the process, several row numbers in the program are removed and replaced by + or ++ — this is how redundant information in the program is removed. The "operation codes" (which are 1 for "print blank," 2 for "print zero," 3 for "print one," 4 for "shift tape left" and 5 for "shift tape right") are omitted from the program; every time the rows are reordered, the op-codes are just carried along. The program used as an example is as follows:

```
row 1    1  9  7
row 2    8  8  8
row 3    9  6  1
row 4    3  2  0
row 5    7  7  8
row 6    6  5  4
row 7    8  6  9
row 8    9  8  1
row 9    9  1  8
```

To prevent confusion later, letters instead of numbers are used in the program:

```
row A      A    I    G
row B      H    H    H
row C      I    F    A
row D      C    B    J
row E      G    G    H
row F      F    E    D
row G      H    F    I
row H      I    H    A
row I      I    A    H
```

Row A is the first row of the table and shall remain so. Replace A by 1 throughout the table:

```
row 1      1    I    G
row B      H    H    H
row C      I    F    1
row D      C    B    J
row E      G    G    H
row F      F    E    D
row G      H    F    I
row H      I    H    1
row I      I    1    H
```

To find to which row of the table to assign the number 2, read across the first row of the table until a letter is reached. Having found an I,

1. replace it by a +,
2. move row I so that it becomes the second row of the table, and
3. replace I by 2 throughout the table:

```
row 1      1    +    G
row 2      2    1    H
row B      H    H    H
row C      2    F    1
row D      C    B    J
row E      G    G    H
row F      F    E    D
row G      H    F    2
row H      2    H    1
```

To find to which row of the table to assign the number 3, read across the second row of the table until a letter is found. Having found an H,

1. replace it by a +,
2. move row H so that it becomes the third row of the table, and
3. replace H by 3 throughout the table:

```
row 1      1    +    G
row 2      2    1    +
row 3      2    3    1
row B      3    3    3
row C      2    F    1
row D      C    B    J
row E      G    G    3
row F      F    E    D
row G      3    F    2
```

To find to which row of the table to assign the number 4, read across the third row of the table until a letter is found. Having failed to find one, read across rows 1, 2 and 3, respectively, until a letter is found. (A letter must be found, for otherwise rows 1, 2 and 3 are the whole program.) Having found a G in row 1,

1. replace it by a ++,
2. move row G so that it becomes the fourth row of the table, and
3. replace G by 4 throughout the table:

row 1	1	+	++
row 2	2	1	+
row 3	2	3	1
row 4	3	F	2
row B	3	3	3
row C	2	F	1
row D	C	B	J
row E	4	4	3
row F	F	E	D

The next two assignments proceed as in the case of rows 2 and 3:

row 1	1	+	++
row 2	2	1	+
row 3	2	3	1
row 4	3	+	2
row 5	5	E	D
row B	3	3	3
row C	2	5	1
row D	C	B	J
row E	4	4	3

row 1	1	+	++
row 2	2	1	+
row 3	2	3	1
row 4	3	+	2
row 5	5	+	D
row 6	4	4	3
row B	3	3	3
row C	2	5	1
row D	C	B	J

To find to which row of the table to assign the number 7, read across the sixth row of the table until a letter is found. Having failed to find one, read across rows 1, 2, 3, 4, 5 and 6, respectively, until a letter is found. (A letter must be found, for otherwise rows 1, 2, 3, 4, 5 and 6 are the whole program.) Having found a D in row 5,

1. replace it by a ++,
2. move row D so that it becomes the seventh row of the table, and
3. replace D by 7 throughout the table:

row 1	1	+	++
row 2	2	1	+
row 3	2	3	1
row 4	3	+	2
row 5	5	+	++
row 6	4	4	3
row 7	C	B	J
row B	3	3	3
row C	2	5	1

After three more assignments the following is finally obtained:

row 1	1	+	++
row 2	2	1	+
row 3	2	3	1
row 4	3	+	2
row 5	5	+	++
row 6	4	4	3
row 7	+	++	++
row 8	2	5	1
row 9	3	3	3
row 10			

This example is atypical in several respects: The state order could have needed a more elaborate scrambling (instead of which the row of the table to which a number was assigned always happened to be the last row of the table at the moment), and the fictitious state used for the purposes of halting (state 0 in the formulation of Section 1.1) could have ended up as any one of the rows of the table except the first row (instead of which it ended up as the last row of the table).

The reader will note, however, that 9 row numbers have been eliminated (and replaced by + or ++) in a program of 9 (actual) rows, and that, in general *this process will eliminate a row number from the program for each row of the program.* Note too that if a program is "linear" (i.e., the machine executes the instruction in storage address 1, then the instruction in storage address 2, then the instruction in storage address 3, etc.), only + will be used; departures from linearity necessitate use of ++.

There follows a description of the redesigned machine. In the formalism of that description the program given above is as follows:

```
10
(1, ,0)   (0, ,1)   (0, ,2)
(2, ,0)   (1, ,0)   (0, ,1)
(2, ,0)   (3, ,0)   (1, ,0)
(3, ,0)   (0, ,1)   (2, ,0)
(5, ,0)   (0, ,1)   (0, ,2)
(4, ,0)   (4, ,0)   (3, ,0)
(0, ,1)   (0, ,2)   (0, ,2)
(2, ,0)   (5, ,0)   (1, ,0)
(3, ,0)   (3, ,0)   (3, ,0)
```

Here the third member of a triple is the number of +'s, the second member is the op-code, and the first member is the number of the next state of the machine if there are no +'s (if there are +'s, the first member of the triple is 0). The number outside the table is the number of the fictitious row of the program used for the purposes of halting.

We define an N-state M-tape-symbol Turing machine by an $(N + 1) \times M$ table and a natural number n ($2 \leq n \leq N + 1$). Each of the $(N + 1)M$ places in this table (with the exception of those in the nth row) must have an entry consisting of an ordered triple (i, j, k) of natural numbers, where k is 0, 1 or 2; j goes from 1 to $M + 2$; and i goes from 1 to $N + 1$ if $k = 0$, $i = 0$ if $k \neq 0$. (Places in the nth row are left blank.) In addition:

(1.3.1)   The entries in which $k = 1$ or $k = 2$ are N in number.

Entries are interpreted as follows:

(1.3.2)   An entry $(i, j, 0)$ the pth row and the mth column of the table means that when the machine is in the pth state and the square of its one-way infinite tape which is being scanned is marked with the mth symbol, then the machine is to go to its ith state if $i \neq n$ (if $i = n$, the machine is instead to halt) after performing the operation of
1.   moving the tape one square to the right if $j = M + 2$,

2.   moving the tape one square to the left if $j = M + 1$, and

3.   marking (overprinting) the square of the tape being scanned with the jth symbol if
     $1 \le j \le M$.

(1.3.3)   An entry $(0, j, 1)$ in the pth row and m th column of the table is to be interpreted in accordance with (1.3.2) as if it were the entry $(p + 1, j, 0)$.

(1.3.4)   For an entry $(0, j, 2)$ in the pth row and m th column of the table the machine proceeds as follows:

    (1.3.4a)   It determines the number p' of entries of the form $(0, , 2)$ in rows of the table preceding the pth row or to the left of the mth column in the pth row.

    (1.3.4b)   It determines the first $p' + 1$ rows of the table which have no entries of the form $(0, , 1)$ or $(0, , 2)$. Suppose the last of these $p' + 1$ rows is the p''th row of the table.

    (1.3.4c)   It interprets the entry in accordance with (1.3.2) as if it were the entry $(p'' + 1, j, 0)$.

1.4. In Section 1.2 it was stated that the programs of the N-state M-tape-symbol Turing machines of Section 1.3 require in order to be specified $(1 - 1/M)$ the number of bits of information required to specify the programs of the N-state M-tape-symbol Turing machines of Section 1.1. (As before, M is regarded to be fixed and N to be large.) This assertion is justified here. In view of (1.3.1), at most

$$N(3(M + 2))^{NM}(N + 1)^{N(M-1)}$$

ways of making entries in the table of an N-state M-tape-symbol Turing machine of Section 1.3 count as programs. Thus only $\log_2$ of this number or asymptotically $N(M - 1) \log_2 N$ bits are required to specify the program of an N-state M-tape-symbol machine of Section 1.3.

Henceforth, in speaking of an N-state M-tape-symbol Turing machine, one of the machines of Section 1.3 will be meant.

1.5. We now define two sets of functions which play a fundamental role in all that follows.

The members $L_M(.)$ of the first set are defined for $M = 3, 4, 5, \ldots$ on the set of all finite binary sequences S as follows: An N-state M-tape-symbol Turing machine can be programmed to calculate S if and only if $N \ge L_M(S)$.

The second set $L_M(C_n)$ $(M = 3, 4, 5, \ldots)$ is defined by

$$L_M(C_n) = \max_S L_M(S),$$

where S is any binary sequence of length n.

Finally, we denote by $_MC_n$ $(M = 3, 4, 5, \ldots)$ the set of all binary sequences S of length n satisfying $L_M(S) = L_M(C_n)$.

1.6. In this section it is shown that for $M = 3, 4, 5, \ldots$,

$$L_M(C_n) \sim \frac{n}{(M - 1) \log_2 n}.$$

We first show that $L_M(C_n)$ is greater than a function of n which is asymptotically equal to $(n/((M - 1) \log_2 n))$. From Section 1.4 it is clear that there are at most

```
Row          Column Number
Number  1              2            3
-----------------------------------------
1       2 , 4        2 , 4        2, 4
2       ..., 2       ..., 3       3, 4
3       ..., 2       ..., 3       4, 4        Section I:
4       ..., 2       ..., 3       5, 4        approximately
5       ..., 2       ..., 3       6, 4        (1 - 1 / log2 N) N
6       ..., 2       ..., 3       7, 4        rows
7       ..., 2       ..., 3       8, 4
8       ..., 2       ..., 3       9, 4
.        .            .            .
.        .            .            .
.        .            .            .
d       d+1, 4       d+1, 4       d+1, 4
d+1     d+2, 4       d+2, 4       d+2, 4
d+2     d+3, 4       d+3, 4       d+3, 4       Section II:
d+3     d+4, 4       d+4, 4       d+4, 4       approximately
d+4     d+5, 4       d+5, 4       d+5, 4       N / log2 N
d+5     d+6, 4       d+6, 4       d+6, 4       rows
d+6     d+7, 4       d+7, 4       d+7, 4
d+7     d+8, 4       d+8, 4       d+8, 4
.        .            .            .
.        .            .            .
.        .            .            .
f       This section is the same
.       (except for the changes
.       in row numbers caused by     Section III:
.       relocation) regardless       a fixed number
.       of the value of N.           of rows
-----------------------------------------
```

Figure 11.　A three section program: This program is in the format of the machines of Section 1.1. There are N rows in this table. The unspecified row numbers in Section I are all in the range from d to $f - 1$, inclusive. The manner in which they are specified determines the finite binary sequence S which the program computes.

$$2^{((1 + \varepsilon_N)N(M-1)\log_2 N)}$$

different programs for an N-state M-tape-symbol Turing machine, where ε_x denotes a (not necessarily positive) function of x and possibly other variables which tends to zero as x goes to infinity with any other variables held fixed. Since a different program is required to calculate each of the 2^n different binary sequences of length n, we see that an N-state M-tape-symbol Turing machine can be programmed to calculate any binary sequence of length n only if

$$(1 + \varepsilon_N)N(M - 1)\log_2 N \geq n$$

or

$$N \geq (1 + \varepsilon_n)\frac{n}{(M - 1)\log_2 n}.$$

It follows from the definition of $L_M(C_n)$ that

$$L_M(C_n) \geq (1 + \varepsilon_n)\frac{n}{(M-1)\log_2 n}.$$

Next we show that $L_M(C_n)$ is less than a function of n which is asymptotically equal to $(n/((M-1)\log_2 n))$. This is done by showing how to construct for any binary sequence S of length not greater than $(1 + \varepsilon_N)N(M-1)\log_2 N$ a program which causes an N-state M-tape-symbol Turing machine to calculate S. The main idea is illustrated in the case where $M = 3$ (see Figure 11 on page 213).

The execution of this program is divided into phases. There are twice as many phases as there are rows in Section I. The current phase is determined by a binary sequence P which is written out starting on the second square of the tape. The nth phase starts in row 1 with the scanner on the first square of the tape and with

$$\begin{cases} P = 111 \ldots 1 \quad (i \text{ 1's}) \text{ if } n = 2i + 1, \\ P = 111 \ldots 10 \ (i \text{ 1's}) \text{ if } n = 2i + 2. \end{cases}$$

Control then passes down column three through the $(i + 1)$-th row of the table, and then control passes to

$$\begin{cases} \text{row } i + 2, \ \text{column 1 if } n = 2i + 1, \\ \text{row } i + 2, \ \text{column 2 if } n = 2i + 2, \end{cases}$$

which

1. changes P to what it must be at the start of the $(n + 1)$-th phase, and
2. transfers control to a row in Section II.

Suppose this row to be the mth row of Section II from the end of Section II.

Once control has passed to the row in Section II, control then passes down Section II until row f is reached. Each row in Section II causes the tape to be shifted one square to the left, so that when row f finally assumes control, the scanner will be on the mth blank square to the right of P. The following diagram shows the way things may look at this point if n is 7 and m happens to be 11:

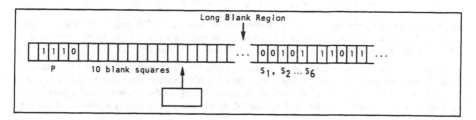

Now control has been passed to Section III. First of all, Section III accumulates in base-two on the tape a count of the number of blank squares between the scanner and P when f assumes control. (This number is $m - 1$.) This base-two count, which is written on the tape, is simply a binary sequence with a 1 at its left end. Section III then removes this 1 from the left end of the binary sequence. The resulting sequence is called S_n.

Note that if the row numbers entered in

$$\begin{cases} \text{row } i + 2, \text{ column 1 if } n = 2i + 1, \\ \text{row } i + 2, \text{ column 2 if } n = 2i + 2, \end{cases}$$

of Section I are suitably specified, this binary sequence S_n can be made any one of the 2^v binary sequences of length $v =$ (the greatest integer not greater than $\log_2 (f - d) - 1$). Finally, Section III writes S_n in a region of the tape far to the right where all the previous S_j $(j = 1, 2, \ldots, n - 1)$ have been written during previous phases, cleans up the tape so that only the sequences P and S_j $(j = 1, 2, 3, \ldots, n)$ remain on it, positions the scanner back on the square at the end of the tape and, as the last act of phase n, passes control back to row 1 again.

The foregoing description of the workings of the program omits some important details for the sake of clarity. These follow.

It must be indicated how Section III knows when the last phase (phase $2(d - 2)$) has occurred. During the nth phase, P is copied just to the right of $S_1, S_2, \ldots, S_n$ (of course a blank square is left between S_n and the copy of P). And during the $(n + 1)$-th phase, Section III checks whether or not P is currently different from what it was during the nth phase when the copy of it was made. If it isn't different, then Section III knows that phasing has in fact stopped and that a termination routine must be executed.

The termination routine first forms the finite binary sequence S^* consisting of

$$S_1, S_2, \ldots, S_{2(d-2)},$$

each immediately following the other. As each of the S_j can be any one of the 2^v binary sequences of length v if the row numbers in the entries in Section I are appropriately specified, it follows that S^* can be any one of the 2^w binary sequences of length $w = 2(d - 2)v$. Note that

$$2(d - 2)(\log_2 (f - d) - 1) \geq w > 2(d - 2)(\log_2 (f - d) - 2),$$

so that

$$w \sim 2\left((1 - \frac{1}{\log_2 N})N \right) \left(\log_2 \frac{N}{\log_2 N} \right) \sim 2N \log_2 N.$$

As we want the program to be able to compute any sequence S of length not greater than $(2 + \varepsilon_N)N \log_2 N$, we have S^* consist of S followed to the right by a single 1 and then a string of 0's, and the termination routine removes the rightmost 0's and first 1 from S^*. Q.E.D.

The result just obtained shows that it is impossible to make further improvement in the logical design of the Turing machine of the kind described in Section 1.2 and actually effected in Section 1.3; if we let the number of tape symbols be fixed and speak asymptotically as the number of states goes to infinity, in our present Turing machines 100 percent of the bits required to specify a program also serve to specify the behavior of the machine.

Note too that the argument presented in the first paragraph of this section in fact establishes that, say, for any fixed s greater than zero, at most $n^{-s} 2^n$ binary sequences S of length n satisfy

$$L_M(S) \leq (1 + \varepsilon_n)\frac{n}{(M - 1) \log_2 n}.$$

Thus we have: For any fixed s greater than zero, at most $n^{-s} 2^n$ binary sequences of length n fail to satisfy the double inequality

ON THE LENGTH OF PROGRAMS FOR COMPUTING FINITE BINARY SEQUENCES

$$(1 + \varepsilon_n)\frac{n}{(M - 1) \log_2 n} \leq L_M(S) \leq (1 + \varepsilon_n')\frac{n}{(M - 1) \log_2 n}.$$

1.7. It may be desirable to have some idea of the "local" as well as the "global" behavior of $L_M(C_n)$. The following program of 8 rows causes an 8-state 3-tape-symbol Turing machine to compute the binary sequence 01100101 of length 8 (this program is in the format of the machines of Section 1.1):

```
1,2   2,4   2,4
2,3   3,4   3,4
3,3   4,4   4,4
4,2   5,4   5,4
5,2   6,4   6,4
6,3   7,4   7,4
7,2   8,4   8,4
8,3   0,4   0,4.
```

And in general:

(1.7.1) $$L_M(C_n) \leq n.$$

From this it is easy to see that for m greater than n:

(1.7.2) $$L_M(C_m) \leq L_M(C_n) + (m - n).$$

Also, for m greater than n:

(1.7.3) $$L_M(C_m) + 1 \geq L_M(C_n).$$

For if one can calculate any binary sequence of length m greater than n with an M-tape-symbol Turing machine having $L_M(C_m)$ states, one can certainly program any M-tape-symbol Turing machine having $L_M(C_m) + 1$ states to calculate the binary sequence consisting of (any particular sequence of length n) followed by a 1 followed by [a sequence of $(m - n - 1)$ 0's], and then — instead of immediately halting — to first erase all the 0's and the first 1 on the right end of the sequence. This last part of the program takes up only a single row of the table; in the format of the machines of Section 1.1 this row r is:

    row r   r,5   r,1   0,1.

Together (1.7.2) and (1.7.3) yield:

(1.7.4) $$\left| L_M(C_{n+1}) - L_M(C_n) \right| \leq 1.$$

From (1.7.1) it is obvious that $L_M(C_1) = 1$, and with (1.7.4) and the fact that $L_M(C_n)$ goes to infinity with n it finally is concluded that:

(1.7.5) For any positive integer p there is at least one solution n of $L_M(C_n) = p$.

1.8. In this section a certain amount of insight is obtained into the properties of finite binary sequences S of length n for which $L_M(S)$ is close to $L_M(C_n)$. M is considered to be fixed throughout this section. There is some connection between the present subject and that of Shannon in [2, Pt. I, especially Th. 9].

The main result is as follows:

(1.8.1) For any $e > 0$ and $d > 1$ one has for all sufficiently large n: If S is any binary sequence of length n satisfying the statement that

(1.8.2) the ratio of the number of 0's in S to n differs from $\frac{1}{2}$ by more than e,

then

$$L_M(S) < L_M(C_{[n\,d\,H(\frac{1}{2}+e,\ \frac{1}{2}-e)]}).$$

Here $H(p, q)$ ($p \geq 0$, $q \geq 0$, $p + q = 1$) is a special case of the entropy function of Boltzmann statistical mechanics and information theory and equals 0 if $p = 0$ or 1, and $-p \log_2 p - q \log_2 q$ otherwise. Also, a real number enclosed in brackets denotes the least integer greater than the enclosed real. The H function comes up because the logarithm to the base-two of the number

$$\sum_{\left|\frac{k}{n} - \frac{1}{2}\right| > e} \binom{n}{k}$$

of binary sequences of length n satisfying (1.8.2) is asymptotic to $n\, H(\frac{1}{2} + e,\ \frac{1}{2} - e)$. This may be shown easily by considering the ratio of successive binomial coefficients and using the fact that $\log(n!) \sim n \log n$.

To prove (1.8.1), first construct a class of effectively computable functions $M_n(.)$ with the natural numbers from 1 to 2^n as range and all binary sequences of length n as domain. $M_n(S)$ is defined to be the ordinal number of the position of S in an ordering of the binary sequences of length n defined as follows:

1. If two binary sequences S and S' have, respectively, m and m' 0's, then S comes before (after) S' according as $|(m/n) - \frac{1}{2}|$ is greater (less) than $|(m'/n) - \frac{1}{2}|$.
2. If 1 does not settle which comes first, take S to come before (after) S' according as S represents (ignoring 0's to the left) a larger (smaller) number in base-two notation than S' represents.

The only essential feature of this ordering is that it gives small ordinal numbers to sequences for which $|(m/n) - \frac{1}{2}|$ has large values. In fact, as there are only

$$2^{(1 + \epsilon_n)\, n\, H(\frac{1}{2}+e,\ \frac{1}{2}-e)}$$

binary sequences S of length n satisfying (1.8.2), it follows that at worst $M_n(S)$ is a number which in base-two notation is represented by a binary sequence of length $\sim n\, H(\frac{1}{2} + e,\ \frac{1}{2} - e)$. Thus in order to obtain a short program for computing an S of length n satisfying (1.8.2), let us just give a program of fixed length r the values of n and $M_n(S)$ and have it compute S ($= M_n^{-1}(M_n(S))$) from this data. The manner in which for n sufficiently large we give the values of n and $M_n(S)$ to the program is to pack them into a single binary sequence of length at most

$$\left[n\left(1 + \frac{d-1}{2}\right) H(\frac{1}{2} + e,\ \frac{1}{2} - e)\right] + 2(1 + [\log_2 n])$$

as follows: Consider (the binary sequence representing $M_n(S)$ in base-two notation) followed by 01 followed by [the binary sequence representing n with each of its bits doubled (e.g., if $n = 43$, this is 110011001111)]. Clearly both n and $M_n(S)$ can be recovered from this sequence. And this sequence can be computed by a program of

$$L_M\left(C_{\left[n\,(1 + \frac{d-1}{2})\,H(\frac{1}{2}+e,\ \frac{1}{2}-e)\right] + 2(1 + [\log_2 n])}\right)$$

ON THE LENGTH OF PROGRAMS FOR COMPUTING FINITE BINARY SEQUENCES

rows. Thus for n sufficiently large this many rows plus r is all that is needed to compute any binary sequence S of length n satisfying (1.8.2). And by the asymptotic formula for $L_M(C_n)$ of Section 1.6, it is seen that the total number of rows of program required is, for n sufficiently large, less than

$$L_M(C_{[n \, d \, H(\frac{1}{2}+e, \, \frac{1}{2}-e)]}).$$

Q.E.D.

From (1.8.1) and the fact that $H(p, q) \leq 1$ with equality if and only if $p = q = \frac{1}{2}$, it follows from $L_M(C_n) \sim (n/((M - 1) \log_2 n))$ that, for example,

(1.8.3) For any $e > 0$, all binary sequences S in ${}_M C_n$, n sufficiently large, violate (1.8.2);

and more generally,

(1.8.4) Let

$$S_{n_1}, \; S_{n_2}, \; S_{n_3}, \; \ldots$$

be any infinite sequence of distinct finite binary sequences of lengths, respectively, $n_1, \; n_2, \; n_3, \; \ldots$ which satisfies

$$L_M(S_{n_k}) \sim L_M(C_{n_k}).$$

Then as k goes to infinity, the ratio of the number of 0's in S_{n_k} to n_k tends to the limit $\frac{1}{2}$.

We now wish to apply (1.8.4) to programs for Turing machines. In order to do this we need to be able to represent the table of entries defining any program as a single binary sequence. A method is sketched here for coding any program $T_{N, M}$ occupying the table of an N-state M-tape-symbol Turing machine into a single binary sequence $C(T_{N, M})$ of length $(1 + \varepsilon_N)N(M - 1) \log_2 N$.

First, write all the members of the ordered triples entered in the table in base-two notation, adding a sufficient number of 0's to the left of the numerals for all numerals to be

1. as long as the base-two numeral for $N + 1$ if they result from the first member of a triple,
2. as long as the base-two numeral for $M + 2$ if they result from the second member, and
3. as long as the base-two numeral for 2 if they result from the third member.

The only exception to this rule is that if the third member of a triple is 1 or 2, then the first member of the triple is not written in base-two notation; no binary sequences are generated from the first members of such triples. Last, all the binary sequences that have just been obtained are joined together, starting with the binary sequence that was generated from the first member of the triple entered at the intersection of row 1 with column 1 of the table, then with the binary sequence generated from the second member of the triple..., ...from the third member..., ...from the first member of the triple entered at the intersection of row 1 with column 2, ...from the second member..., ...from the third member..., and so on across the first row of the table, then across the second row of the table, then the third, ...and finally across the Nth row.

The result of all this is a single binary sequence of length $(1 + \varepsilon_N)N(M - 1) \log_2 N$ (in view of (1.3.1)) from which one can effectively determine the whole table of entries which was coded into it, if only one is given the values of N and M. But it is possible to code in these last pieces of information using only the rightmost

$$2(1 + [\log_2 N]) \; + \; 2(1 + [\log_2 M])$$

bits of a binary sequence consequently of total length

$$(1 + \varepsilon_N)N(M - 1) \log_2 N \; + \; 2(1 + [\log_2 N]) \; + \; 2(1 + [\log_2 M])$$

$$= (1 + \varepsilon_N')N(M - 1) \log_2 N,$$

by employing the same trick that was used to pack two pieces of information into a single binary sequence earlier in this section.

Thus we have a simple procedure for coding the whole table of entries $T_{N,M}$ defining a program of an N-state M-tape-symbol Turing machine and the parameters N and M of the machine into a binary sequence $C(T_{N,M})$ of $(1 + \varepsilon_N)N(M - 1) \log_2 N$ bits.

We now obtain the result:

(1.8.5) Let

$$T_{L_M(S_1),\, M},\ T_{L_M(S_2),\, M},\ \cdots$$

be an infinite sequence of tables of entries which define programs for computing, respectively, the distinct finite binary sequences S_1, S_2, ... Then

$$L_M(C(T_{L_M(S_k),\, M})) \sim L_M(C_{n_k}),$$

where n_k is the length of

$$C(T_{L_M(S_k),\, M}).$$

With (1.8.4) this gives the proposition:

(1.8.6) On the hypothesis of (1.8.5), as k goes to infinity, the ratio of the number of 0's in

$$C(T_{L_M(S_k),\, M})$$

to its length tends to the limit ½.

The proof of (1.8.5) depends on three facts:

(1.8.7a) There is an effective procedure for coding the table of entries $T_{N,M}$ defining the program of an N-state M-tape-symbol Turing machine together with the two parameters N and M into a single binary sequence $C(T_{N,M})$ of length $(1 + \varepsilon_N)N(M - 1) \log_2 N$.

(1.8.7b) Any binary sequence of length not greater than $(1 + \varepsilon_N)N(M - 1) \log_2 N$ can be calculated by a suitably programmed N-state M-tape-symbol Turing machine.

(1.8.7c) From a universal Turing machine program it is possible to construct a program for a Turing machine (with a fixed number r of rows) to take $C(T_{N,M})$ and decode it and to then imitate the calculations of the machine whose table of entries $T_{N,M}$ it then knows, until it finally calculates the finite binary sequence S which the program being imitated calculates, if S exists.

(1.8.7a) has just been demonstrated. (1.8.7b) was shown in Section 1.6. (The concept of a universal program is due to Turing [3].)

The proof of (1.8.5) follows. From (1.8.7a) and (1.8.7b),

$$L_M(C(T_{L_M(S_k),\, M})) \leq (1 + \varepsilon_k)L_M(S_k),$$

and from (1.8.7c) and the hypothesis of (1.8.5),

$$L_M(C(T_{L_M(S_k),\, M})) + r \geq L_M(S_k).$$

It follows that

$$L_M(C(T_{L_M(S_k),\ M})) = (1 + \varepsilon_k)L_M(S_k),$$

which is — since the length of

$$C(T_{L_M(S_k),\ M})$$

is

$$(1 + \varepsilon_k)L_M(S_k)(M - 1) \log_2 L_M(S_k)$$

and

$$L_M(C_{(1+\ \varepsilon_k)L_M(S_k)(M-1)\ \log_2 L_M(S_k)}) = (1 + \varepsilon_k')L_M(S_k)$$

— simply the conclusion of (1.8.5).

1.9. The topic of this section is an application of everything that precedes with the exception of Section 1.7 and the first half of Section 1.8. C.E. Shannon suggests [1, p. 165] that the state-symbol product NM is a good measure of the calculating abilities of an N-state M-tape-symbol Turing machine. If one is interested in *comparing* the calculating abilities of *large Turing machines whose M values vary over a finite range*, the results that follow suggest that $N(M - 1)$ is a good measure of calculating abilities. We have as an application of a slight generalization of the ideas used to prove (1.8.5):

(1.9.1a) Any calculation which an N-state M-tape-symbol Turing machine can be programmed to perform can be imitated by any N'-state M'-tape-symbol Turing machine satisfying

$$(1 + \varepsilon_N)N(M - 1) \log_2 N < (1 + \varepsilon_N'')N'(M' - 1) \log_2 N'$$

if it is suitably programmed.

And directly from the asymptotic formula for $L_M(C_n)$ we have:

(1.9.1b) If

$$(1 + \varepsilon_N)N(M - 1) \log_2 N < (1 + \varepsilon_N'')N'(M' - 1) \log_2 N',$$

then there exist finite binary sequences which an N'-state M'-tape-symbol Turing machine can be programmed to calculate and which it is impossible to program an N -state M-tape-symbol Turing machine to calculate.

As

$$(1 + \varepsilon_N)N(M - 1) \log_2 N = ((1 + \varepsilon_N')N(M - 1)) \log_2 ((1 + \varepsilon_N')N(M - 1))$$

and for x and x' greater than one, $x \log_2 x$ is greater (less) than $x' \log_2 x'$ according as x is greater (less) than x', it follows that the inequalities of (1.9.1a) and (1.9.1b) give the same *ordering* of calculating abilities as do inequalities involving functions of the form $(1 + \varepsilon_N)N(M - 1)$.

Part 2

2.1. In this section we return to the Turing machines of Section 1.1 and add to the conventions (1.1A), (1.1B) and (1.1C),

(2.1D) An entry (i, j) in the pth row of the table of a Turing machine must satisfy $|i - p| \leq b$. In addition, while a fictitious state is used (as before) for the purpose of halting, the row of the table for this fictitious state is now considered to come directly after the actual last row of the program.

Here b is a constant whose value is to be regarded as fixed throughout Part 2. In Section 2.2 it will be shown that b can be chosen sufficiently large that the Turing machines thus defined (which we take the liberty of naming "bounded-transfer Turing machines") have all the calculating capabilities that are basically required of Turing machines for theoretical purposes (e.g., such purposes as defining what one means by "effective process for determining..."), and hence have calculating abilities sufficient for the proofs of Part 2 to be carried out.

(2.1D) may be regarded as a mere convention, but it is more properly considered as a change in the basic philosophy of the logical design of the Turing machine (i.e., the philosophy expressed by A. M. Turing [3, Sec. 9]).

Here in Part 2 there will be little point in considering the general M-tape-symbol machine. It will be understood that we are always speaking of 3-tape-symbol machines.

There is a simple and convenient notational change which can be made at this point; it makes all programs for bounded-transfer Turing machines instantly relocatable (which is convenient if one puts together a program from subroutines) and it saves a great deal of superfluous writing. Entries in the tables of machines will from now on consist of ordered pairs (i', j'), where i' goes from $-b$ to b and j' goes from 1 to 5. A "new" entry (i', j') is to be interpreted in terms of the functioning of the machine in a manner depending on the number p of the row of the table it is in; this entry has the same meaning as the "old" entry $(p + i', j')$ used to have.

Thus, halting is now accomplished by entries of the form (k, j) $(1 \leq k \leq b)$ in the kth row (from the end) of the table. Such an entry causes the machine to halt after performing the operation indicated by j.

2.2. In this section we attempt to give an idea of the versatility of the bounded-transfer Turing machine. It is here shown in two ways that b can be chosen sufficiently large so that any calculation which one of the Turing machines of Section 1.1 can be programmed to perform can be imitated by a suitably programmed bounded-transfer Turing machine.

As the first proof, b is taken to be the number of rows in a 3-tape-symbol universal Turing machine program for the machines of Section 1.1. This universal program (with its format changed to that of the bounded-transfer Turing machines) occupies the last rows of a program for a bounded-transfer Turing machine, a program which is mainly devoted to writing out on the tape the information which will enable the universal program to imitate any calculation which any one of the Turing machines of Section 1.1 can be programmed to perform. One row of the program is used to write out each symbol of this information (as in the program in Section 1.7), and control passes straight through the program row after row until it reaches the universal program.

Now for the second proof. To program a bounded-transfer Turing machine in such a manner that it imitates the calculations performed by a Turing machine of Section 1.1, consider alternate squares on the tape of the bounded-transfer Turing machine to be the squares of the tape of the machine being imitated. Thus

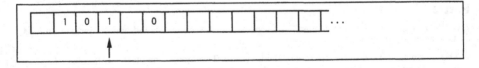

is imitated by

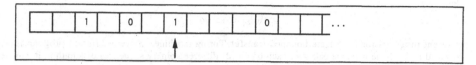

After the operation of a state (i.e., write 0, write 1, write blank, shift tape left, shift tape right) has been imitated, as many 1's as the number of the next state to be imitated are written on the squares of the tape of the bounded-transfer Turing machine which are not used to imitate the squares of the other machine's tape, starting on the square immediately to the right of the one on which is the scanner of the bounded-transfer Turing machine. Thus if in the foregoing situation the next state to be imitated is state number three, then the tape of the bounded-transfer Turing machine becomes

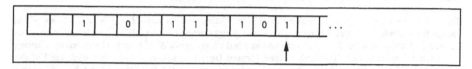

The rows of the table which cause the bounded-transfer Turing machine to do the foregoing (type I rows) are interwoven or braided with two other types of rows. The first of these (type II rows) is used for the sole purpose of putting the bounded-transfer Turing machine back in its initial state (row 1 of the table; this row is a type III row). They appear (as do the other two types of rows) periodically throughout the table, and each of them does nothing but transfer control to the preceding one. The second of these (type III rows) serve to pass control back in the other direction; each time control is about to pass a block of type I rows that imitate a particular state of the other machine while traveling through type III rows, the type III rows erase the rightmost of the 1's used to write out the number of the next state to be imitated. When finally none of these place-marking 1's is left, control is passed to the group of type I rows that was about to be passed, which then proceeds to imitate the appropriate state of the Turing machine of Section 1.1.

Thus the obstacle of the upper bound on the length of transfers in bounded-transfer Turing machines is overcome by passing up and down the table by small jumps, while keeping track of the progress to the desired destination is achieved by subtracting a unit from a count written on the tape just prior to departure.

Although bounded-transfer Turing machines have been shown to be versatile, it is not true that as the number of states goes to infinity, asymptotically 100 percent of the bits required to specify a program also serve to specify the behavior of the bounded-transfer Turing machine.

2.3. In this section the following fundamental result is proved.

(2.3.1) $L(C_n) \sim a^*n$, where a^* is, of course, a positive constant.

First it is shown that there exists an a greater than zero such that:

(2.3.2) $$L(C_n) \geq an.$$

It is clear that there are exactly

$$((5)(2b + 1))^{3N}$$

different ways of making entries in the table of an N-state bounded-transfer Turing machine; that is, there are

$$2^{((3 \log_2 (10b+5))N)}$$

different programs for an N-state bounded-transfer Turing machine. Since a different program is required to have the machine calculate each of the 2^n different binary sequences of length n, it can be seen that an N-state bounded-transfer Turing machine can be programmed to calculate any binary sequence of length n only if

$$(3 \log_2 (10b + 5)) N \geq n \quad \text{or} \quad N \geq (1/(3 \log_2 (10b + 5))) n.$$

Thus one can take $a = (1/(3 \log_2 (10b + 5)))$.

Next it is shown that:

(2.3.3)
$$L(C_n) + L(C_m) \geq L(C_{n+m}).$$

To do this we present a way of making entries in a table with at most $L(C_n) + L(C_m)$ rows which causes the bounded-transfer Turing machine thus programmed to calculate any particular binary sequence S of length $n + m$. S can be expressed as a binary sequence S' of length n followed by a binary sequence S'' of length m. The table is then formed from two sections which are numbered in the order in which they are encountered in reading from row 1 to the last row of the table. Section I consists of at most $L(C_n)$ rows. It is a program which calculates S'. Section II consists of at most $L(C_m)$ rows. It is a program which calculates S''. It follows from this construction and the definitions that (2.3.3) holds.

(2.3.2) and (2.3.3) together imply (2.3.1). This will be shown by a demonstration of the following general proposition:

(2.3.4) Let A_1, A_2, A_3, ... be an infinite sequence of natural numbers satisfying

(2.3.5)
$$A_n + A_m \geq A_{n+m}.$$

Then as n goes to infinity, (A_n/n) tends to a limit from above.

For all n, $A_n \geq 0$, so that $(A_n/n) \geq 0$; that is, $\{(A_n/n)\}$ is a set of reals bounded from below. It is concluded that this set has a greatest lowest bound a^*. We now show that

$$\lim_{n \to \infty} (A_n/n) = a^*.$$

Since a^* is the greatest lower bound of the set $\{(A_n/n)\}$, for any e greater than zero there is a d for which

(2.3.6)
$$(A_d/d) < a^* + e.$$

Every natural number n can be expressed in the form $n = qd + r$, where $0 \leq r < d$. From (2.3.5) it can be seen that for any n_1, n_2, n_3, ... , n_{q+1},

$$\sum_{k=1}^{q+1} A_{n_k} \geq A \left(\sum_{k=1}^{q+1} n_k \right).$$

Taking $n_k = d$ ($k = 1, 2, \ldots , q$) and $n_{q+1} = r$ in this, we obtain

ON THE LENGTH OF PROGRAMS FOR COMPUTING FINITE BINARY SEQUENCES

$$qA_d + A_r \geq A_{qd+r} = A_n,$$

which with (2.3.6) gives

$$qd(a^* + e) = (n - r)(a^* + e) \geq A_n - A_r$$

or

$$(1 - r/n)(a^* + e) \geq (A_n/n) - (A_r/n),$$

which implies

$$a^* + e \geq (A_n/n) + \varepsilon_n$$

or

$$\varlimsup_{n \to \infty} (A_n/n) \leq a^* + e.$$

Since $e > 0$ is arbitrary, it can be concluded that

$$\varlimsup_{n \to \infty} (A_n/n) \leq a^*,$$

which with the fact that $(A_n/n) \geq a^*$ for all n gives

$$\lim_{n \to \infty} (A_n/n) = a^*.$$

2.4. In Section 2.3 an asymptotic formula analogous to a part of Section 1.6 was demonstrated; in this section a result is obtained which completes the analogy. This result is most conveniently stated with the aid of the notation $B(m)$ (where m is a natural number) for the binary sequence which is the numeral representing m in base-two notation (e.g., $B(6) = 110$).

(2.4.1) There exists a constant c such that those binary sequences S of length n satisfying

$$\begin{aligned}
\text{(2.4.2)} \qquad L(S) \leq{}& L(C_n) - L(B(L(C_n))) - \lceil \log_2 L(B(L(C_n))) \rceil \\
&- L(C_m) - \lceil \log_2 L(C_m) \rceil - c
\end{aligned}$$

are less than 2^{n-m} in number.

The proof of (2.4.1) is by contradiction. We suppose that those S of length n satisfying (2.4.2) are 2^{n-m} or more in number and we conclude that for any particular binary sequence $S^\sim$ of length n there is a program of at most $L(C_n) - 1$ rows that causes a bounded-transfer Turing machine to calculate $S^\sim$. This table consists of 11 sections which come one after the other. The first section consists of a single row which moves the tape one square to the left (1,4 1,4 1,4 will certainly do this). The second section consists of exactly $L(B(L(C_n)))$ rows; it is a program for computing $B(L(C_n))$ consisting of the smallest possible number of rows. The third section is merely a repetition of the first section. The fourth section consists of exactly $\lceil \log_2 L(B(L(C_n))) \rceil$ rows. Its function is to write out on the tape the binary sequence which represents the number $L(B(L(C_n)))$ in base-two notation. Since this is a sequence of exactly $\lceil \log_2 L(B(L(C_n))) \rceil$ bits, a simple program exists for calculating it consisting of exactly $\lceil \log_2 L(B(L(C_n))) \rceil$ rows each of which causes the machine to write out a single bit of the sequence and then shift the tape a single square to the left (e.g., 0,2 1,4 1,4 will do for a 0 in the sequence). The fifth section is merely a repetition of the first section. The sixth section consists of at most $L(C_m)$ rows; it is a program consisting of the smallest possible number of rows for computing the sequence S^R of the m rightmost bits of $S^\sim$. The seventh section is merely a repetition of the first section. The eighth section consists of exactly $\lceil \log_2 L(C_m) \rceil$ rows. Its function is to write out on the tape the binary sequence which represents the number $L(C_m)$ in base-two notation. Since this is a sequence of exactly $\lceil \log_2 L(C_m) \rceil$ bits, a simple program exists for calculating it consisting of exactly $\lceil \log_2 L(C_m) \rceil$ rows each of which causes the machine to write out a single bit of the sequence and then shift the tape a single square to the left. The ninth section is merely a repetition of the first section. The tenth section consists of at most as many rows as the expression on the right-hand side of the inequality (2.4.2). It is a program for calculating one (out of not less than 2^{n-m}) of the sequences of length n satisfying (2.4.2) (which one it is depends on $S^\sim$ in a manner which will become clear from the discussion of the eleventh section; for now we merely denote it by S^L).

We now come to the last and crucial eleventh section, which consists *by definition* of $(c - 6)$ rows, and which therefore brings the total number of rows up to at most $1 + L(B(L(C_n))) + 1 + \lceil \log_2 L(B(L(C_n))) \rceil + 1 + L(C_m) + 1 + \lceil \log_2 L(C_m) \rceil + 1 +$ (the expression on the right-hand side of the inequality (2.4.2)) $+ (c - 6) = L(C_n) - 1$. When this section of the program takes over, the numbers and sequences $L(C_n)$, $L(B(L(C_n)))$, S^R, $L(C_m)$, S^L are written — in the above order — on the tape. Note, first of all, that section 11 can:

1. compute the value v of the right-hand side of the inequality (2.4.2) from this data,
2. find the value of n from this data (simply by counting the number of bits in the sequence S^L), and
3. find the value of m from this data (simply by counting the number of bits in S^R).

Using its knowledge of v, m and n, section 11 then computes from the sequence S^L a new sequence $S^{L'}$ which is of length $(n - m)$. The manner in which it does this is discussed in the next paragraph. Finally, section 11 adjoins the sequence S^R to the right of $S^{L'}$, positions this sequence which is in fact $S^\sim$ properly for it to be able to be regarded calculated, cleans up the rest of the tape, and halts scanning the square just to the right of $S^\sim$. $S^\sim$ has been calculated.

To finish the proof of (2.4.1) we must now only indicate how section 11 arrives at $S^{L'}$ (of length $(n - m)$) from v, m, n, and S^L. (And it must be here that it is made clear how the choice of S^L depends on $S^\sim$.) By assumption, S^L satisfies

(2.4.3) $$L(S^L) \leq v \text{ and } S^L \text{ is of length } n.$$

Also by assumption there are at least 2^{n-m} sequences which satisfy (2.4.3). Now *section 11 contains a procedure which when given any one of some particular serially ordered set* ${}_nQ_v$ *of* 2^{n-m} *sequences satisfying (2.4.3), will find the ordinal number of its position in* ${}_nQ_v$. And the number of the position of S^L in ${}_nQ_v$ is the number of the position of $S^{L'}$ in the natural ordering of all binary sequences of length $(n - m)$ (i.e., 000...00, 000...01, 000...10, 000...11, ..., 111...00, 111...01, 111...10, 111...11). In the next and final paragraph of this proof, the foregoing italicized sentence is explained.

It is sufficient to give here a procedure for serially calculating the members of ${}_nQ_v$ in order. (That is, we define a serially ordered ${}_nQ_v$ for which there is a procedure.) By assumption we know that the predicate which is satisfied by all members of ${}_nQ_v$, namely,

$$(L(\ldots) \leq v) \ \& \ (\ldots \text{ is of length } n),$$

is satisfied by at least 2^{n-m} sequences. It should also be clear to the reader on the basis of some background in Turing machine and recursive function theory (see especially Davis [4], where recursive function theory is developed from the concept of the Turing machine) that the set Q of

all natural numbers of the form $2^n 3^v 5^e$, *where e is the natural number represented in base-two notation by a binary sequence S satisfying*

$$(L(S) \leq v) \ \& \ (S \text{ is of length } n)$$

is recursively enumerable. Let T denote some particular Turing machine which is programmed in such a manner that it recursively enumerates (or, to use E. Post's term, generates) Q. The definition of $_nQ_v$ can now be given:

> $_nQ_v$ is the set of binary sequences of length n which represent in base-two notation the exponents of 5 in the prime factorization of the first 2^{n-m} members of Q generated by T whose prime factorizations have 2 with an exponent of n and 3 with an exponent of v, and their order in $_nQ_v$ is the order in which T generates them.

Q.E.D.

It can be proved by contradiction that the set Q is not recursive. For were Q recursive, there would be a program which given any finite binary sequence S would calculate $L(S)$. Hence there would be a program which given any natural number n would calculate the members of C_n. Giving n to this program can be done by a program of length $\lceil \log_2 n \rceil$. Thus there would be a program of length $\lceil \log_2 n \rceil + c$ which would calculate an element of C_n. But we know that the shortest program for calculating an element of C_n is of length $\sim a^*n$, so that we would have for n sufficiently large an impossibility.

It should be emphasized that if $L(C_n)$ is an effectively computable function of n then the method of this section yields the far stronger result: There exists a constant c such that those binary sequences S of length n satisfying $L(S) \leq L(C_n) - L(C_m) - c$ are less than 2^{n-m} in number.

2.5. The purpose of this section is to investigate the behavior of the right-hand side of (2.4.2). We start by showing a result which is stronger for n sufficiently large than the inequality $L(C_n) \leq n$, namely, that the constant a^* in the asymptotic evaluation $L(C_n) \sim a^*n$ of Section 2.3 is less than 1. This is done by deriving:

(2.5.1) For any s there exist n and m such that

$$L(C_s) \leq L(C_n) + L(C_m) + c,$$

and $(n + m)$ is the smallest integral solution x of the inequality

$$s \leq x + \lceil \log_2 x \rceil - 1.$$

From (2.5.1) it will follow immediately that if $e(n)$ denotes the function satisfying $L(C_n) = a^*n + e(n)$ (note that by Section 2.3 $(e(n)/n)$ tends to 0 *from above* as n goes to infinity), then for any s, $L(C_s) \leq L(C_n) + L(C_m) + c$ for some n and m satisfying $(n + m) = s - (1 + \varepsilon_s) \log_2 s$, which implies

$$a^*s \leq a^*(s - (1 + \varepsilon_s) \log_2 s) + e(n) + e(m)$$

or

$$(a^* + \varepsilon_s) \log_2 s \leq e(n) + e(m).$$

Hence as n and m are both less than s and at least one of $e(n)$, $e(m)$ is greater than $\frac{1}{2}(a^* + \varepsilon_s) \log_2 s$, there are an infinity of n for which $e(n) \geq \frac{1}{2}(a^* + \varepsilon_n) \log_2 n$. That is,

(2.5.2)
$$\overline{\lim} \frac{L(C_n) - a^*n}{a^* \log_2 n} \geq \frac{1}{2}.$$

From (2.5.2) with $L(C_n) \leq n$ follows immediately

(2.5.3)
$$a^* < 1.$$

```
Section I:
1,4       1,4        1,4
```

Section II consists of $L(F_L^{-1}(S))$ rows. It is a program
with the smallest possible number of rows for calculating
$F_L^{-1}(S)$.

```
Section III:
1,4       1,4        1,4
```

Section IV consists of $L(F_R^{-1}(S))$ rows. It is a program
with the smallest possible number of rows for calculating
$F_R^{-1}(S)$.

(Should x be of the form 2^h,
another section is added at this point to tell Section V
which of the two possible values s happens to have.
This section consists of two rows; it is either

```
1,4       1,4        1,4
1,2       1,2        1,2
```
or
```
1,4       1,4        1,4
1,3       1,3        1,3.)
```

Section V consists of c - 4 rows, by definition.
It is a program that is able to compute F. It computes

$$F(F_L^{-1}(S),\ F_R^{-1}(S)) = S,$$

positions S properly on the tape,
cleans up the rest of the tape, positions the scanner
on the square just to the right of S, and halts.

Figure 12.   A five section program

The proof of (2.5.1) is presented by examples. The notation $T*U$ is used, where T and U are finite binary sequences for the sequence resulting from adjoining U to the right of T. Suppose it is desired to calculate some finite binary sequence S of length s, say $S = 010110010100110$ and $s = 15$. The smallest integral solution x of $s \leq x + \lceil \log_2 x \rceil - 1$ for this value of s is 12. Then S is expressed as $S'*S^T$ where S' is of length $x = 12$ and S^T is of length $s - x = 15 - 12 = 3$, so that $S' = 010110010100$ and $S^T = 110$. Next S' is expressed as S^L*S^R where the length m of S^L satisfies $A*B(m) = S^T$ for some (possibly null) sequence A consisting entirely of 0's, and the length n of S^R is $x - m$. In this case $A*B(m) = 110$, so that $m = 6$, $S^L = 010110$ and $S^R = 010100$. The final result is that one has obtained the sequences S^L and S^R from the sequence S. And — this is the crucial point — if one is given the S^L and S^R resulting by the foregoing process from some unknown sequence S, one can reverse the procedure and determine S. Thus suppose $S^L = 1110110$ and $S^R = 01110110000$ are given. Then the length m of S^L is 7, the length n of S^R is 11, and the sum x of m and n is $7 + 11 = 18$. Therefore the length s of S must be $s = x + \lceil \log_2 x \rceil - 1 = 18 + 5 - 1 = 22$. Thus $S = S^L*S^R*S^T$, where S^T is of length $s - x = 22 - 18 = 4$, and so from $A*B(m) = S^T$ or $0*B(7) = S^T$ one finds $S^T = 0111$. It is concluded that

$$S = S^L*S^R*S^T = 1110110011101100000111.$$

ON THE LENGTH OF PROGRAMS FOR COMPUTING FINITE BINARY SEQUENCES

(For x of the form 2^h what precedes is not strictly correct. In such cases s may equal the foregoing indicated quantity or the foregoing indicated quantity minus one. It will be indicated later how such cases are to be dealt with.)

Let us now denote by F the function carrying (S^L, S^R) into S, and by F_R^{-1} the function carrying S into S^R, defining F_L^{-1} similarly. Then for any particular binary sequence S of length s the program of Figure 12 on page 227 consists of at most

$$1 + L(F_L^{-1}(S)) + 1 + L(F_R^{-1}(S)) + 2 + (c - 4) \leq L(C_n) + L(C_m) + c$$

rows with $m + n = x$ being the smallest integral solution of $s \leq x + \lceil \log_2 x \rceil - 1$. As this program causes S to be calculated, the proof is easily seen to be complete.

The second result is:

(2.5.4) Let $f(n)$ be any effectively computable function that goes to infinity with n and satisfies $f(n + 1) - f(n) = 0$ or 1. Then there are an infinity of distinct n_k for which $L(B(L(C_{n_k}))) < f(n_k)$.

This is proved from (2.5.5), the proof being identical with that of (1.7.5).

(2.5.5) For any positive integer p there is at least one solution n of $L(C_n) = p$.

Let the n_k satisfy $L(C_{n_k}) = f^{-1}(k)$, where $f^{-1}(k)$ is defined to be the smallest value of j for which $f(j) = k$. Then since $L(C_n) \leq n, f^{-1}(k) \leq n_k$. Noting that f^{-1} is an effectively computable function, it is easily seen that

$$L(B(L(C_{n_k}))) = L(B(f^{-1}(k))) \leq L(B(k)) + c \leq \lceil \log_2 k \rceil + c.$$

Hence, for all sufficiently large k,

$$L(B(L(C_{n_k}))) \leq \lceil \log_2 k \rceil + c < k = f(f^{-1}(k)) \leq f(n_k).$$

Q.E.D.

(2.5.4) and (2.4.1) yield:

(2.5.6) Let $f(n)$ be any effectively computable function that goes to infinity with n and satisfies $f(n + 1) - f(n) = 0$ or 1. Then there are an infinity of distinct n_k for which less than $2^{n_k - f(n_k)}$ binary sequences S of length n_k satisfy $L(S) \leq L(C_{n_k}) - (a^* + \varepsilon_k)f(n_k)$.

Part 3

3.1. Consider a scientist who has been observing a closed system that once every second either emits a ray of light or does not. He summarizes his observations in a sequence of 0's and 1's in which a zero represents "ray not emitted" and a one represents "ray emitted." The sequence may start

$$0110101110 \ldots$$

and continue for a few thousand more bits. The scientist then examines the sequence in the hope of observing some kind of pattern or law. What does he mean by this? It seems plausible that a sequence of 0's and 1's is patternless if there is no better way to calculate it than just by writing it all out at once from a table giving the whole sequence:

My Scientific Theory

```
0
1
1
0
1
0
1
1
1
0
⋮
```

This would not be considered acceptable. On the other hand, if the scientist should hit upon a method by which the whole sequence could be calculated by a computer whose program is short compared with the sequence, he would certainly not consider the sequence to be entirely patternless or random. And the shorter the program, the greater the pattern he might ascribe to the sequence.

There are many genuine parallels between the foregoing and the way scientists actually think. For example, a simple theory that accounts for a set of facts is generally considered better or more likely to be true than one that needs a large number of assumptions. By "simplicity" is *not* meant "ease of use in making predictions." For although General or Extended Relativity is considered to be the simple theory par excellence, very extended calculations are necessary to make predictions from it. Instead, one refers to the number of arbitrary choices which have been made in specifying the theoretical structure. One naturally is suspicious of a theory the number of whose arbitrary elements is of an order of magnitude comparable to the amount of information about reality that it accounts for.

On the basis of these considerations it may perhaps not appear entirely arbitrary to define a patternless or random finite binary sequence as a sequence which in order to be calculated requires, roughly speaking, at least as long a program as any other binary sequence of the same length. A patternless or random infinite binary sequence is then defined to be one whose initial segments are all random. In making these definitions mathematically approachable it is necessary to specify the kind of computer referred to in them. This would seem to involve a rather arbitrary choice, and thus to make our definitions less plausible, but in fact both of the kinds of Turing machines which have been studied by such different methods in Parts 1 and 2 lead to precise mathematical definitions of patternless sequences (namely, the patternless or random finite binary sequences are those sequences S of length n for which $L(S)$ is approximately equal to $L(C_n)$, or, fixing M, those for which $L_M(S)$ is approximately equal to $L_M(C_n)$) whose provable statistical properties start with forms of the law of large numbers. Some of these properties will be established in a paper of the author to appear.[57]

A final word. In scientific research it is generally considered better for a proposed new theory to account for a phenomenon which had not previously been contained in a theoretical structure, before the discovery of that phenomenon rather than after. It may therefore be of some interest to mention that the intuitive considerations of this section antedated the investigations of Parts 1 and 2.

---

[57] The author has subsequently learned of work of P. Martin-Löf ("The Definition of Random Sequences," research report of the Institutionen för Försäkringsmatematik och Matematisk Statistik, Stockholm, Jan. 1966, 21 pp.) establishing statistical properties of sequences defined to be patternless on the basis of a type of machine suggested by A. N. Kolmogorov. Cf. footnote [58].

[58] The author has subsequently learned of the paper of A. N. Kolmogorov, Three approaches to the definition of the concept "amount of information," *Problemy Peredachi Informatsii* [Problems of Information Transmission], *1*, 1 (1965), 3-11 [in Russian], in which essentially the definition offered here is put forth.

3.2. The definition which has just been proposed[58] is one of many attempts which have been made to define what one means by a patternless or random sequence of numbers. One of these was begun by R. von Mises [5] with contributions by A. Wald [6], and was brought to its culmination by A. Church [7]. K. R. Popper [8] criticized this definition. The definition given here deals with the concept of a patternless binary sequence, a concept which corresponds roughly in intuitive intent with the random sequences associated with probability half of Church. However, the author does not follow the basic philosophy of the von Mises-Wald-Church definition; instead, the author is in accord with the opinion of Popper [8, Sec. 57, footnote 1]:

> I come here to the point where I failed to carry out fully my intuitive program — that of analyzing randomness as far as it is possible within the region of *finite* sequences, and of proceeding to *infinite* reference sequences (in which we need *limits* of relative frequencies) only afterwards, with the aim of obtaining a theory in which the existence of frequency limits follows from the random character of the sequence.

Nonetheless the methods given here are similar to those of Church; the concept of effective computability is here made the central one.

A discussion can be given of just how patternless or random the sequences given in this paper appear to be for practical purposes. How do they perform when subjected to statistical tests of randomness? Can they be used in the Monte Carlo method? Here the somewhat tantalizing remark of J. von Neumann [9] should perhaps be mentioned:

> Any one who considers arithmetical methods of producing random digits is, of course, in a state of sin. For, as has been pointed out several times, there is no such thing as a random number — there are only methods to produce random numbers, and a strict arithmetical procedure of course is not such a method. (It is true that a problem that we suspect of being solvable by random methods may be solvable by some rigorously defined sequence, but this is a deeper mathematical question than we can now go into.)

Acknowledgment

The author is indebted to Professor Donald Loveland of New York University, whose constructive criticism enabled this paper to be much clearer than it would have been otherwise.

RECEIVED OCTOBER, 1965; REVISED MARCH, 1966

References

1. SHANNON, C. E. A universal Turing machine with two internal states. In *Automata Studies*, Shannon and McCarthy, Eds., Princeton U. Press, Princeton, N. J., 1956.

2. —. A mathematical theory of communication. *Bell Syst. Tech. J. 27* (1948), 379-423.

3. TURING, A. M. On computable numbers, with an application to the Entscheidungsproblem. *Proc. London Math. Soc.* {2} *42* (1936-37), 230-265; Correction, *ibid.*, *43* (1937), 544-546.

4. DAVIS, M. *Computability and Unsolvability.* McGraw-Hill, New York, 1958.

5. VON MISES, R. *Probability, Statistics. and Truth.* MacMillan, New York, 1939.

6. WALD, A. Die Widerspruchsfreiheit des Kollektivbegriffes der Wahrscheinlichkeitsrechnung. *Ergebnisse eines mathematischen Kolloquiums 8* (1937), 38-72.

7. CHURCH, A. On the concept of a random sequence. *Bull. Amer. Math. Soc. 46* (1940), 130-135.

8.   POPPER, K. R. *The Logic of Scientific Discovery.* U. of Toronto Press, Toronto, 1959.

9.   VON NEUMANN, J. Various techniques used in connection with random digits. In *John von Neumann, Collected Works, Vol. V.* A. H. Taub, Ed., MacMillan, New York, 1963.

10. CHAITIN, G. J. On the length of programs for computing finite binary sequences by bounded-transfer Turing machines. Abstract 66T-26, *Notic. Amer. Math. Soc. 13* (1966), 133.

11. —. On the length of programs for computing finite binary sequences by bounded-transfer Turing machines II. Abstract 631-6, *Notic. Amer. Math. Soc. 13* (1966), 228-229. (Erratum, p. 229, line 5: replace "*P*" by "*L*".)

ON THE LENGTH OF PROGRAMS FOR COMPUTING FINITE BINARY SEQUENCES: STATISTICAL CONSIDERATIONS

Journal of the ACM 16 (1969), pp. 145-159.

GREGORY J. CHAITIN[59]
Buenos Aires, Argentina

Abstract

An attempt is made to carry out a program (outlined in a previous paper) for defining the concept of a random or patternless, finite binary sequence, and for subsequently defining a random or patternless, infinite binary sequence to be a sequence whose initial segments are all random or patternless finite binary sequences. A definition based on the bounded-transfer Turing machine is given detailed study, but insufficient understanding of this computing machine precludes a complete treatment. A computing machine is introduced which avoids these difficulties.

Key Words and Phrases

computational complexity, sequences, random sequences, Turing machines

CR Categories

5.22, 5.5, 5.6

1. Introduction

In this section a definition is presented of the concept of a random or patternless binary sequence based on 3-tape-symbol bounded-transfer Turing machines.[60] These computing machines have been introduced and studied in [1], where a proposal to apply them in this manner is made. The results from [1] which are used in studying the definition are listed for reference at the end of this section.

An N-state, 3-tape-symbol bounded-transfer Turing machine is defined by an N-row, 3-column table. Each of the $3N$ places in this table must contain an ordered pair (i, j) of natural numbers where i takes on values from $-b$ to b, and j from 1 to 5.[61] These entries constitute, when specified, the program of the N-state, 3-tape-symbol bounded-transfer Turing machine and are to be interpreted as follows. An entry (i, j) in the kth row and the pth column of the table means that when the machine is in its kth state, and the square of its one-way infinite tape which is being scanned contains the pth symbol, then if $1 \leq k + i \leq N$, the machine is to go to its $(k + i)$-th state (otherwise, the machine is to halt) after performing one of the following operations:

(a) moving the tape one square to the right if $j = 5$;
(b) moving the tape one square to the left if $j = 4$;
(c) marking (overprinting) the square being scanned with the j th symbol if $1 \leq j \leq 3$.

---

[59]   Address: Mario Bravo 249, Buenos Aires, Argentina.

The first, second, and third symbols are called, respectively, the blank (for unmarked square), 0, and 1.

A bounded-transfer Turing machine may be represented schematically as shown in Figure 13 on page 234. We make the following stipulations: initially the machine is in its first state and scanning the first square of the tape; no bounded-transfer Turing machine may in the course of a calculation scan the end square of the tape and then move the tape one square to the right; initially all squares of the tape are blank; only orders to transfer to state $N + 1$ may be used to halt the machine. A bounded-transfer Turing machine is said to calculate a particular finite binary sequence (e.g. 01111000) if the machine stops with that sequence written at the end of its tape, with all other squares of the tape blank, and with its scanner on the first blank square of the tape. Figure 14 on page 234 illustrates a machine which has calculated the particular sequence mentioned above.

Before proceeding we would like to make a comment from the point of view of the programmer. The logical design of the bounded-transfer Turing machine provides automatically for relocation of programs, and the preceding paragraph establishes linkage conventions for subroutines which calculate finite binary sequences.

Two functions are now defined which play fundamental roles in all that follows. L, the first function,[62] is defined on the set of all finite binary sequences S as follows: An N-state, 3-tape-symbol bounded-transfer Turing machine can be programmed to calculate S if and only if $N \geq L(S)$.

The second function $L(C_n)$ is defined as

$$L(C_n) = \max_{S \text{ of length } n} L(S)$$

where the maximum is taken (as indicated) over all binary sequences S of length n. Also denote by[63] C_n the set of all binary sequences S of length n satisfying $L(S) = L(C_n)$.

An attempt is made in [1, Sec. 3.1] to make it plausible, on the basis of various philosophical considerations, that the patternless or random finite binary sequences of length n are those sequences S for which $L(S)$ is approximately equal to $L(C_n)$. Here an attempt is made to clarify this (somewhat informal) definition and to make it plausible by proving various results concerning what may be termed statistical properties of such finite binary sequences. The set C_∞ of patternless or random, infinite binary sequences is formally defined to be the set of all infinite binary sequences S which satisfy the following inequality for all sufficiently large values of n:

$$L(S_n) > L(C_n) - f(n)$$

where $f(n) = 3 \log_2 n$ and S_n is the sequence of the first n bits of S.

This definition, unlike the first, is quite precise but is also somewhat arbitrary. The failure to state the exact cut-off point at which $L(S)$ becomes too small for S to be considered random or patternless gives to the first definition its informal character. But in the case of finite binary sequences, no gain in clarity is achieved by arbitrarily settling on a cut-off point, while the opposite is true for infinite sequences.

The results from [1] which we need are as follows:

---

[60]   The choice of 3-tape-symbol machines is made merely for the purpose of fixing ideas.

[61]   Here b is a constant whose value is to be regarded as fixed throughout this paper. Its exact value is not important as long as it is not "too small." For an explanation of the meaning of "too small," and proofs that b can be chosen so that it is not too small, see [1, Secs. 2.1 and 2.2]. (b will not be mentioned again.)

[62]   Use of the letter "L" is suggested by the phrase "the Length of program necessary for computing...".

[63]   Use of the letter "C" is suggested by the phrase "the most Complex binary sequences of length...".

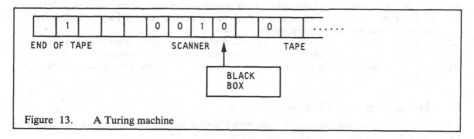

Figure 13.   A Turing machine

$$L(S*S') \leq L(S) + L(S') \tag{1}$$

where S and S' are finite binary sequence and * is the concatenation operation.

$$L(C_{n+m}) \leq L(C_n) + L(C_m) \tag{2}$$

There exists a positive real constant a^* such that (3)

$$(L(C_n)/n) \geq a^* \tag{3a}$$

$$\lim_{n \to \infty} (L(C_n)/n) = a^*. \tag{3b}$$

There exists an integer c such that there are less than 2^{n-m} binary sequences S of length n satisfying the inequality

$$L(S) \leq L(C_n) - \log_2 n - m - c. \tag{4}$$

Inequalities (1), (2), and (3a) are used only in Section 6, and (4) is used only in Section 7. For the proofs of (1), (2), and (3) see [1, Sec. 2.3]. The validity of inequality (4) is easily demonstrated using the method of [1, Sec. 2.4].

The following notational conventions are used throughout this paper:

(a)   * denotes the concatenation operation.
(b)   Let S be a finite binary sequence. S^n denotes the result of concatenating S with itself $n - 1$ times.
(c)   Let m be a positive integer. $B(m)$ denotes the binary sequence which is the numeral representing m in base-two notation; e.g. $B(37) = 100101$. Note that the bit at the left end of $B(m)$ is always 1.

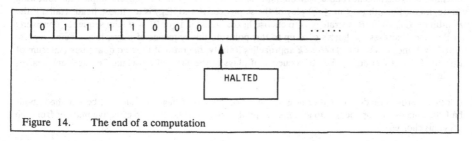

Figure 14.   The end of a computation

(d) Let x be a real number. $[x]$ denotes the least integer greater than the enclosed real x. Note that this is not the usual convention, and that the length of $B(m)$ is equal to $[\log_2 m]$. This last fact will be used but not explicitly mentioned.

(e) ε_x denotes a (not necessarily positive) function of x, and possibly other variables, which approaches zero as x approaches infinity with any other variables held fixed.

(f) Let S be an infinite binary sequence. S_k denotes the binary sequence consisting of the first k bits of S.

2. The Fundamental Theorem

All of our results concerning the statistical properties of random binary sequences will be established by applying the result which is proved in this section.

Theorem 1: Let q be an effective ordering of the finite binary sequences of any given length among themselves; i.e. let q be an effectively computable function with domain consisting of the set of all finite binary sequences and with range consisting of the set of all positive integers, and let the restriction of q to the domain of the set of all binary sequences of length n have the range $\{1, 2, 3, \ldots, 2^n\}$. Then there exists a positive integer c such that for all binary sequences S of length n,

$$L(S) \leq L(C_{[\log_2 q(S)]}) + L(C_{[\log_2 n]}) + c.$$

Proof: The program in Figure 15 on page 236 calculates S and consists of the following number of rows:

$$1 + L(B(q(S))) + 1 + L(B(n)) + (c - 2) \leq L(C_{[\log_2 q(S)]}) + L(C_{[\log_2 n]}) + c.$$

3. An Application: Matching Pennies

The following example of an application of Theorem 1 concerns the game of Matching Pennies.

According to the von Neumann and Morgenstern theory [2] of the mixed strategy for nonstrictly determined, zero-sum two-person games, a rational player will choose heads and tails equiprobably by some "device subject to chance".

Theorem 2: Let $S_1, S_2, S_3, \ldots$ be a sequence of distinct, finite binary sequences of lengths, respectively, $n_1, n_2, n_3, \ldots$ which satisfies $L(S_k) \sim L(C_{n_k})$. Let st be an effectively computable binary function defined on the set of all finite binary sequences and the null sequence. For each positive integer k consider a sequence of n_k plays of the game of Matching Pennies. There are two players: A who attempts to avoid matches and B who attempts to match A's penny. The players employ the following strategies for the mth ($1 \leq m \leq n_k$) play of the sequence. A's strategy is to choose heads (tails) if the mth bit of S_k is 1 (0). B's strategy is to choose heads (tails) if 1 (or 0) $= st$ (the sequence consisting of the $m - 1$ successive choices of A up to this point on this sequence of plays, heads being represented by 1 and tails by 0). Then as k approaches infinity, the ratio of the two quantities (the sum of the payoffs to A (B) during the kth sequence of plays of the game of Matching Pennies) approaches the limit 1.

In other words, a random or patternless sequence of choices of heads or tails will be matched about half the time by an opponent who attempts to predict the next choice in an effective manner from the previous choices.

```
Section I:
1,4       1,4       1,4

Section II consists of L(B(q(S))) rows.  It is a
program for calculating B(q(S)) consisting of the
smallest possible number of rows.

Section III:
1,4       1,4       1,4

Section IV consists of L(B(n)) rows.  It is a
program for calculating B(n) consisting of the
smallest possible number of rows.

Section V consists by definition of c - 2 rows.
It calculates the effectively computable function

q⁻¹(q(S), n) = S;

it finds the two arguments on the tape.
```

Figure 15.   Proof of Theorem 1

The proof of Theorem 2 is similar to the proof of Theorem 3 below, and therefore is omitted.

4. An Application: Simple Normality

In analogy to Borel's concept of the simple normality of a real number r in the base b (see [3, Ch. 9] for a definition of this concept of Borel), let a sequence $S_1, S_2, S_3, \ldots$ of finite b-ary sequences be called simply normal if

$$\lim_{k \to \infty} \frac{\text{the number of occurrences of } a \text{ in } S_k}{\text{the length of } S_k} = \frac{1}{b}$$

for each of the b possible values of a. The application of Theorem 1 given in this section concerns the simple normality of a sequence $S_1', S_2', S_3', \ldots$ of finite b-ary sequences in which each of the S_k' is associated with a binary sequence S_k in a manner defined in the next paragraph. It will turn out that $L(S_k) \sim L(C_{n_k})$, where n_k is the length of S_k, is a sufficient condition for the simple normality of the sequence of associated sequences.

Given a finite binary sequence, we may place a binary point to its left and consider it to be the base-two notation for a nonnegative real number r less than 1. Having done so it is natural to consider, say, the ternary sequence used to represent r to the same degree of precision in base-three notation. Let us define this formally for an arbitrary base b. Suppose that the binary sequence S of length n represents a real number r when a binary point is affixed to its left. Let n' be the smallest positive integer for which $2^n \le b^{n'}$. Now consider the set of all reals written in base-b notation as a "decimal" point followed by any of the $b^{n'}$ b-ary sequences of length n', including those with 0's at the right end. Let r' be the greatest of these reals which is less than or equal to r, and let the b-ary sequence S' be the one used to represent r' in base-b notation. S' is the b-ary sequence which we will associate with the binary sequence S. Note that no two binary sequences of the same length are associated with the same b-ary sequence.

It is now possible to state the principal result of this section.

Theorem 3: Let $S_1, S_2, S_3, \ldots$ be a sequence of distinct, finite binary sequences of lengths, respectively, $n_1, n_2, n_3, \ldots$ which satisfies $L(S_k) \sim L(C_{n_k})$. Then the sequence $S_1', S_2', S_3', \ldots$ of associated b-ary sequences is simply normal.

We first prove a subsidiary result.

Lemma 1: For any real number $e > 0$, any real number $d > 1$, b, and $0 \leq j < b$, for all sufficiently large values of n, if S is a binary sequence of length n whose associated b-ary sequence S' of length n' satisfies the following condition

$$\left| \frac{\text{the number of occurrences of } j \text{ in } S'}{n'} - \frac{1}{b} \right| > e, \tag{5}$$

then

$$L(S) < L(C_{\left[nd^{\dfrac{H\left(\frac{1}{b} - \frac{e}{b-1}, \ldots, \frac{1}{b} - \frac{e}{b-1}, \frac{1}{b} + e\right)}{\log_2 b} \right]}}).$$

Here

$$H(p_1, p_2, \ldots, p_b) \quad (p_1 \geq 0, \; p_2 \geq 0, \; \ldots, \; p_b \geq 0, \; \sum_{i=1}^{b} p_i = 1)$$

is defined to be equal to

$$- \sum_{i=1}^{b} p_i \log_2 p_i$$

where in this sum any terms $0 \log_2 0$ are to be replaced by 0.

The H function occurs because the logarithm to the base two of

$$\sum_{\left| \frac{k}{n'} - \frac{b-1}{b} \right| > e} (b-1)^k \binom{n'}{k},$$

the number of b-ary sequences S' of length n' which satisfy (5) is asymptotic, as n approaches infinity, to

$$n'H\left(\frac{1}{b} - \frac{e}{b-1}, \ldots, \frac{1}{b} - \frac{e}{b-1}, \frac{1}{b} + e \right),$$

which is in turn asymptotic to $nH/\log_2 b$, for $n' \sim n/\log_2 b$. This may be shown by considering the ratio of successive terms of the sum and using Stirling's approximation, $\log (n!) \sim n \log n$ [4, Ch. 6, Sec. 3].

To prove Lemma 1 we first define an ordering q by the following two conditions:

(a) Consider two binary sequences (of length n) S and T whose associated b-ary sequences (of length n') S' and T' contain, respectively, s and t occurrences of j. S comes before (after) T if

$$\left| \frac{s}{n'} - \frac{1}{b} \right| \text{ is greater (less) than } \left| \frac{t}{n'} - \frac{1}{b} \right|.$$

(b) If condition (a) doesn't settle which of the two sequences of length n comes first, take S to come before (after) T if S' represents (ignoring 0's to the left) a larger (smaller) number in base-b notation than T' represents.[64]

Proof: We now apply Theorem 1 to any binary sequence S of length n such that its associated b-ary sequence S' of length n' satisfies (5). Theorem 1 gives us

$$L(S) \leq L(C_{\lceil \log_2 q(S) \rceil}) + L(C_{\lceil \log_2 n \rceil}) + c \tag{6}$$

where, as we know from the paragraph before the last, for all sufficiently large values of n,

$$\log_2 q(S) < (1 + \frac{1}{4}(d - 1))\frac{nH}{\log_2 b}. \tag{7}$$

From (3b) and (7) we obtain for large values of n,

$$L(C_{\lceil \log_2 q(S) \rceil}) < a*(1 + \frac{1}{2}(d - 1))\frac{nH}{\log_2 b}. \tag{8}$$

And eq. (3b) implies that for large values of n,

$$L(C_{\lceil \log_2 n \rceil}) + c < a*\frac{1}{4}(d - 1)\frac{nH}{\log_2 b}. \tag{9}$$

Adding ineqs. (8) and (9), we see that ineq. (6) yields, for large values of n,

$$L(S) < a*(1 + \frac{3}{4}(d - 1))\frac{nH}{\log_2 b}.$$

Applying eq. (3b) to this last inequality, we see that for all sufficiently large values of n,

$$L(S) < L(C_{\lceil nd\frac{H}{\log_2 b} \rceil}),$$

which was to be proved.

Having demonstrated Lemma 1 we need only point out that Theorem 3 follows immediately from Lemma 1, eq. (3b), and the fact that

$$H(p_1, p_2, \ldots, p_b) \leq \log_2 b,$$

with equality if and only if

$$p_1 = p_2 = \cdots = p_b = \frac{1}{b}$$

(for a proof of this inequality, see [5, Sec. 2.2]).

---

[64] This condition was chosen arbitrarily for the sole purpose of "breaking ties."

5. Applications of a von Mises Place Selection V

In this section we consider the finite binary sequence S' resulting from the application of a von Mises place selection V to a finite binary sequence S which is random in the sense of Section 1. For S not to be rejected as random in the sense of von Mises [6] (i.e. in von Mises' terminology, for S not to be rejected as a collective[65]), S' must contain about as many 0's as 1's.

A place selection V is defined to be a binary function (following Church [8], it must be effectively computable) defined on the set of all finite binary sequences and the null sequence. If $S = S'*S''$ is a finite binary sequence, then $V(S') = 0$ (1) is interpreted to mean that the first (i.e. the leftmost) bit of S'' is not (is) selected from S by the place selection V.

By applying Theorem 1 and eq. (3b) we obtain the principal result of this section.

Theorem 4: Let $S_1, S_2, S_3, \ldots$ be a sequence of distinct finite binary sequences of lengths, respectively, $n_1, n_2, n_3, \ldots$ which satisfies $L(S_k) \sim L(C_{n_k})$. Let V be any place selection such that

$$\inf \left(\frac{\text{length of subsequence of } S \text{ selected by } V}{\text{length of } S} \right) > 0 \qquad (10)$$

where the infinum is taken over all finite binary sequences S. Then as k approaches infinity, the ratio of the number of 0's in the subsequence of S_k which is selected by V to the number of 1's in this subsequence tends to the limit 1.

Before proceeding to the proof it should be mentioned that a similar result can be obtained for the generalized place selections due to Loveland [9-11].

The proof of Theorem 4 runs parallel to the proof of Theorem 3. The subsidiary result which is proved by taking in Theorem 1 the ordering q defined below is

Corollary 1: Let e be a real number greater than 0, d be a real number greater than 1, S be a binary sequence of length n, and let V be a place selection which selects from S a subsequence S' of length n'. Suppose that

$$\left| \frac{\text{the number of 0's in } S'}{n'} - \frac{1}{2} \right| > e. \qquad (a)$$

Then for n' greater than N we have

$$L(S) \leq L(C_{[\log_2 q(S)]}) + L(C_{[\log_2 n]}) + c,$$

where

$$\log_2 q(S) < n' d H(\tfrac{1}{2} + e, \tfrac{1}{2} - e) + (n - n').$$

Here N depends only on e and d, and c depends only on V.

Definition[66]: Let S be a binary sequence of length n, let S' of length n' be the subsequence of S selected by the place selection V, and let S'' be the subsequence of S which is not selected by V. Let[67]

$$Q = F(S')*S''*01*B_1^2*B_2^2*B_3^2* \ldots$$

---

[65]   Strictly speaking we cannot employ von Mises' terminology here for von Mises was interested only in infinite sequences. Kolmogorov [7] considers finite sequences.

where each B_i is a single bit and

$$1*B_1*B_2*B_3* \ldots = B(\text{the length of } F(S')).$$

We then define $q(S)$ to be the unique solution of $B(q(S)) = Q$.

Definition: (Let us emphasize that $F(S')$ is never more than about

$$n' \, H(\tfrac{1}{2} + e, \, \tfrac{1}{2} - e)$$

bits long for S' which satisfy supposition (a) of Cor. 1: this is the crux of the proof.) Consider the "padded" numerals for the integers from 0 to $2^{n'} - 1$; padded to a length of n' bits by adding 0's on the left. Arrange these in order of decreasing

$$\left| \frac{m}{n'} - \frac{1}{2} \right|$$

where m is the number of 0's in the padded numeral, and when this does not decide the order, in numerical order (e.g. the list starts

$$0^{n'}, \ 1^{n'}, \ 0^{n'-1}*1, \ 0^{n'-2}*10, \ \ldots \qquad).$$

Suppose that S' is the kth padded numeral in the list. We define $F(S')$ to be equal to $B(k)$. Further details are omitted.

6. Fundamental Properties of the L-Function

In Sections 3–5 the random or patternless finite binary sequences have been studied. Before turning our attention to the random or patternless infinite binary sequences, we would like to show that many fundamental properties of the L-function are simple consequences of the inequality $L(S*S') \leq L(S) + L(S')$ taken in conjunction with the simple normality of sequences of random finite binary sequences.

In Theorem 3 take $b = 2^k$ and let the infinite sequence $S_1, S_2, S_3, \ldots$ consist of all the elements of the various C_n's. We obtain

Corollary 2: For any $e > 0$, k, and for all sufficiently large values of n, consider any element S of C_n to be divided into between $(n/k) - 1$ and (n/k) nonoverlapping binary subsequences of length k with not more than $k - 1$ bits left over at the right end of S. Then the ratio of the number of occurrences of any particular one of the 2^k possible binary subsequences of length k to (n/k) differs from 2^{-k} by less than e.

Keeping in mind the hypothesis of Corollary 2, let S be some element of C_n. Then we have $L(C_n) = L(S)$, and from Corollary 2 with

---

[66]   Strictly speaking, this definition is incorrect. S', reconstructed from $F(S')$ and n, and S'' can be "pieced together" to form S using V to dictate the intermixing, and thus $q(S) = q(T)$ for S and T of the same length only if $S = T$. But $q(S)$ is greater than 2^n for some binary sequences S of length n. To correct this it is necessary to obtain the "real" ordering q' from the ordering q that we define here by "pressing the function q down so as to eliminate gaps in its range." Formally, consider the restriction of q to the domain of all binary sequences of length n. Let the kth element in the range of this restriction of q, ordered according to magnitude, be denoted by r_k. Let S satisfy $q(S) = r_k$. We define $q'(S)$ to be equal to k. As, however, the result of this redefinition is to decrease the value of $q(S)$ for some S, this is a quibble.

[67]   Our superscript notation for concatenation is invoked here for the first time.

$$L(S) = L(S'*S''*S'''* \ldots) \leq L(S') + L(S'') + L(S''') + \cdots$$

(this inequality is an immediate consequence of (1)) this gives us

$$L(C_n) \leq \frac{n}{k}(1 + \varepsilon_n)(2^{-k}\sum L(S)),$$

where the sum is taken over the set of all binary sequences of length k. That is,

$$(L(C_n)/n)k \leq (1 + \varepsilon_n)(2^{-k}\sum L(S)),$$

with which (3a) or $(L(C_n)/n) \geq a^*$ gives

$$a^*k \leq (1 + \varepsilon_n)(2^{-k}\sum L(S)).$$

We conclude from this last inequality the following theorem.

Theorem 5: For all positive integers k,[68]

$$a^*k \leq 2^{-k} \sum_{S \text{ of length } k} L(S).$$

Note that the right-hand side of the inequality of Theorem 5 is merely the expected value of the random variable $L = L(S)$ where the sample space is the set of all binary sequences of length k to which equal probabilities have been assigned. With this probabilistic framework understood, we can denote the right-hand side of the inequality of Theorem 5 by $E\{L\}$ and use the notation $Pr\{\ldots\}$ for the probability of the enclosed event. Recalling eq. (3b) and the definition of $L(C_k)$ as max L, we thus have for any $e > 0$,

$$a^*k \leq E\{L\} = \sum Pr\{S\}L(S)$$
$$\leq Pr\{L \leq (1 - e)a^*k\}\,((1 - e)a^*k) + (1 - Pr\{L \leq (1 - e)a^*k\})\,L(C_k)$$
$$= Pr\{L \leq (1 - e)a^*k\}\,((1 - e)a^*k) + (1 - Pr\{L \leq (1 - e)a^*k\})\,((1 + \varepsilon_k)a^*k),$$

or

$$\varepsilon_k - (e + \varepsilon_k)Pr\{L \leq (1 - e)a^*k\} \geq 0.$$

Thus for any real $e > 0$,

$$\lim_{k \to \infty} Pr\{L \leq (1 - e)a^*k\} = 0. \tag{11}$$

Although eq. (11) is weaker than (4), it is reached by a completely different route. It must be admitted, however, that it is easy to prove Theorem 5 from (4) by taking into account the subadditivity of the right-hand side of the inequality of Theorem 5.

From Theorem 5 we now demonstrate

---

[68]   This statement remains true, as can be proved in several ways, if " < " replaces " ≤ ".

Corollary 3: For all positive integers n, $(L(C_n)/n) > a^*$.

Proof: Since $L(0^n) \leq L(B(n)) + c \leq \lceil \log_2 n \rceil + c$, for large n, $L(S) < a^* n$ for at least one binary sequence of length n, and we therefore may conclude from Theorem 5 that for large n there must be at least one binary sequence S' of length n for which $L(S') > a^* n$; that is, for large n,

$$(L(C_n)/n) > a^*. \tag{12}$$

We now finish the proof of Corollary 3 by contradiction. Suppose that Corollary 3 is false, and there exists an n_0 such that

$$(L(C_{n_0})/n_0) = a^*.$$

$((L(C_{n_0})/n_0) < a^*$ is impossible by (3a).) Then from (2) and (3a) it would follow that for all positive integers k,

$$(L(C_{kn_0})/kn_0) = a^*,$$

which contradicts (12). ∎

The final topic of this section is a derivation of

Theorem 6: $L(C_n) - a^* n$ is unbounded.

Proof: Consider some particular binary sequence S which is a member of C_n. Then from Corollary 2, for large values of n there must certainly be a sequence of k consecutive 0's in S. Suppose that $S = R^* 0^{k*} T$. Then we have

$$L(C_n) = L(S) = L(R^* 0^{k*} T) \leq L(R) + L(0^k) + L(T)$$
$$\leq L(R) + L(B(k)) + c + L(T) \leq L(C_i) + L(C_j) + \lceil \log_2 k \rceil + c$$

where i is the length of R, j is the length of T, and $n - k = i + j$. That is,

Lemma 2: For any positive integer k, for all sufficiently large values of n there exist i and j such that

$$L(C_n) \leq L(C_i) + L(C_j) + \lceil \log_2 k \rceil + c,$$

and $n - k = i + j$.

Theorem 6 follows immediately from Lemma 2 through proof by contradiction. ∎

7. Random or Patternless Infinite Binary Sequences

This section and Section 8 are devoted to a study of the set C_∞ of random or patternless infinite binary sequences defined in Section 1. Two proofs that C_∞ is nonempty, both based on (4), are presented here. The first proof is measure theoretic; the measure space employed may be defined in probabilistic terms as follows: the successive bits of an infinite binary sequence are independent random variables which assume the values 0 and 1 equiprobably. The second proof exhibits an element from C_∞.

```
Start: Set k = 0, set S = null sequence, go to Loop1.

Loop1: Is k ≤ N or L(S) > L(C_k) - 3 log_2 k ?
       If so, set k = k + 1, set S = S*0, go to Loop1.
       If not, go to Loop2.

Loop2: If S = S'*0, set S = S'*1, go to Loop1.
       If S = S'*1, set k = k - 1, set S = S'.
       If k ≠ 0, go to Loop2.
       If k = 0, stop.
```

Figure 16.   "Calculating" an infinite random sequence

Theorem 7: C_∞ is nonempty.

First Proof: From (4) and the Borel-Cantelli lemma, it follows immediately that

$$C_\infty \text{ is a set of measure } 1. \tag{13}$$

Second Proof: It is easy to see from (4) that we can find an N so large that

$$\sum_{k>N} N_k 2^{-k} < 1 \tag{14}$$

where N_k is the number of binary sequences S of length k for which

$$L(S) \leq L(C_k) - 3 \log_2 k.$$

Consider Figure 16, the flowchart for a process which never terminates (for that would contradict (14)). Then from Dirichlet's box principle (if an infinity of letters is placed in a finite number of pigeonholes, then there is always a pigeonhole which receives an infinite number of letters) it is clear that from some point on, the first bit of S will remain fixed; from some point on, the first two bits of S will remain fixed; ... ; from some point on (depending on n), the first n bits of S will remain fixed; ... Let us denote by $S_{\lim}$ the infinite binary sequence whose nth bit is 0 (1) if from some point on, the nth bit of S remains 0 (1). It is clear that $S_{\lim}$ is in C_∞.

Remark: When C_∞ was defined in Section 1, we pointed out that this definition contains an arbitrary element, i.e. the choice of 3 $\log_2 n$ as the function $f(n)$. In defining C_∞ it is desirable to choose an f which goes to infinity as slowly as possible and which results in a C_∞ of measure 1. We will call such f's "suitable." From results in [1, Secs. 2.4 and 2.5], which are more powerful than (4), it follows that there is an f which is suitable and satisfies the equations

$$\begin{cases} \overline{\lim} \, (f(n)/\log_2 n) = 2a^*, \\ \underline{\lim} \, (f(n)/\log_2 n) = a^*. \end{cases}$$

The question of obtaining lower bounds on the growth of an f which is suitable will be considered in Section 10, but there a different computing machine is used as the basis for the definition of random or patternless infinite binary sequence.

8. Statistical Properties of Infinite, Random or Patternless Binary Sequences

Results concerning the statistical properties of infinite, random or patternless binary sequences follow from the corresponding results for finite sequences. Thus Theorem 8 is an immediate consequence of Theorem 3, and Corollary 1 and eq. (3b) yield Theorem 9.

Theorem 8: Real numbers whose binary expansions are sequences in C_∞ are simply normal in every base.[69]

Theorem 9: Any infinite binary sequence in C_∞ is a collective with respect to the set of place selections[70] which are effectively computable and satisfy the following condition: For any infinite binary sequence S,

$$\lim \frac{\text{the number of bits in } S_k \text{ which are selected by } V}{k} > 0.$$

9. A General Formulation: Binary Computing Machines

Throughout the study of random or patternless binary sequences which has been attempted in the preceding sections, there has been a recurring difficulty. Theorem 1 and the relationship $L(C_n) \sim a*n$ have been used as the cornerstones of our treatment, but the assumption that $L(C_n) \sim a*n$ does not ensure that $L(C_n)$ behaves sufficiently smoothly to make really effective use of Theorem 1. Indeed it is conceivable that greater understanding of the bounded-transfer Turing machine would reveal that $L(C_n)$ behaves rather roughly and irregularly. Therefore a new computing machine is now introduced.[71]

To understand the logical design of this computing machine, it is helpful to provide a general formulation of computing machines for calculating finite binary sequences whose programs are also finite binary sequences. We call these *binary computing machines*. Formally, a binary computing machine is a partial recursive function M of the finite binary sequences which is finite binary sequence valued. The argument of M is the program, and the partial recursive function gives the output (if any) resulting from that program. $L_M(S)$ and $L_M(C_n)$ (if the computing machine is understood, the subscript will be omitted) are defined as follows:

$$L_M(S) = \begin{cases} \min_{M(P)=S} (\text{length of } P), \\ \infty \quad \text{if there are no such } P, \end{cases}$$

$$L_M(C_n) = \max_{S \text{ of length } n} L_M(S).$$

---

[69] It is known from probability theory that a real r which is simply normal in every base has the following property. Let b be a base, and denote by a_n the nth "digit" in the base-b expansion of r. Consider a b-ary sequence $c_1, c_2, \ldots, c_m$. As n approaches infinity the ratio of (the number of those positive integers k less than n which satisfy $a_k = c_1, a_{k+1} = c_2, \ldots, a_{k+m-1} = c_m$) to n tends to the limit b^{-m}.

[70] Wald [12] introduced the notion of a collective with respect to a set of place selections; von Mises had originally permitted "all place selections which depend only on elements of the sequence previous to the one being considered for selection."

[71] The author has subsequently learned of Kolmogorov [13], in which a similar kind of computing machine is used in essentially the same manner for the purpose of defining a finite random sequence. Martin-Löf [14-15] studies the statistical properties of these random sequences and puts forth a definition of an infinite random sequence.

In this general setting the program for the definition of a random or patternless binary sequence assumes the following form: The patternless or random finite binary sequences of length n are those sequences S for which $L(S)$ is approximately equal to $L(C_n)$. The patternless or random infinite binary sequences S are those whose truncations S_n are all patternless or random finite sequences. That is, it is necessary that for large values of n, $L(S_n) > L(C_n) - f(n)$ where f approaches infinity slowly.

We define below a binary computing machine M^* which has, as is easily seen, the following very convenient properties.

(a) $L(C_n) = n + 1$.
(b) Those binary sequences S of length n for which $L(S) < L(C_n) - m$ are less than 2^{n-m} in number.
(c) For any binary computer M there exists a constant c such that for all finite binary sequences S,
$$L_{M^*}(S) \leq L_M(S) + c.$$

The computing machine M^* is constructed from the two-argument partial recursive function $U(P, M')$, a universal binary computing machine. That is, U is characterized (following Turing) by the property that for any binary computer M there exists a finite binary sequence M' such that for all programs P, $U(P, M') = M(P)$ where both sides of the equation are undefined whenever one of them is.

Definition: If possible[72] let $P = P'*B$ where B is a single bit. If $B = 1$ then we define $M^*(P)$ to be equal to P'. If $B = 0$ then let the following equation be examined for ·a solution: $P' = S*T*01*B_1^{2*}B_2^{2*}B_3^{2*} \ldots$ where each B_i is a single bit, $1*B_1*B_2*B_3* \ldots = B(n)$, and T is of length n. If this equation has a solution then the solution must be unique, and we define $M^*(P)$ to be equal to $U(S, T)$.

10. Bounds on Suitable Functions

In Section 7 we promised to provide bounds on any f which is suitable (i.e. suitable for defining a C_∞ of measure 1). We prove here that $\overline{\lim} f(k)/\log_2 k \geq 1$, the constant being best possible.

We use the result [4 (1950 ed.), p. 163, prob. 4] that the set # of those infinite binary sequences S for which $r(S_k) > \lceil \log_2 k \rceil$ infinitely often is of measure 1; here r denotes the length of the run of 0's at the right end of the sequence.[73] As # and C_∞ are both of measure 1, they have an element S in common. Then for infinitely many values of k,
$$\begin{cases} L(S_k) > L(C_k) - f(k), \\ r(S_k) > \lceil \log_2 k \rceil. \end{cases}$$

But taking into account property (c) of M^*, we see that $S_k = \cdots *0^{\lceil \log_2 k \rceil}$ implies that $L(S_k) \leq L(C_{k-\lceil \log_2 k \rceil}) + c$. Thus for infinitely many values of k,
$$L(C_{k-\lceil \log_2 k \rceil}) + c \geq L(S_k) > L(C_k) - f(k)$$

or

$$k - \lceil \log_2 k \rceil + 1 + c \geq L(S_k) > k + 1 - f(k),$$

which implies that $f(k) > \lceil \log_2 k \rceil - c$. Hence $\overline{\lim} f(k)/\log_2 k$ must be greater than or equal to 1.

---

[72] That is, if P is a single bit this is not possible. $M^*(P)$ is therefore undefined.
[73] We are indebted to Professor Leonard Cohen of the City University of New York for pointing out to us the existence of such results.

ON THE LENGTH OF PROGRAMS FOR COMPUTING FINITE BINARY SEQUENCES

Now it is necessary to show that the constant is the best possible. From the Borel-Cantelli lemma and property (b) of M^*, we see at once that for f to be suitable, it is sufficient that

$$\sum_{k=1}^{\infty} 2^{-f(k)}$$

converges. Thus $f(k) = \log_2 (k(\log k)^2)$ is suitable, and this $f(k)$ is asymptotic to $\log_2 k$.

11. Two Analogues to the Fundamental Theorem

To study the statistical properties of binary sequences which are defined to be random or patternless on the basis of the computing machine M^*, it is necessary to have, in addition to properties (a) and (b) of M^*, an analogue to Theorem 1. We state two, the second of which is just a refinement of the first. Both are proved using property (c) of M^*.

Theorem 10: On the hypothesis of Theorem 1, for all binary sequences S of length n,

$$L(S) \leq L(B(q(S))^*B(n)^*01^*B_1^2{}^*B_2^2{}^*B_3^2{}^* \ldots) + c$$

where each B_i is a single bit and $1^*B_1{}^*B_2{}^*B_3{}^* \ldots = B([\log_2 n])$. Thus

$$L(S) \leq L(C_{g(q(S), n)}) + c \leq g(q(S), n) + c'$$

where

$$g(q(S), n) = [\log_2 q(S)] + [\log_2 n] + 2[\log_2 [\log_2 n]].$$

Theorem 11: On the hypothesis of Theorem 1, for all binary sequences S of length n,

$$L(S) \leq L(B(q(S))^*B(n + 2 - [\log_2 q(S)])^*01^*B_1^2{}^*B_2^2{}^*B_3^2{}^* \ldots) + c$$

where each B_i is a single bit, $1^*B_1{}^*B_2{}^*B_3{}^* \ldots = B([\log_2 g(q(S), n)])$, and

$$g(q(S), n) = n + 2 - [\log_2 q(S)].$$

Thus

$$L(S) \leq L(C_{h(q(S), n)}) + c \leq h(q(S), n) + c'$$

where

$$h(q(S), n) = [\log_2 q(S)] + [\log_2 g(q(S), n)] + 2[\log_2 [\log_2 g(q(S), n)]].$$

On comparing property (a) of M^*, property (b) of M^*, and Theorem 11 with, respectively, (3b), (4), and Theorem 1, we see that they are analogous but far more powerful. It therefore follows that Sections 3-5, 7, and 8 can be applied almost *verbatim* to the present computing machine. In particular, Theorem 2, Lemma 1, Theorem 4, (13), Theorem 8, and Theorem 9 hold, without any change whatsoever, for the random sequences defined on the basis of M^*. In all cases, however, much stronger assertions can be made. For example, in place of Theorem 9 we can state that

Theorem 12[74]: The set C_∞ of all infinite binary sequences S which have the property that for all sufficiently large values of k, $L(S_k) > L(C_k) - \log_2 (k(\log k)^2)$, is of measure 1, and each element of

C_∞ is a collective with respect to the set of place selections V which are effectively computable and satisfy the following condition:[75] For any infinite binary sequence S,

$$\lim_{k \to \infty} \frac{\text{the number of bits in } S_k \text{ which are selected by } V}{\log_2 k} = \infty.$$

References

1. CHAITIN, G. J. On the length of programs for computing finite binary sequences. *J. ACM 13,* 4 (Oct. 1966), 547-569.

2. VON NEUMANN, J., AND MORGENSTERN, O. *Theory of Games and Economic Behavior.* Princeton U. Press, Princeton, N. J., 1953.

3. HARDY, G. H., AND WRIGHT, E. M. *An Introduction to the Theory of Numbers.* Oxford U. Press, Oxford, 1962.

4. FELLER, W. *An Introduction to Probability Theory and Its Applications, Vol. I.* Wiley, New York, 1964.

5. FEINSTEIN, A. *Foundations of Information Theory.* McGraw-Hill, New York, 1958.

6. VON MISES, R. *Probability, Statistics, and Truth.* Macmillan, New York, 1939.

7. KOLMOGOROV, A. N. On tables of random numbers. *Sankhya* {A}, 25 (1963), 369-376.

8. CHURCH, A. On the concept of a random sequence. *Bull. Amer. Math. Soc. 46* (1940), 130-135.

9. LOVELAND, D. W. *Recursively Random Sequences.* Ph.D. Diss., N.Y.U., June 1964.

10. —. The Kleene hierarchy classification of recursively random sequences. *Trans. Amer. Math. Soc. 125* (1966), 487-510.

11. —. A new interpretation of the von Mises concept of random sequence. *Z. Math. Logik Grundlagen Math. 12* (1966), 279-294.

12. WALD, A. Die Widerspruchsfreiheit des Kollectivbegriffes der Wahrsheinlichkeitsrechnung. *Ergebnisse eines mathematischen Kolloquiums 8* (1937), 38-72.

13. KOLMOGOROV, A. N. Three approaches to the definition of the concept "quantity of information." *Problemy Peredachi Informatsii 1* (1965), 3-11. (in Russian)

14. MARTIN-LÖF, P. The definition of random sequences. Res. Rep., Inst. Math. Statist., U. of Stockholm, Stockholm, 1966, 21 pp.

15. —. The definition of random sequences. *Inform. Contr. 9* (1966), 602-619.

16. LÖFGREN, L. Recognition of order and evolutionary systems. In *Computer and Information Sciences-II,* Academic Press, New York, 1967, pp. 165-175.

17. LEVIN, M., MINSKY, M., AND SILVER, R. On the problem of the effective definition of "random sequence". Memo 36 (revised), RLE and MIT Comput. Center, 1962, 10 pp.

---

[74] Compare the last paragraph of Section 10.

[75] In view of Section 10, it apparently is not possible by the methods of this paper to replace the "$\log_2 k$" here by a significantly smaller function.

RECEIVED NOVEMBER, 1965; REVISED NOVEMBER, 1966

ON THE SIMPLICITY AND SPEED OF PROGRAMS FOR COMPUTING INFINITE SETS OF NATURAL NUMBERS

Journal of the ACM 16 (1969), pp. 407-422.

GREGORY J. CHAITIN[76]
Buenos Aires, Argentina

Abstract

It is suggested that there are infinite computable sets of natural numbers with the property that no infinite subset can be computed more simply or more quickly than the whole set. Attempts to establish this without restricting in any way the computer involved in the calculations are not entirely successful. A hypothesis concerning the computer makes it possible to exhibit sets without simpler subsets. A second and analogous hypothesis then makes it possible to prove that these sets are also without subsets which can be computed more rapidly than the whole set. It is then demonstrated that there are computers which satisfy both hypotheses. The general theory is momentarily set aside and a particular Turing machine is studied. Lastly, it is shown that the second hypothesis is more restrictive then requiring the computer to be capable of calculating all infinite computable sets of natural numbers.

Key Words and Phrases

computational complexity, computable set, recursive set, Turing machine, constructive ordinal, partially ordered set, lattice

CR Categories

5.22

Introduction

Call a set of natural numbers *perfect* if there is no way to compute infinitely many of its members essentially better (i.e. simpler or quicker) than computing the whole set. The thesis of this paper is that perfect sets exist. This thesis was suggested by the following vague and imprecise considerations.

One of the most profound problems of the theory of numbers is that of calculating large primes. While the sieve of Eratosthenes appears to be as simple and as quick an algorithm for calculating all the primes as is possible, in recent times hope has centered on calculating large primes by calculating a subset of the primes, those that are Mersenne numbers. Lucas's test is simple and can test whether or not a Mersenne number is a prime with rapidity far greater than is furnished by the sieve method. If there are an infinity of Mersenne primes, then it appears that Lucas has achieved a decisive advance in this classical problem of the theory of numbers.[77]

---

[76]   Address: Mario Bravo 249, Buenos Aires, Argentina.
[77]   For Lucas's test, cf. Hardy and Wright [1, Sec. 15.5]. For a history of number theory, cf. Dantzig [2], especially Sections 3.12 and B.8.

An opposing point of view is that there is no way to calculate large primes essentially better than to calculate them all. If this is the case it apparently follows that there must be only finitely many primes.

1. General Considerations

The notation and terminology of this paper are largely taken from Davis [3].

Definition 1: A computing machine Σ is defined by a 2-ary nonvanishing computable function σ in the following manner. The natural number n is part of the output $\Sigma(p, t)$ of the computer Σ at time t resulting from the program p if and only if the nth prime[78] [79] divides $\sigma(p, t)$. The infinite set $\Sigma(p)$ of natural numbers which the program p causes the computing machine Σ to calculate is defined to be

$$\bigcup_t \Sigma(p, t)$$

if infinitely many numbers are put out by the computer in numerical order and without any repetition. Otherwise, $\Sigma(p)$ is undefined.

Definition 2: A program complexity measure Π is a computable 1-ary function with the property that only finitely many programs p have the same complexity $\Pi(p)$.

Definition 3: The complexity $\Pi_\Sigma(S)$ of an infinite computable set S of natural numbers as computed by the computer Σ under the complexity measure Π is defined to be equal to

$$\begin{cases} \min_{\Sigma(p)=S} \Pi(p), & \text{if there are such } p, \\ \infty, & \text{otherwise.} \end{cases}$$

I.e. $\Pi_\Sigma(S)$ is the complexity of the simplest program which causes the computer to calculate S, and if there is no such program,[80] the complexity is infinite.[81]

In this section we do not see any compelling reason for regarding any particular computing machine and program complexity measure as most closely representing the state of affairs with which number theorists are confronted in their attempts to compute large primes as simply and as quickly as possible.[82] The four theorems of this section and their extensions hold for any computer Σ and any program complexity measure Π. Thus, although we don't know which computer and complexity measure to select, as this section holds true for all of them, we are covered.

Theorem 1: For any natural number n, there exists an infinite computable set S of natural numbers which has the following properties:

(a) $\Pi_\Sigma(S) > n$.

---

[78] The 0th prime is 2, the 1st prime is 3, etc.
[79] The primes are, of course, used here only for the sake of convenience.
[80] This possibility can never arise for the simple-program computers or the quick-program computers introduced later; such computers can be programmed to compute any infinite computable set of natural numbers.
[81] A more formal definition would perhaps use ω, the first transfinite ordinal, instead of ∞.
[82] In Sections 2 and 3 the point of view is different; some computing machines are dismissed as degenerate cases and an explicit choice of program complexity function is suggested.

(b) For any infinite computable set R of natural numbers, $R \subset S$ implies $\Pi_\Sigma(R) \geq \Pi_\Sigma(S)$.

Proof: We first prove the existence of an infinite computable set A of natural numbers having no infinite computable subset B such that $\Pi_\Sigma(B) \leq n$. The infinite computable sets C of natural numbers for which $\Pi_\Sigma(C) \leq n$ are finite in number. Each such C has a smallest element c. Let the (finite) set of all these c be denoted by D. We take $A = \overline{D}$.

Now let $A_0, A_1, A_2, \ldots$ be the infinite computable subsets of A. Consider the following set:

$$E = \{\Pi_\Sigma(A_0), \Pi_\Sigma(A_1), \Pi_\Sigma(A_2), \ldots\}.$$

From the manner in which A was constructed, we know that each member of E is greater than n. And as the natural numbers are well-ordered, we also know that E has a smallest element r. There exists a natural number s such that $\Pi_\Sigma(A_s) = r$. We take $S = A_s$, and we are finished. Q.E.D.

Theorem 2: For any natural number n and any infinite computable set T of natural numbers with infinite complement, there exists a computable set S of natural numbers which has the following property: $T \subset S$ and $\Pi_\Sigma(S) > n$.

Proof: There are infinitely many computable sets of natural numbers which have T as a subset, but the infinite computable sets F of natural numbers for which $\Pi_\Sigma(F) \leq n$ are finite in number. Q.E.D.

Theorem 3: For any 1-ary computable function f, there exists an infinite computable set S of natural numbers which has the following property: $\Sigma(p) \subset S$ implies the existence of a t_0 such that for $t > t_0$, $n \in \Sigma(p, t)$ only if $t > f(n)$.

Proof: We describe a procedure for computing S in successive stages (each stage being divided into two successive steps); during the k th stage it is determined in the following manner whether or nor $k \in S$. Two subsets of the computing machine programs p such that $p < \lceil k/4 \rceil$ are considered: set A, consisting of those programs which have been "eliminated" during some stage previous to the kth; and set B, consisting of those programs not in A which cause Σ to output the natural number k during the first $f(k)$ time units of calculation.

STEP 1. Put k in S if and only if B is empty.

STEP 2. Eliminate all programs in B (i.e. during all future stages they will be in A).

The above constructs S. That S contains infinitely many natural numbers follows from the fact that up to the kth stage at most $k/4$ programs have been eliminated, and thus at most $k/4$ natural numbers less than or equal to k can fail to be in S.[83]

It remains to show that $\Sigma(p) \subset S$ implies the existence of a t_0 such that for $t > t_0$, $n \in \Sigma(p, t)$ only if $t > f(n)$. Note that for $n \geq 4p + 4$, $n \in \Sigma(p, t)$ only if $t > f(n)$. For a value of n for which this failed to be the case would assure p's being in A, which is impossible. Thus given a program p such that $\Sigma(p) \subset S$, we can calculate a point at which the program has become slow and will remain so; i.e. we can calculate a permissible value for t_0. In fact, $t_0(p) = \max_{j < 4p+4} f(j)$. Q.E.D.

The following theorem and the type of diagonal process used in its proof are similar in some ways to Blum's exposition of a theorem of Rabin in [5, pp. 241-242].

---

[83]   I.e. the Schnirelman density $d(S)$ of S is greater than or equal to $3/4$. It follows from $d(S) \geq 3/4$ that S is a basis of the second order; i.e. every natural number can be expressed as the sum of two elements of S. Cf. Gelfond and Linnik [4, Sec. 1.1]. We conclude that the mere fact that a set is a basis of the second order for the natural numbers does not provide a quick means for computing infinitely many of its members.

COMPUTING INFINITE SETS OF NATURAL NUMBERS

Theorem 4: For any 1-ary computable function f and any infinite computable set T of natural numbers with infinite complement, there exists an infinite computable set S of natural numbers which is a superset of T and which has the following property: $\Sigma(p) = S$ implies the existence of a t_0 such that for $t > t_0$, $n \in \Sigma(p, t)$ only if $t > f(n)$.

Proof: First we define three functions: $a(n)$ is equal to the nth natural number in $\overline{T}$; $b(n)$ is equal to the smallest natural number j greater than or equal to n such that $j \in T$ and $j + 1 \notin T$; and $c(n)$ is equal to $\max_{n \le k \le b(n)} f(k)$. As proof, we give a process for computing $S \cap \overline{T}$ in successive stages; during the kth stage it is determined in the following manner whether or not $a(k) \in S$. Consider the computing machine programs 0, 1, 2, ..., k to fall into two mutually exclusive sets: set A, consisting of those programs which have been eliminated during some stage previous to the kth; and set B, consisting of all others.

STEP 1. Determine the set C consisting of the programs in B which cause the computing machine Σ to output during the first $c(a(k))$ time units of calculation any natural numbers greater than or equal to $a(k)$ and less then or equal to $b(a(k))$.

STEP 2. Check whether C is empty. Should $C = \emptyset$, we neither eliminate programs nor put $a(k)$ in S; we merely proceed to the next (the $(k + 1)$-th) stage. Should $C = \emptyset$, however, we proceed to step 3.

STEP 3. We determine p_0, the smallest natural number in C.

STEP 4. We ask, "Does the program p_0 cause Σ to output the number $a(k)$ during the first $c(a(k))$ time units of calculation?" According as the answer is "no" or "yes" we do or don't put $a(k)$ in S.

STEP 5. Eliminate p_0 (i.e. during future stages p_0 will be in A).

The above constructs S. We leave to the reader the verification that the constructed S has the desired properties. Q.E.D.

We now make a number of remarks.

Remark 1: We have actually proved somewhat more. Let U be any infinite computable set of natural numbers. Theorems 1 and 3 hold even if it is required that the set S whose existence is asserted be a subset of U. And if in Theorems 2 and 4 we make the additional assumption that T is a subset of U, and $U \cap \overline{T}$ is infinite, then we can also require that S be a subset of U.

The above proofs can practically be taken word for word (with obvious changes which may loosely be summarized by the command "ignore natural numbers not in U") as proofs for these extended theorems. It is only necessary to keep in mind the essential point, which in the case of Theorem 3 assumes the following form. If during the kth stage of the diagonal process used to construct S we decide whether to put in S the kth element of U, we are still sure that $\Sigma(p) \subset S$ is impossible for all the p which were eliminated before. For if $\Sigma(p) \subset U$, then p is eliminated as before; while if $\Sigma(p)$ has elements not in U, then it is clear that $\Sigma(p) \subset S$ is impossible, for S is a subset of U.

Remark 2: In Theorems 1 and 2 we see two possible extremes for S. In Theorem 1 we contemplate an arbitrarily complex infinite computable set of natural numbers that has the property that there is no way to compute infinitely many of its members which is simpler than computing the whole set. On the other hand, in Theorem 2 we contemplate an infinite computable set of natural numbers that has the property that there is a way to compute infinitely many of its members which is very much simpler than computing the whole set. Theorems 3 and 4 are analogous to Theorems 1 and 2, but Theorem 3 does not go as far as Theorem 1. Although Theorem 3 asserts the existence of infinite computable sets of natural numbers which have no infinite subsets which can be computed quickly, it does not establish that no infinite subset can be computed more quickly than the whole set. In this generality we are unable to demonstrate a Theorem 3 truly analogous to Theorem 1, although an attempt to do so is made in Remark 5.

Remark 3: The restriction in the conclusions of Theorems 3 and 4 that t be greater than t_0 is necessary. For as Arbib remarks in [6, p. 8], in some computers Σ any finite part of S can be computed very quickly by a table look-up procedure.

Remark 4: The 1-ary computable function f of Theorems 3 and 4 can go to infinity very quickly indeed with increasing values of its argument. For example, let $f_0(n) = 2^n, f_{k+1}(n) = f_k(f_k(n))$. For each k, $f_{k+1}(n)$ is greater than $f_k(n)$ for all but a finite number of values of n. We may now proceed from finite ordinal subscripts to the first transfinite ordinal by a diagonal process: $f_\omega(n) = \max_{k \le n} f_k(n)$. We choose to continue the process up to ω^2 in the following manner, which is a natural way to proceed (i.e. the fundamental sequences can be computed by simple programs) but which is by no means the only way to get to ω^2. i and j denote finite ordinals.

$$f_{\omega i + j + 1}(n) = f_{\omega i + j}(f_{\omega i + j}(n)),$$

$$f_{\omega(i+1)}(n) = \max_{k \le n} f_{\omega i + k}(n),$$

$$f_{\omega^2}(n) = \max_{k \le n} f_{\omega k}(n).$$

Taking $f = f_{\omega^2}$ in Theorem 3 yields an S such that any attempt to compute infinitely many of its elements requires an amount of time which increases almost incomprehensibly quickly with the size of the elements computed.

More generally, the above process may be continued through to any constructive ordinal.[84] For example, there are more or less natural manners to reach ε_0, the first epsilon-number; the territory up to it is very well charted.[85]

The above is essentially a constructive version of remarks by Borel [8] in an appendix on a theorem of P. du Bois-Reymond. These remarks are partly reproduced in Hardy [9].

Remark 5: Remark 4 suggests the following approach to the speed of programs. For any constructive ordinal α there is a computable 2-ary function f (by no means unique) with the property that the set of 1-ary functions f_k defined by $f_k(n) = f(k,n)$ is a representative of α when ordered in such a manner that a function g comes before a function h if and only if $g(n) < h(n)$ holds for all but a finite number of values of n. We now associate an ordinal $\mathrm{Ord}_\Sigma(S)$ with each infinite computable set S of natural numbers in accordance with the following rules:

(a) $\mathrm{Ord}_\Sigma(S)$ equals the smallest ordinal $\beta < \alpha$ such that f_{k_0}, the βth element of the set of functions f_k, has the following property: There exists a program p and a time t_0 such that $\Sigma(p) = S$ and for $t > t_0$, $n \in \Sigma(p, t)$ only if $t \le f_{k_0}(n)$.

(b) If (a) fails to define $\mathrm{Ord}_\Sigma(S)$ (i.e. if the set of ordinals β is empty), then $\mathrm{Ord}_\Sigma(S) = \alpha$.

Then for any constructive ordinal α we have the following analogue to Theorem 1.

---

[84]   Cf. Davis [3, Sec. 11.4] for a definition of the concept of a constructive ordinal number.
[85]   Cf. Fraenkel [7, pp. 207–208].

Theorem 1′: Any infinite computable set T of natural numbers has an infinite computable subset S with the following properties:

(a) $\text{Ord}_\Sigma(S) \le \text{Ord}_\Sigma(T)$.

(b) For any infinite computable set R of natural numbers, $R \subset S$ implies $\text{Ord}_\Sigma(S) \le \text{Ord}_\Sigma(R)$.

Proof: Let $T_0, T_1, T_2, \ldots$ be the infinite computable subsets of T. Consider the following set of ordinal numbers less than or equal to α:

$$\{\text{Ord}_\Sigma(T_0), \text{Ord}_\Sigma(T_1), \text{Ord}_\Sigma(T_2), \ldots \}.$$

As the ordinal numbers less than or equal to α are well-ordered, this set has a smallest element β. There exists a natural number s such that $\text{Ord}_\Sigma(T_s) = \beta$. We take $S = T_s$. Q.E.D.

However, we must admit that this approach to the speed of programs does not seem to be a convincing support for the thesis of this paper.

2. Connected Sets, Simple-Program Computers, and Quick-Program Computers

The principal results of subsections 2.A and 2.B, namely, Theorems 6 and 8, hold only for certain computers, but we argue that all other computing machines are degenerate cases of computers which in view of their unnecessarily restricted capabilities do not merit consideration.

In this section and the next we attempt to make plausible the contention that some connected sets (defined below) may well be considered to be perfect sets. In subsection 2.A we study the complexity of subsets of connected sets, and in subsection 2.B we study the speed of programs for computing subsets of connected sets. The treatments are analogous but we find the second more convincing, because in the first treatment one explicit choice is made for the program complexity measure Π. $\Pi(p)$ is taken to be $[\log_2 (p + 1)]$.

The concept of a connected set is analogous to the concept of a retraceable set, cf. Dekker and Myhill [10].

Definition 4: A connecting function γ is a one-to-one onto mapping carrying the set of all finite sets of natural numbers onto the set of all natural numbers. The monotonicity conditions $\gamma(V \cup W) \ge \gamma(W)$ must be satisfied and there must be a 1-ary computable function g such that $\gamma(W) = g(\Pi_{n \in W} p_n)$, where p_n denotes the n th prime.[78] [79] Let $S = \{s_0, s_1, s_2, \ldots \}$ ($s_0 < s_1 < s_2 < \cdots$) be a computable set of natural numbers with m members ($0 \le m \le \aleph_0$). From a connecting function γ we define a secondary connecting function Γ as follows:[86]

$$\Gamma(S) = \bigcup_{k<m} \{\gamma(\bigcup_{j\le k} \{s_j\})\}.$$

A γ-connected set is defined to be any infinite computable set of natural numbers which is in the range of Γ.

Remark 6: Consider a connecting function γ. Note that any two γ-connected sets which have an infinite intersection must be identical. In fact, two γ-connected sets which have an element in common must be identical up to that element.

---

[86]   Thus $\Gamma(S)$ always has the same number of elements as S, be S empty, finite or infinite.

The following important results concerning γ-connected sets are established by the methods of Section 1 and thus hold for any computer Σ and complexity measure Π. For any natural number n there exists a γ-connected set S such that $\Pi_\Sigma(S) > n$. This follows from the fact that there are infinitely many γ-connected sets, while the infinite computable sets H of natural numbers such that $\Pi_\Sigma(H) \leq n$ are only finite in number. Theorem 3 remains true if we require that the set S whose existence is asserted be a γ-connected set. S may be constructed by a procedure similar to that of the proof of Theorem 3; during the k th stage instead of deciding whether or not $k \in S$, it is decided whether or not $k \in \Gamma^{-1}(S)$. These two results should be kept in mind while appraising the extent to which the theorems of this section and the next corroborate the thesis of this paper.

2.A. Simplicity

In this subsection we make one explicit choice for the program complexity measure Π. We consider programs to be finite binary sequences as well as natural numbers:

PROGRAMS
Binary Sequence Λ 0 1 00 01 10 11 000 001 010 011 100 ...
Natural Number 0 1 2 3 4 5 6 7 8 9 10 11 ...

Henceforth, when we denote a program by a lowercase (uppercase) Latin letter, we are referring to the program considered as a natural number (binary sequence). Next we define the complexity of a program P to be the number of bits in P (i.e. its length). I.e. the complexity $\Pi(p)$ of a program p is equal to $[\log_2 (p + 1)]$, the greatest integer not greater than the base-2 logarithm of $p + 1$.

We now introduce the simple-program computers. Computers similar to them have been used in Solomonoff [11], Kolmogorov [12], and in [13].

Definition 5: A simple-program computer Σ has the following property: For any computer Ξ, there exists a natural number $_\Sigma c_\Xi$ such that $\Pi_\Sigma(S) \leq \Pi_\Xi(S) + {}_\Sigma c_\Xi$ for all infinite computable sets S of natural numbers.

To the extent that it is plausible to consider all computer programs to be binary sequences, it seems plausible to consider all computers which are not simple-program computers as unnecessarily awkward degenerate cases which are unworthy of attention.

Remark 7: Note that if Σ and Ξ are two simple-program computers, then there exists a natural number $c_{\Sigma\Xi}$ which has the following property: $| \Pi_\Sigma(S) - \Pi_\Xi(S) | \leq c_{\Sigma\Xi}$ for all infinite computable sets S of natural numbers. In fact we can take

$$c_{\Sigma\Xi} = \max({}_\Sigma c_\Xi, {}_\Xi c_\Sigma).$$

Theorem 5: For any connecting function γ, there exists a simple-program computer Σ^γ which has the following property: For any γ-connected set S and any infinite computable subset R of S,

$$\Pi_{\Sigma^\gamma}(S) \leq \Pi_{\Sigma^\gamma}(R).$$

Proof: Taking for granted the existence of a simple program computer Σ^* (cf. Theorem 9), we construct the computer Σ^γ from it as follows:

$$\Sigma^\gamma(\Lambda, t) = \emptyset,$$

$$\Sigma^\gamma(P0, t) = \Sigma^*(P, t), \tag{1}$$

$$\Sigma^{\gamma}(P1, t) = \bigcap_{t' < t} \overline{\Sigma^{\gamma}(P1, t')} \ \cap \ \Gamma(\bigcup_{n \in \Sigma^*(P, t)} \gamma^{-1}(n)).$$

As Σ^* is a simple-program computer, so is Σ^{γ}, for $\Sigma^{\gamma}(P0, t) = \Sigma^*(P, t)$. Σ^{γ} also has the following very important property: For all programs $P0$ for which $\Sigma^{\gamma}(P0)$ is a subset of some γ-connected set S, $\Sigma^{\gamma}(P1) = S$. Moreover, $\Sigma^{\gamma}(P1)$ cannot be a proper subset of any γ-connected set. In summary, given a P such that $\Sigma^{\gamma}(P)$ is a proper subset of a γ-connected set S, then by changing the rightmost bit of P to a 1 we get a program P' with the property that $\Sigma^{\gamma}(P') = S$. This implies that for any infinite computable subset R of a γ-connected set S,

$$\Pi_{\Sigma^{\gamma}}(S) \le \Pi_{\Sigma^{\gamma}}(R).$$

Q.E.D.

In view of Remark 7, the following theorem is merely a corollary to Theorem 5.

Theorem 6: Consider a simple-program computer Σ. For any connecting function γ, there exists a natural number c_{γ} which has the following property: For any γ-connected set S and any infinite computable subset R of S, $\Pi_{\Sigma}(S) \le \Pi_{\Sigma}(R) + c_{\gamma}$. In fact, we can take[87]

$$c_{\gamma} = 2 \max({}_{\Sigma}c_{\Sigma^{\gamma}}, \ {}_{\Sigma^{\gamma}}c_{\Sigma}).$$

2.B. Speed

This treatment runs parallel to that of subsection 2.A.

Definition 6: A quick-program computer Σ has the following property: For any computer Ξ, there exists a 1-ary computable function ${}_{\Sigma}s_{\Xi}$ such that for all programs p for which $\Xi(p)$ is defined, there exists a program p' such that $\Sigma(p') = \Xi(p)$ and

$$\bigcup_{t' \le t} \Xi(p, t') \ \subset \ \bigcup_{t' \le {}_{\Sigma}s_{\Xi}(t)} \Sigma(p', t')$$

for all but a finite number of values of t.

Theorem 7: For any connecting function γ, there exists a quick-program computer Σ^{γ} which has the following property: For any program P such that $\Sigma^{\gamma}(P)$ is a proper subset of a γ-connected set S, there exists a program P' such that $\Sigma^{\gamma}(P') = S$ and $\Sigma^{\gamma}(P, t) \subset \Sigma^{\gamma}(P', t)$ for all t. In fact, P' is just P with the 0 at its right end changed to a 1, as the reader has no doubt guessed.

Proof: Taking for granted the existence of a quick-program computer Σ^* (cf. Theorem 9), we construct Σ^{γ} from it exactly as in the proof of Theorem 5. I.e. Σ^{γ} is defined, as before, by eqs. (1). The remainder of the proof parallels the proof of Theorem 5. Q.E.D.

---

[87] That

$$c_{\gamma} = {}_{\Sigma}c_{\Sigma^{\gamma}} + {}_{\Sigma^{\gamma}}c_{\Sigma}$$

will do follows upon taking a slightly closer look at the matter.

Theorem 7 yields the following corollary in a manner analogous to the manner in which Theorem 5 yields Theorem 6.

Theorem 8: Consider a quick-program computer Σ. For any connecting function γ there exists a 1-ary computable function s_γ which has the following property: For any program p such that $\Sigma(p)$ is a subset of a γ-connected set S, there exists a program p' such that

$$\Sigma(p') = S,$$

$$\bigcup_{t' \le t} \Sigma(p, t') \ \subset \ \bigcup_{t' \le s_\gamma(t)} \Sigma(p', t')$$

for all but a finite number of values of t. In fact we can take

$$s_\gamma(n) = {}_\Sigma s_{\Sigma^\gamma}({}_{\Sigma^\gamma} s_\Sigma(n)).$$

Remark 8: Arbib and Blum [14] base their treatment of program speed upon the idea that if two computers can imitate act by act the computations of the other, and not take too many time units of calculation to imitate the first several time units of calculation of the other, then these computers are essentially equivalent. The idea used to derive Theorem 8 from Theorem 7 is similar: Any two quick-program computers (and in particular Σ^γ and Σ) can imitate act by act each other's computations and are thus in a sense equivalent.

In order to clarify the above, let us formally define within the framework of Arbib and Blum a concept analogous to that of the quick-program computer. In what remains of this remark we use the notation and terminology of Arbib and Blum, not that of this paper. However, in order to prove that this analogous concept is not vacuous, it is necessary to make explicit an assumption which is implicit in their framework. For any machine M there exists a total recursive function m such that $m(i, x, t) = 2y$ if and only if ${}^M\phi_i(x) = y$ and ${}^M\Phi_i(x) = t$.

Definition AB: A quick-program machine M is a machine with the following property. Consider any machine N. There exists a total recursive function f_{NM} increasing in both its variables such that $N \ge_{(f_{NM})} M$; i.e. M is at least as complex as N (modulo (f_{NM})). Here, a two-variable function enclosed in parentheses denotes the monoid with multiplication * and identity $e(x, y) = y$, which is generated by the function.

Then by (ii) of Theorem 2 [14] we have

Theorem AB: Consider two quick-program machines M and N. There exists a total recursive function g_{NM} increasing in both of its variables such that $N \equiv_{(g_{NM})} M$; i.e. N and M are (g_{NM})-equivalent.

Remark 9: In an effort to make this subsection more comprehensible, we now cast it into the framework of lattice theory, cf. Birkhoff [15].

Definition L1: Let Σ_1 and Σ_2 be computing machines. Σ_1 **im** Σ_2 (Σ_1 can be imitated by Σ_2) if and only if there exists a 1-ary computable function f which has the following property: For any program p for which $\Sigma_1(p)$ is defined, there exists a program p' such that $\Sigma_2(p') = \Sigma_1(p)$ and

$$\bigcup_{t' \le t} \Sigma_1(p, t') \ \subset \ \bigcup_{t' \le f(t)} \Sigma_2(p', t')$$

for all but a finite number of values of t.

Lemma L1: The binary relation **im** is reflexive and transitive.

Definition L2: Let Σ_1 and Σ_2 be computing machines. Σ_1 **eq** Σ_2 if and only if Σ_1 **im** Σ_2 and Σ_2 **im** Σ_1.

Lemma L2: The binary relation **eq** is an equivalence relation.

Definition L3: L is the set of equivalence classes induced by the equivalence relation **eq**. For any computer Σ, (Σ) is the equivalence class of Σ, i.e. the set of all computers Σ' such that Σ' **eq** Σ. For any $(\Sigma_1),(\Sigma_2) \in L$, $(\Sigma_1) \leq (\Sigma_2)$ if and only if Σ_1 **im** Σ_2.

Lemma L3: L is partially ordered by the binary relation $\leq$.

Lemma L4: Consider a computer which cannot be programmed to compute any infinite set of natural numbers, e.g. the computer Σ_0 defined by $\Sigma_0(p, t) = \emptyset$. Denote by 0 the equivalence class of this computer; i.e. denote by 0 the computers which compute no infinite sets of natural numbers. 0 bounds L from below; i.e. $0 \leq A$ for all $A \in L$.

Lemma L5: Consider a quick-program computer, e.g. the computer Σ^* of Theorem 9. Denote by 1 the equivalence class of this computer; i.e. denote by 1 the quick-program computers. 1 bounds L from above; i.e. $A \leq 1$ for all $A \in L$.

Lemma L6: Let Σ_1 and Σ_2 be computers. Define the computer Σ_3 as follows: $\Sigma_3(\Lambda, t) = \emptyset$, $\Sigma_3(P0, t) = \Sigma_1(P, t)$, $\Sigma_3(P1, t) = \Sigma_2(P, t)$. (Σ_3) is the l.u.b. of (Σ_1) and (Σ_2).

Lemma L7: Let Σ_1 and Σ_2 be computers. Define the computer Σ_3 as follows: Consider the sets

$$S_1 = \bigcup_{t' \leq t} \Sigma_1(K(p), t'),$$

$$S_2 = \bigcup_{t' \leq t} \Sigma_2(L(p), t'),$$

where $(K(p), L(p))$ is the pth ordered pair in an effective enumeration of the ordered pairs of natural numbers (cf. Davis [3, pp. 43-45]). If Σ_1 and Σ_2 output in size order and without repetitions the elements of, respectively, S_1 and S_2, and $S_1 \subset S_2$ or $S_2 \subset S_1$, then

$$\Sigma_3(p, t) = S_1 \cap S_2 \cap \overline{\bigcup_{t' < t} \Sigma_3(p, t')}.$$

Otherwise, $\Sigma_3(p, t) = \emptyset$. (Σ_3) is the g.l.b. of (Σ_1) and (Σ_2).

Theorem L: L is a denumerable, distributive lattice with zero element and one element.

We may describe the g.l.b. and l.u.b. operations of this lattice as follows. The l.u.b. of two computers is the slowest computer which is faster than both of them, and the g.l.b. of two computers is the fastest computer which is slower than both of them.

3. A Simple, Quick-Program Computer

This section is the culmination of this paper. A computer is constructed which is both a simple-program computer and a quick-program computer.

If it is believed that programs are essentially binary sequences and that the only natural measure of the complexity of a program considered as a binary sequence is its length, then apparently the conclusion would have to be drawn that only simple, quick-program computers are worthy of attention, all other computers being degenerate cases.

It would seem to follow that the connected sets indeed corroborate this paper's thesis. For there is a simple, quick-program computer which best represents mathematically the possibilities open to number theorists in their attempts to calculate large primes. We do not know which it may happen to be, but we do know (cf. Remark 6) that there are connected sets which are very complex and which must be computed very slowly when one is using this computer. In view of Theorems 6 and 8 it would seem to be appropriate to consider these connected sets to be perfect sets. Thus our quest for perfect sets comes to a close.

Theorem 9: There exists a simple, quick-program computer, namely Σ^*.

Proof: We take it for granted that there is a computer $\Sigma^\$$ which can compute every 2-ary computable function f in the following sense: There exists a binary sequence P_f and a 2-ary computable function $\#_f$ increasing in its second argument such that

$$\{f(n, m)\} = \Sigma^\$(B(n)P_f, \#_f(n, m))$$

for all natural numbers n and m. Moreover, $\Sigma^\$(B(n)P_f, t)$ is nonempty only if there exists an m such that $t = \#_f(n, m)$. Here B is the function carrying each natural number into its associated binary sequence, as in Section 2.

From $\Sigma^\$$ we now construct the computer Σ^*: $n \in \Sigma^*(p, t)$ if and only if $\Sigma^\$(p, t)$ has only a single element, this element is not zero, and the nth prime divides it.[78][79]

We now verify that Σ^* is a simple, quick-program computer. Consider a computer Ξ. We give the natural number $_\Sigma \cdot c_\Xi$ explicitly: $_\Sigma \cdot c_\Xi$ is the length of P_ξ. We also give the 1-ary computable function $_\Sigma \cdot s_\Xi$ explicitly:

$$_\Sigma \cdot s_\Xi(n) = \max_{k \le n} \#_\xi(k, n).$$

Here ξ is, of course, the 2-ary computable function which defines the computer Ξ as in Definition 1. Q.E.D.

Appendix A. A Turing Machine

The contents of this appendix have yet to be fitted into the general framework which we attempted to develop in Sections 1-3.

Definition A: Δ is a Turing machine. Δ's "tape" is a quarter-plane or quadrant divided into squares. It has a single scanner which scans one of the squares. If the scanner runs off the quadrant, Δ halts. Δ can perform any one of the following operations: quadrant one square left (L), right (R), up (U), or down (D); or the scanner can overprint a 0 (0), a 1 (1), or erase (E) the square of the quadrant being scanned. The programs of Δ are tables with three columns headed "blank," "0," and "1," and consecutively numbered rows. Each place in the table must have an ordered pair; the first member of the pair gives the operation to be performed and the second member gives the number of the next row of the table to be obeyed. As program complexity measure Ω, we take the number of rows in the program's table. One operation (L, R, U, D, 0, 1, or E) is performed per unit time. The computing machine Δ begins calculating with its scanner on the corner square, with the quadrant completely erased, and obeying the last row of its program's table. The Turing machine outputs a natural number n when the binary sequence which represents n in base-2 notation appears at the bottom of the quadrant, starting in the corner square, ending in the square being scanned, and with Δ obeying the next to last row of its program's table.

Theorem A1: For any connecting function γ there exists a natural number c_γ and a 1-ary computable function s_γ which have the following property: For any program p for which $\Delta(p)$ is a subset of a γ-connected set S, there is a program p' such that

(a) $\Delta(p') = S,\ \Omega(p') = \Omega(p) + c_\gamma$; [88] and
(b)   for all natural numbers t,

$$\bigcup_{t' \le t} \Delta(p, t') \ \subset\ \bigcup_{t' \le t + s_\gamma(n)} \Delta(p', t'),$$

where n stands for the largest element of the left-hand side of the relation, if this set is not empty (otherwise, n stands for 0).

Proof: p' is obtained from p in the following manner. c_γ rows are added to the table defining the program p. All transfers to the next to the last row in the program p are replaced by transfers to the first row of the added section. The new rows of the table use the program p as a subroutine. They make the program p think that it is working as usual, but actually p is using neither the quadrant's three edge rows nor the three edge columns; p has been fooled into thinking that these squares do not exist because the new rows moved the scanner to the fourth square on the diagonal of the quadrant before turning control over to p for the first time by transferring to the last row of p. This protected region is used by the new rows to do its scratch-work, and also to keep permanent records of all natural numbers which it causes Δ to output.

Every time the subroutine thinks it is making Δ output a natural number n, it actually only passes n and control to the new rows. These proceed to find out which natural numbers are in $\Gamma(\gamma^{-1}(n))$. Then the new rows eliminate those elements of $\Gamma(\gamma^{-1}(n))$ which Δ put out previously. Finally, they make Δ output those elements which remain, move the scanner back to what the subroutine last thought was its position, and return control to the subroutine. Q.E.D.

Remark A: Assuming that only the computer Δ and program complexity measure Ω are of interest, it appears that we have before us some connected sets which are in a very strong sense perfect sets. For, as was mentioned in Remark 6, there are γ-connected sets which Δ must compute very slowly. For such sets, the term $s_\gamma(n)$ in (b) above is negligible compared with t.

---

[88]   This implies

$$\Omega_\Delta(S) \le \Omega_\Delta(\Delta(p)) + c_\gamma.$$

I.e.

$$\Omega_\Delta(S) \le \Omega_\Delta(R) + c_\gamma$$

for any infinite computable subset R of the γ-connected set S.

Theorem A2: Consider a simple-program computer Σ and the program complexity measure $\Pi(p) = \lceil \log_2(p+1) \rceil$. Let $S_0, S_1, S_2, \ldots$ be a sequence of distinct infinite computable sets of natural numbers. Then we may conclude that

$$\lim_{k \to \infty} (\Pi_\Sigma(S_k))/(2\,\Omega_\Delta(S_k) \log_2 \Omega_\Delta(S_k))$$

exists and is in fact unity.

Proof: Apply the technique of [16, Pt. 1].

Of course,

Theorem A3: Δ is a quick-program computer.

Appendix B. A Lattice of Computer Speeds

The purpose of this appendix is to study L^*, the lattice of speeds of computers which calculate all infinite computable sets of natural numbers. L^* is a sublattice (in fact, a filter) of the lattice L of Remark 9. It will be shown that L^* has a rich structure: every countable partially ordered set is imbeddable in L^*.[89] Thus to require a computer to be a quick-program computer is more than to require that it be able to compute all infinite computable sets of natural numbers.

Definition B1: L^* is the sublattice of L consisting of the (Σ) such that Σ can be programmed to compute all infinite computable sets of natural numbers.

In several respects the following theorem is quite similar to Theorem 9 of Hartmanis and Stearns [18] and to Theorem 8 of Blum [19]. The diagonal process of the proof of Theorem 3 is built into a computer's circuits.

Theorem B1: There exists a quick-program computer Σ_1 with the property that for any 1-ary computable function f and any infinite computable set U of natural numbers, there exists a 1-ary computable function g and an infinite computable set S of natural numbers such that

(a) $S \subset U$,
(b) $g(n) > f(n)$ for all but a finite number of values of n,
(c) there exists a program p such that $\Sigma_1(p) = S$ and $n \in \Sigma_1(p, g(n) + 1)$ for all $n \in S$,
(d) for all programs p' such that $\Sigma_1(p') \subset S$, $n \in \Sigma_1(p', t)$ only if $t > g(n)$, with the possible exception of a finite number of values of n.

Proof: Let Σ be a quick-program computer. We construct Σ_1 from it. $\Sigma_1(\Lambda, t) = \emptyset$, $\Sigma_1(P0, t) = \Sigma(P, t)$, $\Sigma_1(P1, 0) = \emptyset$, and $\Sigma_1(P1, t + 1)$ is a subset of $\Sigma(P, t)$. For each element $n^\#$ of $\Sigma(P, t)$, it is determined in the following manner whether or not $n^\# \in \Sigma_1(P1, t + 1)$. Define m, n_k $(0 \le k \le m)$, m', and t_k $(0 \le k \le m)$ as follows:

$$\bigcup_{t' \le t} \Sigma(P, t') = \{n_0, n_1, n_2, \ldots, n_m\} \quad (n_0 < n_1 < n_2 \cdots < n_m),$$

---

[89] An analogous result is due to Sacks [17, p. 53]. If P is a countable partially ordered set, then P is imbeddable in the upper semilattice of degrees of recursively enumerable sets. Cf. also Sacks [17, p. 21].

COMPUTING INFINITE SETS OF NATURAL NUMBERS

$$n^{\#} = n_{m'},$$

$$n_k \in \Sigma(P, t_k) \quad (0 \le k \le m).$$

Define $A(i, j)$ (the predicate "the program j is eliminated during the ith stage"),[90] A (the set of programs eliminated before the m'th stage), and A' (the set of programs eliminated before or during the m'th stage) as follows:

$$A(i, j) \text{ iff } j < [i/4] \text{ and } n_i \in \bigcup_{t' \le t_i} \Sigma_1(j, t'),$$

$$A = \{j \mid A(i, j) \text{ for some } i < m'\},$$

$$A' = \{j \mid A(t, j) \text{ for some } t \le m'\}.$$

$n^{\#} \in \Sigma_1(P1, t + 1)$ iff $A' = A$.

That the above indeed constructs Σ_1 follows from the fact that each of the t_k is less than $t + 1$, and thus $\Sigma_1(P1, t + 1)$ is defined only in terms of $\Sigma_1(p', t')$, for which t' is less than $t + 1$. I.e. that $\Sigma_1(p, t)$ is defined follows by induction on t. Also, Σ_1 is a quick-program computer, for $\Sigma_1(P0, t) = \Sigma(P, t)$.

We now define the function g and the set S, whose existence is asserted by the theorem. By one of the extensions of Theorem 3, there exists a program P which has the following properties:

1. $\Sigma(P) \subset U$.

2. For all but a finite number of values of n, $n \in \Sigma(P, t)$ only if $t > f(n)$.

$S = \Sigma_1(P1)$. That S is infinite follows from the fact that at most $k/4$ of the first k elements of $\Sigma(P)$ fail to be in S. $g(n)$ is defined for all $n \in \Sigma(P)$ by $n \in \Sigma(P, g(n))$. It is irrelevant how $g(n)$ is defined for $n \notin \Sigma(P)$, as long as $g(n) > f(n)$.

Part (a) of the conclusion follows from the fact that $\Sigma_1(P1, t + 1) \subset \Sigma(P, t) \subset U$ for all t. Part (c) follows from the fact that if $n \in \Sigma_1(P1)$, then

$$n \in \Sigma_1(P1, g(n) + 1).$$

Part (d) follows from the fact that if $\Sigma_1(p')$ is defined and n is the first element of $\Sigma_1(p') \cap \Sigma(p)$ which is greater than or equal to the $(4p' + 4)$-th element[91] of $\Sigma(P)$ and which is contained in a $\Sigma_1(p', t)$ such that $t \le g(n)$, then n is not an element of S. Q.E.D.

Corollary B1: On the hypothesis of Theorem B1, not only do the g and S whose existence is asserted have the properties (a) to (d), they also, as follows immediately from (c) and (d), have the property that for any quick-program computer Σ:

(e) There exists a program p_2 such that $\Sigma(p_2) = S$ and $n \in \Sigma(p_2, t)$ with $t \le {}_{\Sigma}s_{\Sigma_1}(g(n) + 1)$ for all but a finite number of values of $n \in S$.

(f) For all programs p_3 such that $\Sigma(p_3) \subset S$, $n \in \Sigma(p_3, t)$ only if ${}_{\Sigma_1}s_{\Sigma}(t) > g(n)$, with the possible exception of a finite number of values of n.

---

[90]   During the ith stage of this diagonal process it is decided whether or not the ith element of $\Sigma(P)$ is in $\Sigma_1(P1)$.

[91]   I.e. it is greater than or equal to n_{4p+4}.

Remark B: Theorem B1 is a "no speed-up" theorem; i.e. it contrasts with Blum's speed-up theorem (cf. [5, 6, 19]). Each S whose existence is asserted by Theorem B1 has a program for Σ_1 to compute it which is as fast as possible. I.e. no other program for Σ_1 to compute S can output more than a finite number of elements more quickly. Thus it is not possible to speed up every program for computing S by the computer Σ_1. And, as is pointed out by Corollary B1, this also holds for any other quick-program computer, but with the slight "fogginess" that always results in passing from a statement about one particular quick-program computer to a statement about another.

Definition B2: Let S be a computable set of natural numbers and let Σ be a computer. Σ^S denotes the computer which can compute only subsets of S, but which is otherwise identical to Σ. I.e.

$$\Sigma^S(p, t) = \begin{cases} \Sigma(p, t), & \text{if } \bigcup_{t' \leq t} \Sigma(p, t') \subset S, \\ \emptyset, & \text{otherwise.} \end{cases}$$

Theorem B2: There is a computer Σ_0 such that $(\Sigma_0) \in L^*$ and $(\Sigma_0) < 1$. Moreover, for any computable sets T and R of natural numbers,

(a) if $T - R$ and $R - T$ are both infinite, then l.u.b. $(\Sigma_0),(\Sigma_1^T)$ and l.u.b. $(\Sigma_0),(\Sigma_1^R)$ are incomparable members of L^*, and

(b) if $T \subset R$ and $R - T$ is infinite, then the first of these two members of L^* is less than the second.

Proof: Σ_0 is constructed from the computer Σ_1 of Theorem B1 as follows. $n \in \Sigma_0(p, t)$ if and only if there exist t' and t'' with $\max(t', t'') = t$ such that

$$n \in \Sigma_1(p, t'), \quad s_{t'} \in \Sigma_1(p, t'')$$

where

$$\bigcup_{t_3 \leq t} \Sigma_1(p, t_3) = \{s_0, s_1, s_2, \dots\} \quad (s_0 < s_1 < s_2 < \cdots),$$

and for no $n_1 \geq n_2, t_1 < t_2 \leq t$ is it simultaneously the case that $n_1 \in \Sigma_1(p, t_1)$ and $n_2 \in \Sigma_1(p, t_2)$. Note that for all p, $\Sigma_1(p) = \Sigma_0(p)$, both sides of the equation being undefined if one of them is.

Sets S whose existence is asserted by Theorem B1 which must be computed very slowly by Σ_1 must be computed very much more slowly indeed by Σ_0, and thus Σ_1 im Σ_0 cannot be the case. Moreover, within any infinite computable set U of natural numbers, there are such sets S.

We now show in greater detail that $(\Sigma_0) < (\Sigma_1) = 1$ by a reductio ad absurdum of Σ_1 im Σ_0. Suppose Σ_1 im Σ_0. Then by definition there exists a 1-ary computable function h such that for any program p for which $\Sigma_1(p)$ is defined, there exists a program p' such that $\Sigma_0(p') = \Sigma_1(p)$ and

$$\bigcup_{t' \leq t} \Sigma_1(p, t') \subset \bigcup_{t' \leq h(t)} \Sigma_0(p', t')$$

for all but a finite number of values of t.

In Theorem B1 we now take $f(n) = \max(n, \max_{k \leq n} h(k))$. We obtain g and S satisfying

1. $g(n) > n$,

2. $g(n) > \max_{k \le n} h(k)$ for all but finitely many n. From (1) it follows that for all but a finite number of values of $n \in S$, the fastest program for Σ_1 to compute S outputs n at time $t' = g(n) + 1$, while the fastest program for Σ_0 to compute S outputs n at time

$$t'' = g(s_{g(n)+1}) + 1 = g(s_{t'}) + 1 > s_{t'} + 1 \ge t' + 1.$$

Here, as usual, $S = \{s_0, s_1, s_2, \ldots\}$ $(s_0 < s_1 < s_2 < \cdots)$. Note that s_k, the kth element of S, must be greater than or equal to k:

3. $s_k \ge k$.

By the definition of (Σ_1) **im** (Σ_0) we must have

$$h(g(n) + 1) \ge g(s_{g(n)+1}) + 1,$$

for all but finitely many $n \in S$. By (2) this implies

$$h(g(n) + 1) > \max_{k \le s_{g(n)+1}} h(k) + 1;$$

hence $g(n) + 1 > s_{g(n)+1}$ for all but finitely many $n \in S$. Invoking (3) we obtain $g(n) + 1 > s_{g(n)+1} \ge g(n) + 1$, which is impossible. Q.E.D.

A slightly different way of obtaining the following theorem was announced in [20].

Theorem B3: Any countable partially ordered set is order-isomorphic with a subset of L^*. That is, L^* is a "universal" countable partially ordered set.

Proof: We show that an example of a universal partially ordered set is C, the computable sets of natural numbers ordered by set inclusion. Thus the theorem is established if we can find in L^* an isomorphic image of C. This isomorphic image is obtained in the following manner. Let S be a computable set of natural numbers. Let S' be the set of all odd multiples of 2^n, where n ranges over all elements of S. The isomorphic image of the element S of C is the element l.u.b. $(\Sigma_0),(\Sigma_1^{S'})$ of L^*. Here Σ_0 is the computer of Theorem B2, Σ_1 is the computer of Theorem B1, and "$\Sigma_1^{S'}$" is written in accordance with the notational convention of Definition B2.

It only remains to prove that C is a universal partially ordered set. Sacks [17, p. 53] attributes to Mostowski [21] the following result: There is a universal countable partially ordered set $A = \{a_0, a_1, a_2, \ldots\}$ with the property that the predicate $a_n \le a_m$ is computable. We finish the proof by constructing in C an isomorphic image $A' = \{A_0, A_1, A_2, \ldots\}$ of A [as follows:

$$A_i = \{k \mid a_k \le a_i\}.$$

It is easy to see that $A_i \subset A_j$ if and only if $a_i \le a_j$.] Q.E.D.

Corollary B2: L^* has exactly $\aleph_0$ elements.

References

1. HARDY, G. H., AND WRIGHT, E. M. *An Introduction to the Theory of Numbers.* Clarendon Press, Oxford, 1962.

2. DANTZIG, T. *Number, the Language of Science.* Macmillan, New York, 1954.

3. DAVIS, M. *Computability and Unsolvability.* McGraw-Hill, New York, 1958.

4. GELFOND, A. O., AND LINNIK, YU. V. *Elementary Methods in Analytic Number Theory*. Rand McNally, Chicago, 1965.

5. BLUM, M. Measures on the computation speed of partial recursive functions. Quart. Prog. Rep. 72, Res. Lab. Electronics, MIT, Cambridge, Mass., Jan. 1964, pp. 237-253.

6. ARBIB, M. A. Speed-up theorems and incompleteness theorems. In *Automata Theory*, E. R. Cainiello (Ed.), Academic Press, New York, 1966, pp. 6-24.

7. FRAENKEL, A. A. *Abstract Set Theory*. North-Holland, Amsterdam, The Netherlands, 1961.

8. BOREL, É. *Leçons sur la Theorie des Fonctions*. Gauthier-Villars, Paris, 1914.

9. HARDY, G. H. *Orders of Infinity*. Cambridge Math. Tracts, No. 12, U. of Cambridge, Cambridge, Eng., 1924.

10. DEKKER, J. C. E., AND MYHILL, J. Retraceable sets. *Canadian J. Math. 10* (1958), 357-373.

11. SOLOMONOFF, R. J. A formal theory of inductive inference, Pt. I. *Inform. Contr. 7* (1964), 1-22.

12. KOLMOGOROV, A. N. Three approaches to the definition of the concept "amount of information." *Problemy Peredachi Informatsii 1* (1965), 3-11. (Russian)

13. CHAITIN, G. J. On the length of programs for computing finite binary sequences: statistical considerations. *J. ACM 16*, 1 (Jan. 1969), 145-159.

14. ARBIB, M. A., AND BLUM, M. Machine dependence of degrees of difficulty. *Proc. Amer. Math. Soc. 16* (1965), 442-447.

15. BIRKHOFF, G. *Lattice Theory*. Amer. Math. Soc. Colloq. Publ. Vol. 25, Amer. Math. Soc., Providence, R. I., 1967.

16. CHAITIN, G. J. On the length of programs for computing finite binary sequences. *J. ACM 13*, 4 (Oct. 1966), 547-569.

17. SACKS, G. E. *Degrees of Unsolvability*. No. 55, Annals of Math. Studies, Princeton U. Press, Princeton, N. J., 1963.

18. HARTMANIS, J., AND STEARNS, R. E. On the computational complexity of algorithms. *Trans. Amer. Math. Soc. 117* (1965), 285-306.

19. BLUM, M. A machine-independent theory of the complexity of recursive functions. *J. ACM 14*, 2 (Apr. 1967), 322-336.

20. CHAITIN, G. J. A lattice of computer speeds. Abstract 67T-397, *Notices Amer. Math. Soc. 14* (1967), 538.

21. MOSTOWSKI, A. Über gewisse universelle Relationen. *Ann. Soc. Polon. Math. 17* (1938), 117-118.

22. BLUM, M. On the size of machines. *Inform. Contr. 11* (1967), 257-265.

23. CHAITIN, G. J. On the difficulty of computations. Panamerican Symp. of Appl. Math., Buenos Aires, Argentina, Aug. 10, 1968. (to be published)

RECEIVED OCTOBER, 1966; REVISED DECEMBER, 1968